PRAISE FOR

The Red Queen

"Called the Red Queen theory by biologists, after the chess piece in Lewis Carroll's *Through the Looking-Glass* which runs but stays in the same place, this hypothesis is just one of the controversial ideas put forth in this witty, elegantly written inquiry."

—*Publishers Weekly*

"Extensively researched, clearly written: one of the best introductions to its fascinating and controversial subject."

—*Kirkus Reviews*

"Ridley's *Red Queen* is . . . literary and witty. He has the tone of a jolly schoolmaster who wants to share with you the neat things he's read." —*Boston Sunday Globe*

"A fascinating book filled with lucid prose and seductive reasoning."

—*Library Journal*

Pryde Brown

About the Author

MATT RIDLEY is the author of *Nature Via Nurture: Genes, Experience, and What Makes Us Human;* the critically acclaimed national bestseller *Genome: The Autobiography of a Species in 23 Chapters; The Origins of Virtue: Human Instincts and the Evolution of Cooperation;* and the *New York Times* Notable Book *The Red Queen: Sex and the Evolution of Human Nature.* His books have been short-listed for six literary awards, including the Los Angeles Times Book Prize. Formerly a scientist, journalist, and a national newspaper columnist, he is a visiting professor at Cold Spring Harbor Laboratory in New York and the chairman of the International Centre for Life in Newcastle, England.

THE RED QUEEN

*Sex and the Evolution of
Human Nature*

MATT RIDLEY

For Matthew

HARPER ● PERENNIAL

This book was first published in Great Britain in 1993 by Penguin Books Ltd. It is here reprinted by arrangement with Penguin Putnam.

HarperCollins books may be purchased for educational, business, or sales promotional use. For information please write: Special Markets Department, HarperCollins Publishers Inc., 10 East 53rd Street, New York, NY 10022.

First Perennial edition published 2003.

Library of Congress Cataloging-in-Publication Data
Ridley, Matt.
    The red queen : sex and the evolution of human nature / Matt
    Ridley.—1st Perennial ed.
        p. cm.
    Originally published: London: Viking, 1993.
    Includes bibliographical references and index.
    ISBN 0-06-055657-9   ISBN 978-0-06-055657-0
    1. Human evolution. 2. Social evolution. 3. Sex I. Title.

GN365.9.R53 2003
599.93'8—dc21   2003043356

---

07 08 09 **RRD** 20 19 18 17 16

CONTENTS

ACKNOWLEDGMENTS

This book is crammed with original ideas—very few of them my own. Science writers become accustomed to the feeling that they are intellectual plagiarists, raiding the minds of those who are too busy to tell the world about their discoveries. There are scores of people who could have written each chapter of my book better than I. My consolation is that few could have written all the chapters. My role has been to connect the patches of others' research together into a quilt.

But I remain deeply indebted and grateful to all those whose minds I raided. I have interviewed more than sixty people in the course of researching this book and have never met with anything but courtesy, patience, and infectious curiosity about the world. Many became friends. I am especially grateful to those whom I interviewed repeatedly and at length until I had almost picked their minds clean: Laura Betzig, Napoleon Chagnon, Leda Cosmides, Helena Cronin, Bill Hamilton, Laurence Hurst, Bobbi Low, Andrew Pomiankowski, Don Symons, and John Tooby.

Among those who agreed to interviews in person or by telephone, I would like to thank Richard Alexander, Michael Bailey, Alexandra Basolo, Graham Bell, Paul Bloom, Monique Borgehoff Mulder, Don Brown, Jim Bull, Austin Burt, David Buss, Tim Clutton-Brock, Bruce Ellis, John Endler, Bart Gledhill, David Goldstein, Alan Grafen, Tim Guilford, David Haig, Dean Hamer, Kristen Hawkes, Elizabeth Hill, Kim Hill, Sarah Hrdy, William Irons, William James, Charles Keckler, Mark Kirkpatrick, Jochen Kumm, Curtis Lively, John Maynard Smith, Matthew Meselson, Geoffrey

Miller, Anders Moller, Atholl McLachlan, Jeremy Nathans, Magnus Nordborg, Elinor Ostrom, Sarah Otto, Kenneth Oye, Margie Profet, Tom Ray, Michael Ryan, Dev Singh, Robert Smuts, Randy Thornhill, Robert Trivers, Leigh Van Valen, Fred Whitam, George Williams, Margo Wilson, Richard Wrangham, and Marlene Zuk.

My sincere thanks also to those who corresponded with me or sent me their papers and books: Christopher Badcock, Robert Foley, Stephen Frank, Valerie Grant, Toshikazu Hasegawa, Doug Jones, Egbert Leigh, Daniel Perusse, Felicia Pratto, and Edward Tenner.

Other minds I raided more subtly, even surreptitiously. Among those who have given advice or helped to clear my thoughts in many conversations are Alun Anderson, Robin Baker, Horace Barlow, Jack Beckstrom, Rosa Beddington, Mark Bellis, Roger Bingham, Mark Boyce, John Browning, Stephen Budiansky, Edward Carr, Geoffrey Carr, Jeremy Cherfas, Alice Clarke, Nico Colchester, Charles Crawford, Francis Crick, Martin Daly, Kurt Darwin, Marian Dawkins, Richard Dawkins, Andrew Dobson, Emma Duncan, Peter Garson, Anthony Gottlieb, John Hartung, Peter Hudson, Anya Hurlbert, Mark Flinn, Archie Fraser, Steven Gaulin, Charles Godfray, Joel Heinen, Nigella Hillgarth, Michael Kinsley, Richard Ladle, Richard Machalek, Seth Masters, Patrick McKim, Graeme Mitchison, Oliver Morton, Randolph Nesse, Paul Neuburg, Paul Newton, Linda Partridge, Marion Petrie, Steve Pinker, Mike Polioudakis, Jeanne Regalski, Peter Richerson, Mark Ridley (being mistaken for whom has been a great benefit to me), Alan Rogers, Vincent Sarich, Terry Sejnowski, Miranda Seymour, Rachel Smolker, Beverly Strassmann, Jeremy Taylor, Nancy Thornhill, David Wilson, Edward Wilson, Adrian Wooldridge, and Bob Wright.

Several people helped even further by reading drafts of chapters and commenting on them. Their advice was time-consuming to them but immensely valuable to me: Laura Betzig, Mark Boyce, Helena Cronin, Richard Dawkins, Laurence Hurst, Geoffrey Miller, and Andrew Pomiankowski. I owe a special debt to Bill Hamilton, to whom I returned again and again for inspiration at the early stages of this project.

My agents, Felicity Bryan and Peter Ginsberg, were unfailingly encouraging and constructive at all stages. My editors at Penguin and Macmillan, Ravi Mirchandani, Judith Flanders, Bill Rosen, and especially Carrie Chase, were efficient, kind, and inspired.

My wife, Anya Hurlbert, read the entire book, and her advice and support throughout have been invaluable.

Lastly, thanks to the red squirrel that sometimes scratched at my window while I wrote. I still don't know which sex it was.

Chapter 1

HUMAN NATURE

The most curious part of the thing was, that the trees and the other things round them never changed their places at all: however fast they went, they never seemed to pass anything. "I wonder if all the things move along with us?" thought poor puzzled Alice. And the Queen seemed to guess her thoughts, for she cried, "Faster! Don't try to talk!"

—Lewis Carroll, *Through the Looking-Glass*

When a surgeon cuts into a body, he knows what he will find inside. If he is seeking the patient's stomach, for example, he does not expect to find it in a different place in every patient. All people have stomachs, all human stomachs are roughly the same shape, and all are found in the same place. There are differences, no doubt. Some people have unhealthy stomachs; some have small stomachs; some have slightly misshapen stomachs. But the differences are tiny compared with the similarities. A vet or a butcher could teach the surgeon about a much greater variety of different stomachs: big, multichambered cow stomachs; tiny mouse stomachs; somewhat human looking pig stomachs. There is, it is safe to say, such a thing as the typical human stomach, and it is different from a non-human stomach.

It is the assumption of this book that there is also, in the same way, a typical human nature. It is the aim of this book to seek it. Like the stomach surgeon, a psychiatrist can make all sorts of basic assumptions when a patient lies down on the couch. He can assume that the patient knows what it means to love, to envy, to trust, to think, to speak, to fear, to smile, to bargain, to covet, to dream, to remember, to sing, to quarrel, to lie. Even if the person were from a newly discovered continent, all sorts of assumptions about his or her mind and nature would still be valid. When, in the 1930s, contact was made with New Guinea tribes hitherto cut off from the outside world and ignorant of its existence, they were found to smile and frown as unambiguously as any Westerner, despite 100,000 years of separation since they last shared a common ancestor. The "smile" of a baboon is a threat; the smile of a man is a sign of pleasure: It is human nature the world over.

That is not to deny the fact of culture shock. Sheeps' eyeball soup, a shake of the head that means yes, Western privacy, circumcision rituals, afternoon siestas, religions, languages, the difference in smiling frequency between a Russian and an American waiter in a restaurant—there are myriad human particulars as well as human universals. Indeed, there is a whole discipline, cultural anthropology, that devotes itself to the study of human cultural differences. But it is easy to take for granted the bedrock of similarity that underlies the human race—the shared peculiarities of being human.

This book is an inquiry into the nature of that human nature. Its theme is that it is impossible to understand human nature without understanding how it evolved, and it is impossible to understand how it evolved without understanding how human sexuality evolved. For the central theme of our evolution has been sexual.

Why sex? Surely there are features of human nature other than this one overexposed and troublesome procreative pastime. True enough, but reproduction is the sole goal for which human beings are designed; everything else is a means to that end. Human beings inherit tendencies to survive, to eat, to think, to speak, and so on. But above all they inherit a tendency to reproduce. Those of their predecessors that reproduced passed on their characteristics to their offspring; those that remained barren did not. Therefore, anything that increased the chances of a person reproducing successfully was passed on at the expense of anything else. We can confidently assert that there is nothing in our natures that was not carefully "chosen" in this way for its ability to contribute to eventual reproductive success.

This seems an astonishingly hubristic claim. It seems to deny free will, ignore those who choose chastity, and portray human beings as programmed robots bent only on procreation. It seems to imply that Mozart and Shakespeare were motivated only by sex. Yet I know of no other way that human nature can have developed except by evolution, and there is now overwhelming evidence that there is no other way for evolution to work except by competitive reproduction. Those strains that reproduce persist; those that do not reproduce die out. The ability to reproduce is what makes living things different from rocks. Besides, there is nothing inconsistent with free will or even chastity in this view of life. Human beings, I believe, thrive according to their ability to take initiatives and exercise individual talent. But free will was not created for fun; there was a reason that evolution handed our ancestors the ability to take initiatives, and the reason was that free will and initiative are means to satisfy ambition, to compete with fellow human beings, to deal with life's emergencies, and so eventually to be in a better position to reproduce and rear children than human beings who do not reproduce. Therefore, free will itself is any good only to the extent that it contributes to eventual reproduction.

Look at it another way: If a student is brilliant but terrible in examinations—if, say, she simply collapses with nervousness at the very thought of an exam—then her brilliance will count for nothing in a course that is tested by a single examination at the end of the term. Likewise, if an animal is brilliant at survival, has an efficient metabolism, resists all diseases, learns faster than its competitors, and lives to a ripe old age, but is infertile, then its superior genes are simply not available to its descendants. Everything can be inherited except sterility. None of your direct ancestors died childless. Consequently, if we are to understand how human nature evolved, the very core of our inquiry must be reproduction, for reproductive success is the examination that all human genes must pass if they are not to be squeezed out by natural selection. Hence I am going to argue that there are very few features of the human psyche and nature that can be understood without reference to reproduction. I begin with sexuality itself. Reproduction is not synonymous with sex; there are many asexual ways to reproduce. But reproducing sexually must improve an individual's reproductive success or else sex would not persist. I end with intelligence, the most human of all features. It is increasingly hard to understand how human beings came to be so clever without considering sexual competition.

What was the secret that the serpent told Eve? That she could eat a certain fruit? Pah. That was a euphemism. The fruit was carnal knowledge, and everybody from Thomas Aquinas to Milton knew it. How did they know it? Nowhere in Genesis is there even the merest hint of the equation: Forbidden fruit equals sin equals sex. We know it to be true because there *can* only be one thing so central to mankind. Sex.

OF NATURE AND NURTURE

The idea that we were designed by our past was the principal insight of Charles Darwin. He was the first to realize that you can abandon divine creation of species without abandoning the argument from design. Every living thing is "designed" quite unconsciously by the selective reproduction of its own ancestors to suit a particular life-style. Human nature was as carefully designed by natural selection for the use of a social, bipedal, originally African ape as human stomachs were designed for the use of an omnivorous African ape with a taste for meat.

That starting point will already have irritated two kinds of people. To those who believe that the world was made in seven days by a man with a long beard and that therefore human nature cannot have been designed by selection but by an Intelligence, I merely bid a respectful good day. We have little common ground on which to argue because I share few of your assumptions. As for those who protest that human nature did not evolve, but was invented *de novo* by something called "culture," I have more hope. I think I can persuade you that our views are compatible. Human nature is a product of culture, but culture is also a product of human nature, and both are the products of evolution. This does not mean that I am going to argue that it is "all in our genes." Far from it. I am vigorously going to challenge the notion that anything psychological is purely genetic, and equally vigorously challenge the assumption that anything universally human is untainted by genes. But our "culture" does not have to be the way it is. Human culture could be very much more varied and surprising than it is. Our closest relatives, the chimpanzees, live in promiscuous societies in which females seek as many sexual partners as possible and a male will kill the infants of strange females with whom he has not mated. There is no human society that remotely resembles this particular pattern. Why not? Because human nature is different from chimp nature.

If this is so, then the study of human nature must have profound implications for the study of history, sociology, psychology, anthropology, and politics. Each of those disciplines is an attempt to understand human behavior, and if the underlying universals of human behavior are the product of evolution, then it is vitally important to understand what the evolutionary pressures were. Yet I have gradually come to realize that almost all of social science proceeds as if 1859, the year of the publication of the *Origin of Species*, had never happened; it does so quite deliberately, for it insists that human culture is a product of our own free will and invention. Society is not the product of human psychology, it asserts, but vice versa.

That sounds reasonable enough, and it would be splendid for those who believe in social engineering if it were true, but it is simply not true. Humanity is, of course, morally free to make and remake itself infinitely, but we do not do so. We stick to the same monotonously human pattern of organizing our affairs. If we were more adventurous, there would be societies without love, without ambition, without sexual desire, without marriage, without art, without grammar, without music, without smiles—and with as many unimaginable novelties as are in that list. There would be societies in which women killed each other more often than men, in which old people were considered more beautiful than twenty-year-olds, in which wealth did not purchase power over others, in which people did not discriminate in favor of their own friends and against strangers, in which parents did not love their own children.

I am not saying, like those who cry, "You can't change human nature, you know," that it is futile to attempt to outlaw, say, racial persecution because it is in human nature. Laws against racism do have an effect because one of the more appealing aspects of human nature is that people calculate the consequences of their actions. But I am saying that even after a thousand years of strictly enforced laws against racism, we will not one day suddenly be able to declare the problem of racism solved and abolish the laws secure in the knowledge that racial prejudice is a thing of the past. We assume, and rightly, that a Russian is just as human after two generations of oppressive totalitarianism as his grandfather was before him. But why, then, does social science proceed as if it were not the case, as if people's natures are the products of their societies?

It is a mistake that biologists used to make, too. They believed that evolution proceeded by accumulating the changes that individuals gathered during their lives. The idea was most clearly formulated by Jean-Baptiste Lamarck, but Charles Darwin sometimes used it, too. The classic example is a blacksmith's son supposedly inheriting his father's acquired muscles at birth. We now know that Lamarckism cannot work because bodies are built from cakelike recipes, not architectural blueprints, and it is simply impossible to feed information back into the recipe by changing the cake.[1] But the first coherent challenge to Lamarckism was the work of a German follower of Darwin named August Weismann, who began to publish his ideas in the 1880s.[2] Weismann noticed something peculiar about most sexual creatures: Their sex cells— eggs and sperm—remained segregated from the rest of the body from the moment of their birth. He wrote: "I believe that heredity depends upon the fact that a small portion of the effective substance of the germ, the germ-plasm, remains unchanged during the development of the ovum into an organism, and that this part of the germ-plasm serves as a foundation from which germ-cells of the new organism are produced. There is, therefore, continuity of the germ-plasm from one generation to another."[3]

In other words, you are descended not from your mother but from her ovary. Nothing that happened to her body or her mind in her life could affect your nature (though it could affect your nurture, of course—an extreme example being that her addiction to drugs or alcohol might leave you damaged in some nongenetic way at birth). You are born free of sin. Weismann was much ridiculed for this in his lifetime and little believed. But the discovery of the gene and of the DNA from which it is made and of the cipher in which DNA's message is written have absolutely confirmed his suspicion. The germ-plasm is kept separate from the body.

Not until the 1970s were the full implications of this realized. Then Richard Dawkins of Oxford University effectively invented the notion that because bodies do not replicate themselves but are grown, whereas genes do replicate themselves, it inevitably follows that the body is merely an evolutionary vehicle for the gene, rather than vice versa. If genes make their bodies do things that perpetuate the genes (such as eat, survive, have sex, and help rear children), then the genes themselves will be perpetuated. So other kinds of bodies will disappear. Only bodies that suit the survival and perpetuation of genes will remain.

Since then, the ideas of which Dawkins was an early champion have changed biology beyond recognition. What was still—despite Darwin—essentially a descriptive science has become a study of function. The difference is crucial. Just as no engineer would dream of describing a car engine without reference to its function (to turn wheels), so no physiologist would dream of describing a stomach without reference to its function (to digest food). But before, say, 1970, most students of animal behavior and virtually all students of human behavior were content to describe what they found without reference to a function. The gene-centered view of the world changed this for good. By 1980 no detail of animal courtship mattered unless it could be explained in terms of the selective competition of genes. And by 1990 the notion that human beings were the only animals exempt from this logic was beginning to look ever more absurd. If man has evolved the ability to override his evolutionary imperatives, then there must have been an advantage to his genes in doing so. Therefore, even the emancipation from evolution that we so fondly imagine we have achieved must itself have evolved because it suited the replication of genes.

Inside my skull is a brain that was designed to exploit the

conditions of an African savanna between 3 million and 100,000 years ago. When my ancestors moved into Europe (I am a white European by descent) about 100,000 years ago, they quickly evolved a set of physiological features to suit the sunless climate of northern latitudes: pale skin to prevent rickets, male beards, and a circulation relatively resistant to frostbite. But little else changed: Skull size, body proportions, and teeth are all much the same in me as they were in my ancestors 100,000 years ago and are much the same as they are in a San tribesman from southern Africa. And there is little reason to believe that the gray matter inside the skull changed much, either. For a start, 100,000 years is only three thousand generations, a mere eye blink in evolution, equivalent to a day and a half in the life of bacteria. Moreover, until very recently the life of a European was essentially the same as that of an African. Both hunted meat and gathered plants. Both lived in social groups. Both had children dependent on their parents until their late teens. Both used stone, bone, wood, and fiber to make tools. Both passed wisdom down with complex language. Such evolutionary novelties as agriculture, metal, and writing arrived less than three hundred generations ago, far too recently to have left much imprint on my mind.

There is, therefore, such a thing as a universal human nature, common to all peoples. If there were descendants of *Homo erectus* still living in China, as there were a million years ago, and those people were as intelligent as we are, then truly they could be said to have different but still human natures.[4] They might perhaps have no lasting pair bonds of the kind we call marriage, no concept of romantic love, and no involvement of fathers in parental care. We could have some very interesting discussions with them about such matters. But there are no such people. We are all one close family, one small race of the modern *Homo sapiens* people who lived in Africa until 100,000 years ago, and we all share the nature of that beast.

Just as human nature is the same everywhere, so it is recognizably the same as it was in the past. A Shakespeare play is about motives and predicaments and feelings and personalities that are instantly familiar. Falstaff's bombast, Iago's cunning, Leontes's jealousy, Rosalind's strength, and Malvolio's embarrassment have not changed in four hundred years. Shakespeare was writing about the same human nature that we know today. Only his vocabulary (which is nurture, not nature) has aged. When I watch *Anthony and Cleopatra*, I am seeing a four-hundred-year-old interpretation of a two-thousand-year-old history. Yet it never even occurs to me that love was any different then from what it is now. It is not necessary to explain to me why Anthony falls under the spell of a beautiful woman. Across time just as much as across space, the fundamentals of our nature are universally and idiosyncratically human.

THE INDIVIDUAL IN SOCIETY

Having argued that all human beings are the same, that this book is about their shared human nature, I shall now seem to argue the opposite. But I am not being inconsistent.

Human beings are individuals. All individuals are slightly different. Societies that treat their constituent members as identical pawns soon run into trouble. Economists and sociologists who believe that individuals will usually act in their collective rather than their particular interests ("From each according to his ability, to each according to his needs"[5] versus "Devil take the hindmost") are soon confounded. Society is composed of competing individuals as surely as markets are composed of competing merchants; the focus of economic and social theory is, and must be, the individual. Just as genes are the only things that replicate, so individuals, not societies, are the vehicles for genes. And the most formidable threats to reproductive destiny that a human individual faces come from other human individuals.

It is one of the remarkable things about the human race that no two people are identical. No father is exactly recast in his son; no daughter is exactly like her mother; no man is his brother's double, and no woman is a carbon copy of her sister—unless they are that rarity, a pair of identical twins. Every idiot can be father or mother to a genius—and vice versa. Every face and every set of fingerprints is effectively unique. Indeed, this uniqueness goes further in human beings than in any other animal. Whereas every deer or every sparrow is self-reliant and does everything every other deer or sparrow does, the same is not true of a man or a woman, and has not been for thousands of years. Every individual is a specialist of some sort, whether he or she is a welder, a housewife, a playwright, or a prostitute. In behavior, as in appearance, every human individual is unique.

How can this be? How can there be a universal, species-specific human nature when every human being is unique? The solution to this paradox lies in the process known as sex. For it is sex that mixes together the genes of two people and discards half of the mixture, thereby ensuring that no child is exactly like either of its parents. And it is also sex that causes all genes to be contributed eventually to the pool of the whole species by such mixing. Sex causes the differences between individuals but ensures that those differences never diverge far from a golden mean for the whole species.

A simple calculation will clarify the point. Every human being has two parents, four grandparents, eight great-grandparents, sixteen great-great-grandparents, and so on. A mere thirty generations back—in, roughly, A.D. 1066—you had more than a billion direct ancestors in the same generation (2 to the power of 30). Since there were fewer than a billion people alive at that time in the whole world, many of them were your ancestors two or three times over. If, like me, you are of British descent, the chances are that almost all of the few million Britons alive in 1066, including King Harold, William the Conqueror, a random serving wench, and the meanest vassal (but excluding all well-behaved monks and nuns), are your direct ancestors. This makes you a distant cousin many times over of every other Briton alive today except the children of recent immigrants. All Britons are descended from the same set of people a mere thirty generations ago. No wonder there is a certain uniformity about the human (and every other sexual) species. Sex imposes it by its perpetual insistence on the sharing of genes.

If you go back further still, the different human races soon merge. Little more than three thousand generations back, all of our ancestors lived in Africa, a few million simple hunter-gatherers, completely modern in physiology and psychology.[6] As a result, the genetic differences between the average members of different races are actually tiny and are mostly confined to a few genes that affect skin color, physiognomy, or physique. Yet the differences between any two individuals, of the same race or of different races, can still be large. According to one estimate, only 7 percent of the genetic differences between two individuals can be attributed to the fact that they are of different race; 85 percent of the genetic differences are attributable to mere individual variation, and the rest is tribal or national. In the words of one pair of scientists: "What this means is that the average genetic difference between one Peruvian farmer and his neighbor, or one Swiss villager and his neighbor, is twelve times greater than the difference between the 'average genotype' of the Swiss population and the 'average genotype' of the Peruvian population."[7]

It is no harder to explain than a game of cards. There are aces and kings and twos and threes in any deck of cards. A lucky player is dealt a high-scoring hand, but none of his cards is unique. Elsewhere in the room are others with the same kinds of cards in their hands. But even with just thirteen kinds of cards, every hand is different and some are spectacularly better than others. Sex is merely the dealer, generating unique hands from the same monotonous deck of genetic cards shared by the whole species.

But the uniqueness of the individual is only the first of the implications of sex for human nature. Another is that there are, in fact, two human natures: male and female. The basic asymmetry of gender leads inevitably to different natures for the two genders, natures that suit the particular role of each gender. For example, males usually compete for access to females, rather than vice versa. There are good evolutionary reasons for this, and there are clear evolutionary consequences, too; for instance, men are more aggressive than women.

A third implication of sex for human natures is that every other human being alive today is a potential source of genes for your children. And we are descended from only those people who sought the best genes, a habit we inherited from them. Therefore, if you spot somebody with good genes, it is your inherited habit to seek to buy some of those genes; or, put more prosaically, people are attracted to people of high reproductive and genetic potential—the healthy, the fit, and the powerful. The consequences of this fact, which goes under the name of sexual selection, are bizarre in the extreme, as will become clear in the rest of this book.

OURS TO REASON WHY?

To speak of the "purpose" of sex or of the function of a particular human behavior is shorthand. I do not imply some teleological goal-seeking or the existence of a great designer with an aim in mind. Still less will I be implying foresight or consciousness on the part of "sex" itself or of mankind. I merely refer to the astonishing power of adaptation, so well appreciated by Charles Darwin and so little understood by his modern critics. For I must confess at once that I am an "adaptationist," which is a rude word for somebody who believes that animals and plants, their body parts and their behaviors, consist largely of designs to solve particular problems.[8]

Let me explain. The human eye is "designed" to form an image of the visual world on its retina; the human stomach is "designed" to digest food; it is perverse to deny such facts. The only question is how they came to be "designed" for their jobs. And the only answer that has stood the test of time and scrutiny is that there was no designer. Modern people are descended mainly from those people whose eyes and stomachs were better at those jobs than other people's. Small, random improvements in the ability of stomachs to digest and of eyes to see were thus inherited, and small diminishments in ability were not inherited because the owners, equipped with poor digestion or poor vision, did not live so long or breed so well.

We human beings find the notion of engineering design quite easy to grasp and have little difficulty seeing the analogy with the design of an eye. But we seem to find it harder to grasp the idea of "designed" behavior, mainly because we assume that purposeful behavior is evidence of conscious choice. An example might help to clarify what I mean. There is a little wasp that injects its eggs into whitefly aphids, where they grow into new wasps by eating the whitefly from the inside out. Distressing but true. If one of these wasps, upon poking its tail into a whitefly, discovers that the aphid is already occupied by a young wasp, then she does something that seems remarkably intelligent: She withholds sperm from the egg she is about to lay and lays an unfertilized egg inside the wasp larva that is inside the whitefly. (It is a peculiarity of wasps and ants that unfertilized eggs develop into males, while all fertilized ones develop into females.) The "intelligent" thing that the mother wasp has done is to recognize that there is less to eat inside an already-occupied whitefly than in virgin territory. Her egg will therefore grow into a small, stunted wasp. And in her species, males are small, females large. So it was "clever" of her to "choose" to make her offspring male when she "knew" it was going to be small.

But of course this is nonsense. She was not "clever"; she did not "choose" and she "knew" not what she did. She was a minuscule wasp with a handful of brain cells and absolutely no possibility of conscious thought. She was an automaton, carrying out the simple instructions of her neural program: *If whitefly occupied, withhold sperm.* Her program had been designed by natural selection over millions of years: Wasps that inherited a tendency to withhold sperm when they found their prey already occupied had more successful offspring than those that did not. Yet in exactly the same way that natural selection had "designed" an eye, as if for the "purpose" of seeing, so natural selection had produced behavior that seemed designed to suit the wasp's purposes.[9]

This "powerful illusion of deliberate design"[10] is so fundamental a notion and yet so simple that it hardly seems necessary to repeat it. It has been much more fully explored and explained by Richard Dawkins in his wonderful book *The Blind Watchmaker*.[11] Throughout this book I will assume that the greater the degree of complexity there is in a behavior pattern, genetic mechanism, or psychological attitude, the more it implies a design for a function. Just as the complexity of the eye forces us to admit that it is designed to see, so the complexity of sexual attraction implies that it is designed for genetic trade.

In other words, I believe that it is always worth asking the question why. Most of science is the dry business of discovering how the universe works, how the sun shines, or how plants grow. Most scientists live their lives steeped in *how* questions, not *why* questions. But consider for a moment the difference between the question "Why do men fall in love?" and the question "How do men fall in love?" The answer to the second will surely turn out to be merely a matter of plumbing. Men fall in love through the effects of hormones on brain cells and vice versa, or some such physiological effect. One day some scientist will know exactly how the brain of a young man becomes obsessed with the image of a particular young woman, molecule by molecule. But the why question is to me more interesting because the answer gets to the heart of how human nature came to be what it is.

Why has that man fallen in love with that woman? Because she's pretty. Why does pretty matter? Because human beings are a mainly monogamous species and so males are choosy about their mates (as male chimpanzees are not); prettiness is an indication of youth and health, which are indications of fertility. Why does that man care about fertility in his mate? Because if he did not, his genes would be eclipsed by those of men who did. Why does he care about that? He does not, but his genes act as if they do. Those who choose infertile mates leave no descendants. Therefore, everybody is descended from men who preferred fertile women, and every person inherits from those ancestors the same preference. Why is that man a slave to his genes? He is not. He has free will. But you just said he's in love because it is good for his genes. He's free to ignore the dictates of his genes. Why do his genes want to get together with her genes anyway? Because that's the only way they can get into the next generation; human beings have two sexes that must breed by mixing their genes. Why do human beings have two sexes? Because in mobile animals hermaphrodites are less good at doing two things at once than males and females are at each doing his or her own thing. Therefore, ancestral hermaphroditic animals were outcompeted by ancestral sexed animals. But why only two sexes? Because that was the only way to settle a long-running genetic dispute between sets of genes. What? I'll explain later. But why does she need him? Why don't her genes just go ahead and make babies without waiting for his input? That is the most fundamental why question of all, and the one with which the next chapter begins.

In physics, there is no great difference between a why question and a how question. How does the earth go around the sun? By gravitational attraction. Why does the earth go around the sun? Because of gravity. Evolution, however, causes biology to be a very different game because it includes contingent history. As anthropologist Lionel Tiger has put it, "We are perforce in some sense constrained, goaded, or at least affected by the accumulated impact of selective decisions made over thousands of generations."[14] Gravity is gravity however history deals its dice. A peacock is a showy peacock because at some point in history ancestral peahens stopped picking their mates according to mundane utilitarian criteria and instead began to follow a fashion for preferring elaborate display. Every living creature is a product of its past. When a neo-Darwinian asks, "Why?" he is really asking, "How did this come about?" He is a historian.

OF CONFLICT AND COOPERATION

One of the peculiar features of history is that time always erodes advantage. Every invention sooner or later leads to a counterinvention. Every success contains the seeds of its own overthrow. Every hegemony comes to an end. Evolutionary history is no different. Progress and success are always relative. When the land was unoccupied by animals, the first amphibian to emerge from the sea could get away with being slow, lumbering, and fishlike, for it had no enemies and no competitors. But if a fish were to take to the land today, it would be gobbled up by a passing fox as surely as a Mongol horde would be wiped out by machine guns. In history and in evolution, progress is always a futile, Sisyphean struggle to stay in the same relative place by getting ever better at things. Cars move through the congested streets of London no faster than horse-drawn carriages did a century ago. Computers have no effect on productivity because people learn to complicate and repeat tasks that have been made easier.[13]

This concept, that all progress is relative, has come to be known in biology by the name of the Red Queen, after a chess piece that Alice meets in *Through the Looking-Glass*, who perpetually runs without getting very far because the landscape moves with her. It is an increasingly influential idea in evolutionary theory, and one that will recur throughout the book. The faster you run, the more the world moves with you and the less you make progress. Life is a chess tournament in which if you win a game, you start the next game with the handicap of a missing pawn.

The Red Queen is not present at all evolutionary events. Take the example of a polar bear, which is equipped with a thick coat of white fur. The coat is thick because ancestral polar bears better survived to breed if they did not feel the cold. There was a relatively simple evolutionary progression: thicker and thicker fur, warmer and warmer bears. The cold did not get worse just because the bear's insulation got thicker. But the polar bear's fur is white for a different reason: camouflage. White bears can creep up on seals much more easily than brown bears can. Presumably, once upon a time, it was easy to creep up on Arctic seals because they feared no enemies on the ice, just as present-day Antarctic seals are entirely fearless on the ice. In those days, proto–polar bears had an easy time catching seals. But soon nervous, timid seals tended to live longer than trusting ones, so gradually seals grew more and more wary. Life grew harder for bears. They had to creep up on the seals stealthily, but the seals could easily see them coming—until one day (it may not have been so sudden, but the principle is the same) by chance mutation a bear had cubs that were white instead of brown. They thrived and multiplied because the seals did not see them coming. The seal's evolutionary effort was for nothing; they were back where they started. The Red Queen was at work.

In the world of the Red Queen, any evolutionary progress will be relative as long as your foe is animate and depends heavily on you or suffers heavily if you thrive, like the seals and the bears. Thus the Red Queen will be especially hard at work among predators and their prey, parasites and their hosts, and males and females of the same species. Every creature on earth is in a Red Queen chess tournament with its parasites (or hosts), its predators (or prey), and, above all, with its mate.

Just as parasites depend on their hosts and yet make them suffer, and just as animals exploit their mates and yet need them, so the Red Queen never appears without another theme being sounded: the theme of intermingled cooperation and conflict. The relationship between a mother and her child is fairly straightforward: Both are seeking roughly the same goal—the welfare of themselves and each other. The relationship between a man and his wife's lover or between a woman and her rival for a promotion is also fairly straightforward: Both want the worst for each other. One relationship is all about cooperation, the other all about conflict. But what is the relationship between a woman and her husband? It is cooperation in the sense that both want the best for the other. But why? In order to exploit each other. A man uses his wife to produce children for him. A woman uses her husband to make and help rear her children. Marriage teeters on the line between a cooperative venture and a form of mutual exploitation—ask any divorce lawyer. Successful marriages so submerge the costs under mutual benefits that the cooperation can predominate; unsuccessful ones do not.

This is one of the great recurring themes of human history, the balance between cooperation and conflict. It is the obsession of governments and families, of lovers and rivals. It is the key to economics. It is, as we shall see, one of the oldest themes in the history of life, for it is repeated right down to the level of the gene itself. And the principal cause of it is sex. Sex, like marriage, is a cooperative venture between two rival sets of genes. Your body is the scene of this uneasy coexistence.

TO CHOOSE

One of Charles Darwin's more obscure ideas was that animals'
mates can act like horse breeders, consistently selecting certain
types and so changing the race. This theory, known as sexual selection, was ignored for many years after Darwin's death and has only
recently come back into vogue. Its principal insight is that the goal
of an animal is not just to survive but to breed. Indeed, where
breeding and survival come into conflict, it is breeding that takes
precedence; for example, salmon starve to death while breeding.
And breeding, in sexual species, consists of finding an appropriate
partner and persuading it to part with a package of genes. This goal
is so central to life that it has influenced the design not only of the
body but of the psyche. Simply put, anything that increases reproductive success will spread at the expense of anything that does
not—even if it threatens survival.

Sexual selection produces the appearance of purposeful
"design" as surely as natural selection does. Just as a stag is
designed by sexual selection for battle with sexual rivals and a peacock is designed for seduction, so a man's psychology is designed
to do things that put his survival at risk but increase his chances of
acquiring or retaining one or more high-quality mates. Testosterone itself, the very elixir of masculinity, increases susceptibility
to infectious disease. The more competitive nature of men is a consequence of sexual selection. Men have evolved to live dangerously
because success in competition or battle used to lead to more or
better sexual conquests and more surviving children. Women who
live dangerously merely put at risk those children they already have.
Likewise, the intimate connection between female beauty and
female reproductive potential (beautiful women are almost by definition young and healthy; compared with older women, they are
therefore both more fertile and have a longer reproductive life
ahead of them) is a consequence of sexual selection acting on both
men's psyches and women's bodies. Each sex shapes the other.
Women have hourglass-shaped bodies because men have preferred
them that way. Men have an aggressive nature because women have

preferred them that way (or have allowed aggressive men to defeat other men in contests over women—it amounts to the same thing). Indeed, this book will end with the astonishing theory that the human intellect itself is a product of sexual rather than natural selection, for most evolutionary anthropologists now believe that big brains contributed to reproductive success either by enabling men to outwit and outscheme other men (and women to outwit and outscheme other women) or because big brains were originally used to court and seduce members of the other sex.

Discovering and describing human nature and how it differs from the nature of other animals is as interesting a task as any that science has faced; it is on a par with the quest for the atom, the gene, and the origin of the universe. Yet science has consistently shied away from the task. The greatest "experts" our species has produced on the subject of human nature were people like Buddha and Shakespeare, not scientists or philosophers. The biologists stick to animals; those who try to cross the line (as Harvard's Edward Wilson did in his book *Sociobiology* in 1975) are vilified with accusations of political motives.[14] Meanwhile, human scientists proclaim that animals are irrelevant to the study of human beings and that there is no such thing as a universal human nature. The consequence is that science, so coldly successful at dissecting the Big Bang and DNA, has proved spectacularly inept at tackling what the philosopher David Hume called the greatest question of all: Why is human nature what it is?

Chapter 2

THE ENIGMA

Birth after birth the line unchanging runs,

And fathers live transmitted in their sons;

Each passing year beholds the unvarying kinds,

The same their manners, and the same their minds.

Till, as erelong successive buds decay,

And insect-shoals successive pass away,

Increasing wants the pregnant parent vex

With the fond wish to form a softer sex. . . .

—Erasmus Darwin, "The Temple of Nature"

Zog the Martian steered her craft carefully into its new orbit and prepared to reenter the hole in the back of the planet, the one that had never been seen from Earth. She had done it many times before and was not so much nervous as impatient to be home. It had been a long stay on Earth, longer than most Martians made, and she looked forward to a long argon bath and a glass of cold chlorine. It would be good to see her colleagues again. And her children. And her husband—she caught herself and laughed. She had been on Earth so long she had even begun to think like an earthling. Husband indeed! Every Martian knew that no Martian had a husband. There was no such thing as sex on Mars. Zog thought with pride of the report in her knapsack: "Life on Earth: The Reproduction Enigma Solved." It was the finest thing she had ever done; promotion could not be denied her now, whatever Big Zag said.

A week later, Big Zag opened the door of the Earthstudy Inc. committee room and asked the secretary to send Zog in. Zog entered and sat in the seat assigned to her. Big Zag avoided her eyes as she cleared her throat and began.

"Zog, this committee has read your report carefully, and we are all, I think I can say, impressed with its thoroughness. You have certainly made an exhaustive survey of reproduction on Earth. Moreover, with the possible exception of Miss Zeeg here, we are all agreed that you have made an overwhelming case for your hypothesis. I consider it now beyond doubt that life on Earth reproduces in the way you describe, using this strange device called 'sex.' Some of the committee are less happy with your conclusion that many of the peculiar facets of the earthling species known as human beings are a consequence of this sex thing: jealous love, a sense of beauty, male aggression, even what they laughingly call intelligence." The committee chuckled sycophantically at this old joke. "But," said Big Zag suddenly and loudly, looking up from the paper in front of her, "we have one major difficulty with your report. We believe you have entirely failed to address the most interesting issue of all. It is a three-letter question of great simplicity." Big Zag's voice dripped sarcasm. *"Why?"*

Zog stammered: "What do you mean, why?"

"I mean why do earthlings have sex? Why don't they just clone themselves as we do? Why do they need two creatures to have one baby? Why do males exist? Why? Why? Why?"

"Oh," said Zog quickly, "I tried to answer that question, but I got nowhere. I asked some human beings, people who had studied the subject for years, and they did not know. They had a few suggestions, but each person's suggestion was different. Some said sex was a historical accident. Some said it was a way of fending off disease. Some said it was about adapting to change and evolving faster. Others said it was a way of repairing genes. But basically they did not know."

"Did not know?" Big Zag burst out. "Did not know? The most essential peculiarity in their whole existence, the most intriguing scientific question anybody has ever asked about life on Earth, and they *don't know*. Zod save us!"

What is the purpose of sex? At first glance the answer seems obvious to the point of banality. But a second glance brings a different thought. Why must a baby be the product of two people? Why not three, or one? Need there be a reason at all?

About twenty years ago a small group of influential biologists changed their ideas about sex. From considering it logical, inevitable, and sensible as a means of reproduction, they switched almost overnight to the conclusion that it was impossible to explain why it had not disappeared altogether. Sex seemed to make no sense at all. Ever since, the purpose of sex has been an open question, and it has been called the queen of evolutionary problems.[1]

But dimly, through the confusion, a wonderful answer is taking shape. To understand it requires you to enter a looking-glass world, where nothing is what it seems. Sex is not about reproduction, gender is not about males and females, courtship is not about persuasion, fashion is not about beauty, and love is not about affection. Below the surface of every banality and cliché there lies irony, cynicism, and profundity.

In 1858, the year Charles Darwin and Alfred Russel Wallace gave the first plausible account of a mechanism for evolution, the Victorian brand of optimism known as "progress" was in its prime. It is hardly surprising that Darwin and Wallace were immediately interpreted as having given succor to the god of progress. Evolution's immediate popularity (and it was popular) owed much to the fact that it was misunderstood as a theory of steady progress from amoeba to man, a ladder of self-improvement.

As the end of the second millennium approaches, mankind is in a different mood. Progress, we think, is about to hit the buffers of overpopulation, the greenhouse effect, and the exhaustion of resources. However fast we run, we never seem to get anywhere. Has the industrial revolution made the average inhabitant of the world healthier, wealthier, and wiser? Yes, if he is German. No, if he is Bangladeshi. Uncannily (or, a philosopher would have us believe, predictably), evolutionary science is ready to suit the mood. The fashion in evolutionary science now is to scoff at progress; evolution is a treadmill, not a ladder.

PREGNANT VIRGINS

For people, sex is the only way to have babies, and that, plainly enough, is its purpose. It was only in the last half of the nineteenth century that anybody saw a problem with this. The problem was that there seemed to be all sorts of better ways of reproducing. Microscopic animals split in two. Willow trees grow from cuttings.

Dandelions produce seeds that are clones of themselves. Virgin greenfly give birth to virgin young that are already pregnant with other virgins. August Weismann saw this clearly in 1889. "The significance of amphimixis [sex]," he wrote, "cannot be that of making multiplication possible, for multiplication may be effected without amphimixis in the most diverse ways—by division of the organism into two or more, by budding, and even by the production of unicellular germs."[2]

Weismann started a grand tradition. From that day to this, at regular intervals, the evolutionists have declared that sex is a "problem," a luxury that should not exist. There is a story about an early meeting of the Royal Society in London, attended by the king, at which an earnest discussion began about why a bowl of water weighed the same with a goldfish in it as it did without. All sorts of explanations were proffered and rejected. The debate became quite heated. Then the king suddenly said, "I doubt your premise." He sent for a bowl of water and a fish and a balance. The experiment was done. The bowl was put on the balance, and the fish was added; the bowl's weight increased by exactly the weight of the fish. Of course.

The tale is no doubt apocryphal, and it is not fair to suggest that the scientists you will meet in these pages are quite such idiots as to assume a problem exists when it does not. But there is a small similarity. When a group of scientists suddenly said that they could not explain why sex existed and they found the existing explanations unsatisfactory, other scientists found this intellectual sensitivity absurd. Sex exists, they pointed out; it must confer some kind of advantage. Like engineers telling bumblebees they could not fly, biologists were telling animals and plants they would be better off breeding asexually. "A problem for this argument," wrote Lisa Brooks of Brown University, "is that many sexual organisms seem to be unaware of the conclusion."[3] There might be a few holes in existing theories, said the cynics, but do not expect us to give you a Nobel Prize for plugging them. Besides, why must sex have a purpose? Maybe it is just an evolutionary accident that reproduction happens that way, like driving on one side of the road.

Yet lots of creatures do not have sex at all or do it in some generations and not others. The virgin greenfly's great-great-granddaughter, at the end of the summer, will be sexual: She will mate with a male greenfly and have young that are mixtures of their parents. Why does she bother? For an accident, sex seems to have hung on with remarkable tenacity. The debate has refused to die. Every year produces a new crop of explanations, a new collection of essays, experiments, and simulations. Survey the scientists involved now and virtually all will agree that the problem has been solved; but none will agree on the solution. One man insists on hypothesis A, another on hypothesis B, a third on C, a fourth on all of the above. Could there be a different explanation altogether? I asked John Maynard Smith, one of the first people to pose the question "Why sex?," whether he still thought some new explanation was needed. "No. We have the answers. We cannot agree on them, that is all."[4]

OF SEX AND FREE TRADE

A brief genetic glossary is necessary before we proceed. Genes are biochemical recipes written in a four-letter alphabet called DNA, recipes for how to make and run a body. A normal human being has two copies of each of 30,000 genes in every cell in his or her body. The total complement of 60,000 human genes is called the "genome," and the genes live on twenty-three pairs of ribbonlike objects called "chromosomes." When a man impregnates a woman, each one of his sperm contains one copy of each gene, 30,000 in all, on twenty-three chromosomes. These are added to the 30,000 single genes on twenty-three chromosomes in the woman's egg to make a complete human embryo with 30,000 pairs of genes and twenty-three pairs of chromosomes.

A few more technical terms are essential, and then we can discard the whole jargon-ridden dictionary of genetics. The first word is "meiosis," which is simply the procedure by which the male selects the genes that will go into a sperm or the female selects the genes that will go into an egg. The man may choose either the 30,000 genes he inherited from his father or the seventy-five thousand he inherited from his mother or more likely, a mixture. During meiosis something peculiar happens. Each of the 23 pairs of chromosomes is laid alongside it opposite number. Chunks of one set are swapped with chunks of the other in a procedure called "recombination." One whole set is then passed on to the offspring to be married with a set from the other parent—a procedure known as "outcrossing."

Sex is recombination plus outcrossing; this mixing of genes is its principal feature. The consequence is that the baby gets a thorough mixture of its four grandparents' genes (because of recombination) and its two parents' genes (because of outcrossing). Between them, recombination and outcrossing are the essential procedures of sex. Everything else about it—gender, mate choice, incest avoidance, polygamy, love, jealousy—are ways of doing outcrossing and recombination more effectively or carefully.

Put this way, sex immediately becomes detached from reproduction. A creature could borrow another's genes at any stage in its life. Indeed, that is exactly what bacteria do. They simply hook up with each other like refueling bombers, pass a few genes through the pipe, and go their separate ways. Reproduction they do later, by splitting in half.[5]

So sex equals genetic mixing. The disagreement comes when you try to understand why genetic mixing is a good idea. For the past century or so, traditional orthodoxy held that genetic mixing is good for evolution because it helps create variety, from which natural selection can choose. It does not change genes—even Weismann, who did not know about genes and referred vaguely to "ids," realized that—but it throws together new combinations of genes. Sex is a sort of free trade in good genetic inventions and thus greatly increases the chances that they will spread through a species and the species will evolve. "A source of individual variability furnishing material for the operation of natural selection," Weismann called sex.[6] It speeds up evolution.

Graham Bell, an English biologist working in Montreal, has

dubbed this traditional theory the "Vicar of Bray" hypothesis after a fictional sixteenth-century cleric who was quick to adapt to the prevailing religious winds, switching between Protestant and Catholic rites as the ruling monarch changed. Like the flexible vicar, sexual animals are said to be adaptable and quick to change. The Vicar of Bray orthodoxy survived for almost a century; it still survives in biology textbooks. The precise moment when it was first questioned is hard to pin down for sure. There were doubts as far back as the 1920s. Only gradually did it dawn on modern biologists that the Weismann logic was profoundly flawed. It seems to treat evolution as some kind of imperative, as if evolving were what species exist to do—as if evolving were a goal imposed on existence.[7]

This is, of course, nonsense. Evolution is something that happens to organisms. It is a directionless process that sometimes makes an animal's descendants more complicated, sometimes simpler, and sometimes changes them not at all. We are so steeped in notions of progress and self-improvement that we find it strangely hard to accept this. But nobody has told the coelacanth, a fish that lives off Madagascar and looks exactly like its ancestors of 300 million years ago, that it has broken some law by not "evolving." The notion that evolution simply cannot go fast enough, and its corollary that a coelacanth is a failure because it did not become a human being, is easily refuted. As Darwin noticed, mankind has intervened dramatically to speed up evolution, producing hundreds of breeds of dogs, from chihuahuas to St. Bernards, in an evolutionary eye blink. That alone is evidence that evolution was not going as fast as it could. Indeed, the coelacanth, far from being a flop, is rather a success. It has stayed the same—a design that persists without innovation, like a Volkswagen beetle. Evolving is not a goal but a means to solving a problem.

Nonetheless, Weismann's followers, and especially Sir Ronald Fisher and Hermann Muller, could escape the teleology trap by arguing that evolution, if not preordained, was at least essential. Asexual species were at a disadvantage and would fail in competition with sexual species. By incorporating the concept of the gene into Weismann's argument, Fisher's book in 1930[8] and Muller's in 1932[9] laid out a seemingly watertight argument for the advantages of sex, and Muller even went as far as to declare the problem emphatically solved by the new science of genetics. Sexual species shared their newly invented genes among all individuals; asexual ones did not. So sexual species were like groups of inventors pooling their resources. If one man invented a steam engine and another a railway, then the two could come together. Asexual ones behaved like groups of jealous inventors who never shared their knowledge, so that steam locomotives were used on roads and horses dragged carts along railways.

In 1965, James Crow and Motoo Kimura modernized the Fisher-Muller logic by demonstrating with mathematical models how rare mutations could come together in sexual species but not in asexual ones. The sexual species does not have to wait for two rare events in the same individual but can combine them from different individuals. This, they said, would grant the sexual species an advantage over the asexual ones as long as there were at least one thousand individuals in the sexual ones. All was hunky-dory. Sex was explained, as an aid to evolution, and modern mathematics was adding new precision. The case could be considered closed.[10]

MANKIND'S GREATEST RIVAL IS MANKIND

So it might have remained were it not for a voluminous and influential publication by a Scottish biologist named V. C. Wynne Edwards that had appeared a few years before, in 1962. Wynne Edwards did biology an enormous service by exposing a gigantic fallacy that had systematically infected the very heart of evolutionary theory since Darwin's day. He exposed the fallacy not to demolish it but because he believed it to be true and important. But in so doing he made it explicit for the first time.[11]

The fallacy persists in the way many laymen speak of evolution. We talk blithely among ourselves about evolution being a question of the "survival of the species." We imply that species compete with each other, that Darwin's "struggle for existence" is between dinosaurs and mammals, or between rabbits and foxes, or between men and neanderthals. We borrow the imagery of nation-states and football teams: Germany against France, the home team against its rivals.

Charles Darwin, too, slipped occasionally into this way of thinking. The very subtitle of *On the Origin of Species* refers to the "preservation of favored races."[12] But his main focus was on the individual, not the species. Every creature differs from every other; some survive or thrive more readily than others and leave more young behind; if those changes are heritable, gradual change is inevitable. Darwin's ideas were later fused with the discoveries of Gregor Mendel, who had proved that heritable features came in discrete packages, which became known as genes, forming a theory that was able to explain how new mutations in genes spread through a whole species.

But there lay buried beneath this theory an unexamined dichotomy. When the fittest are struggling to survive, with whom are they competing? With other members of their species or with members of other species?

A gazelle on the African savanna is trying not to be eaten by cheetahs, but it is also trying to outrun other gazelles when a cheetah attacks. What matters to the gazelle is being faster than other gazelles, not being faster than cheetahs. (There is an old story of a philosopher who runs when a bear charges him and his friend. "It's no good, you'll never outrun a bear," says the logical friend. "I don't have to," replies the philosopher. "I only have to outrun you.") In the same way, psychologists sometimes wonder why people are endowed with the ability to learn the part of Hamlet or understand calculus when neither skill was of much use to mankind in the primitive conditions where his intellect was shaped. Einstein would probably have been as hopeless as anybody in working out how to catch a woolly rhinoceros. Nicholas Humphrey, a Cambridge psychologist, was the first to see clearly the solution to this puzzle. We use our intellects not to solve practical problems but to outwit each other. Deceiving people, detecting deceit, understanding people's motives, manipulating people—these are what the intellect is used for. So what matters is not how clever and crafty you are but how much more clever and craftier you are than other people. The value of intellect is infinite. Selection within the species is always going to be more important than selection between the species.[13]

Now this may seem a false dichotomy. After all, the best thing an individual animal can do for its species is to survive and breed. Often, however, the two imperatives will be in conflict. Suppose the individual is a tigress whose territory has recently been invaded by another tigress. Does she welcome the intruder and discuss how best they can cohabit the territory, sharing prey? No, she fights her to the death, which from the point of view of the species is unhelpful. Or suppose the individual is an eaglet of a rare species anxiously watched by conservationists in its nest. Eaglets often kill their younger brothers and sisters in the nest. Good for the individual, bad for the species.

Throughout the world of animals, individuals are fighting individuals, whether of the same species or of another. And indeed, the closest competitor a creature is ever likely to meet is a member of its own species. Natural selection is not going to pick genes that help gazelles survive as a species but hurt the chances of individuals—because such genes will be wiped out long before they can show their benefits. Species are not fighting species as nations battle other nations.

Wynne Edwards believed fervently that animals often did things for the species, or at least for the group in which they lived. For example, he thought that seabirds chose not to breed when their numbers were high in order to prevent too much pressure on the food supply. The result of Wynne Edwards's book was that two factions formed: the group selectionists, who argued that much of animal behavior was informed by the interests of the group, not the individual, and the individual selectionists, who argued that individual interests always triumphed. The group selectionist argument is inherently appealing—we are immersed in the ethic of team spirit and charity. It also seemed to explain animal altruism. Bees die as they sting, trying to save the hive; birds warn each other of predators or help to feed their young siblings; even human beings are prepared to die in acts of selfless heroism to save others' lives. But as we shall see, the appearance is misleading. Animal altruism is a myth. Even in the most spectacular cases of selflessness it turns out that animals are serving the selfish interests of their own genes—if sometimes being careless with their bodies.

THE REDISCOVERY OF THE INDIVIDUAL

If you attend a meeting of evolutionary biologists somewhere in America, you might be lucky and spot a tall, gray-whiskered, smiling man bearing a striking resemblance to Abraham Lincoln, standing rather diffidently at the back of the crowd. He will probably be surrounded by a knot of admirers, hanging on his every word—for he is a man of few words. A whisper will go around the room: "George is here." You will sense from people's reactions the presence of greatness.

The man in question is George Williams, who has been a quiet, bookish professor of biology at the State University of New York at Stony Brook on Long Island for most of his career. He has done no memorable experiments and has made no startling discovery. Yet he is the progenitor of a revolution in evolutionary biology almost as profound as Darwin's. In 1966, irritated by Wynne Edwards and other exponents of group selection, he spent a summer vacation writing a book about how he thought evolution worked. Called *Adaptation and Natural Selection*, that book still towers over biology like a Himalayan peak. It did for biology what Adam Smith had done for economics: It explained how collective effects could flow from the actions of self-interested individuals.[14]

In the book Williams exposed the logical flaws in group selection with unanswerable simplicity. The few evolutionists who had stuck to individual selection all along, such as Ronald Fisher, J. B. S. Haldane, and Sewall Wright, were vindicated.[15] The ones who had confused species and individual, such as Julian Huxley, were eclipsed.[16] Within a few years of Williams's book, Wynne Edwards was effectively defeated, and almost all biologists agreed that no creature could ever evolve the ability to help its species at the expense of itself. Only when the two interests coincided would it act selflessly.

This was disturbing. It seemed at first to be a very cruel and heartless conclusion to reach, particularly in a decade when economists were tentatively celebrating the discovery that the ideal of helping society could persuade people to pay high taxes to support welfare. Society, they said, need not be based on tempering the greed of individuals but on appealing to their better natures. And here were biologists coming to exactly the opposite conclusion about animals, depicting a harsh world in which no animal ever sacrificed its own ambition to the need of the team or the group. Crocodiles would eat one another's babies even on the brink of extinction.

Yet that was not what Williams said. He knew full well that individual animals often cooperate and that human society is not a ruthless free-for-all. But he also saw that cooperation is nearly always between close relatives—mothers and children, sister worker bees—or that it is practiced where it directly or eventually benefits the individual. The exceptions are few indeed. This is because where selfishness brings higher rewards than altruism, selfish individuals leave more descendants, so altruists inevitably become extinct. But where altruists help their relatives, they are helping those who share some of their genes, including whatever genes had caused them to be altruistic. So without any conscious intention on the part of individuals, such genes spread.[17]

But Williams realized that there was one troubling exception to this pattern: sex. The traditional explanation for sex, the Vicar of Bray theory, was essentially group selectionist. It demanded that an individual altruistically share its genes with those of another individual when breeding because if it did not, the species would not innovate and would, a few hundred thousand years later, be outcompeted by other species that did. Sexual species, it said, were better off than asexual species.

But were sexual *individuals* better off than asexual ones? If not, sex could not be explained by the Williams "selfish" school of thought. Therefore, either there was something wrong with the selfish theories and true altruism could indeed emerge, or the traditional explanation of sex was wrong. And the more Williams and his allies looked, the less sense sex seemed to make for the individual as opposed to the species.

Michael Ghiselin of the California Academy of Sciences in San Francisco was at the time engaged in a study of Darwin's work and was struck by Darwin's own insistence on the primacy of the struggle between individuals rather than the struggle between groups. But Ghiselin, too, began thinking about how sex seemed such an exception to this. He posed the following question: How could a gene for sexual reproduction spread at the expense of an asexual gene? Suppose all members of a species were asexual but one day one pair of them invented sex. What benefit would it bring? And if it brought no benefit, why would it spread? And if it could not spread, why were so many species sexual? Ghiselin could not see how the new sexual individuals could possibly leave behind more offspring than the old asexual ones. Indeed, surely they would leave fewer because, unlike their rivals, they had to waste time finding each other, and one of them, the male, would not produce babies at all.[18]

John Maynard Smith, an engineer-turned-geneticist at the University of Sussex in England, with a penetrating and somewhat playful mind that had been trained by the great neo-Darwinist J. B. S. Haldane, answered Ghiselin's question without solving his dilemma. He said that a sexual gene could spread only if it doubled the number of offspring an individual could have, which seemed absurd. Suppose, he said, turning Ghiselin's thought around, that in a sexual species one day a creature decides to forgo sex and put all of its genes into its own offspring, taking none from its mate. It would then have passed twice as many genes on to the next generation as its rivals had. Surely it would be at a huge advantage. It would contribute twice as much to the next generation and would soon be left in sole possession of the genetic patrimony of the species.[19]

Imagine a Stone Age cave inhabited by two men and two

women, one of them a virgin. One day the virgin gives birth "asexually" to a baby girl that is essentially her identical twin. (She becomes, in the jargon, a "parthenogen.") It could happen in several ways—for example, by a process called "automixis," in which an egg is, roughly speaking, fertilized by another egg. The cave woman has another daughter two years later by the same means. Her sister, meanwhile, has had a son and a daughter by the normal method. There are now eight people in the cave. Next, the three young girls each have two children and the first generation dies off. Now there are ten people in the cave, but five of them are parthenogens. In two generations the gene for parthenogenesis has spread from one-quarter to one-half of the population. It will not be long before men are extinct.

This is what Williams called the cost of meiosis and Maynard Smith called the cost of males. For what dooms the sexual cave people is simply that half of them are men, and men do not produce babies. It is true that men do occasionally help in child rearing, killing woolly rhinos for dinner or whatever, but even that does not really explain why men are necessary. Suppose that the asexual women at first gave birth only when they had intercourse. Again there are precedents. There are grasses that only set seed when fertilized by pollen from a related species, but the seed inherits no genes from the pollen. It is called "pseudogamy."[20] In this case the men in the cave would have no idea that they are being genetically excluded and would treat the asexual babies as their own, serving woolly rhino meat to them just as they would to their own children.

This thought-experiment illustrates the numerically huge advantage a gene that makes its owner asexual has. Logic such as this set Maynard Smith, Ghiselin, and Williams to wondering what compensating advantage of sex there must be, given that every mammal and bird, most invertebrate animals, most plants and fungi, and many protozoa are sexual.

For those who think that to talk about the "cost of sex" is merely to illustrate how absurdly pecuniary we have become, and who reject the whole logic of this argument as specious, I offer the following challenge. Explain hummingbirds—not how they work but why they exist at all. If sex had no cost, hummingbirds would not exist. Hummingbirds eat nectar, which is produced by flowers to lure pollinating insects and birds. Nectar is a pure gift by the plant of its hard-won sugar to the hummingbird, a gift given only because the hummingbird will then carry pollen to another plant. To have sex with another plant, the first plant must bribe the pollen carrier with nectar. Nectar is therefore a pure, unadulterated cost incurred by the plant in its quest for sex. If sex had no cost, there would be no hummingbirds.[21]

Williams was inclined to conclude that perhaps his logic was good, but for animals like us the practical problems were simply insurmountable. In other words, getting from being sexual to being asexual would indeed confer advantages, but it would be just too difficult to achieve. About this time sociobiologists were beginning to fall into a trap of being too readily enamored of "adaptationist" arguments—just-so stories, as Stephen Jay Gould of Harvard called them. Sometimes, he pointed out, things were the way they were for accidental reasons. Gould's own example is of the triangular space between two cathedral arches at right angles, known as a spandrel, which has no function but is simply the by-product of putting a dome on four arches. The spandrels between the arches on St. Mark's Basilica in Venice were not there because somebody wanted spandrels. They were there because there is no way to put two arches next to each other without producing a space in between. The human chin may be such a spandrel; it has no function but is the inevitable result of having jaws. Likewise the fact that blood is red is surely a photochemical accident, not a design feature. Perhaps sex was a spandrel, an evolutionary relic of a time when it served a purpose. Like chins or little toes or appendixes, it no longer served a purpose but was not easily got rid of.[22]

Yet this argument for sex is pretty unconvincing because quite a few animals and plants have abandoned sex or have it only occasionally. Take the average lawn. The grass in it never has sex—unless you forget to cut it, at which point it grows flower heads. And what about water fleas? For many generations in a row water fleas are asexual: They are all female, they give birth to other females, they never mate. Then as the pond fills up with water fleas, some start to give birth to males, which mate with other females to produce "winter" eggs that lie on the bottom of the pond and regenerate when the pond is flooded again. Water fleas can turn sex on and off again, which seems to prove that it has some immediate purpose beyond helping evolution to happen. It is worth an individual water flea's while to have sex at least in certain seasons.

So we are left with an enigma. Sex serves the species but at the expense of the individual. Individuals could abandon sex and rapidly outcompete their sexual rivals. But they do not. Sex must therefore in some mysterious manner "pay its way" for the individual as well as for the species. How?

PROVOCATION BY IGNORANCE

Until the mid-1970s the debate that Williams had started remained an arcane and obscure one. And the protagonists sounded fairly confident in their attempts to resolve the dilemma. But in the mid-1970s two crucial books changed that forever by throwing down a gauntlet that other biologists could not resist picking up. One book was by Williams himself, the other by Maynard Smith.[23] "There is a kind of crisis at hand in evolutionary biology," wrote Williams melodramatically. But whereas Williams's book, *Sex and Evolution*, was an ingenious account of several possible theories of sex—an attempt to defuse the crisis—Maynard Smith's book, *The Evolution of Sex*, was very different. It was a counsel of despair and bafflement. Again and again Maynard Smith came back to the enormous price of sex: the twofold disadvantage—two parthenogenetic virgins can have twice as many babies as one woman and one man. Again and again he declared it insurmountable by current theories. "I fear the reader may find these models insubstantial and unsatisfactory," he wrote. "But they are the best we have." And in a separate paper: "One is left with the feeling that some essential feature of the situation is being overlooked."[24] By insisting that the problem was emphatically not solved, Maynard Smith's book had an electrifying impact. It was an unusually humble and honest gesture.

Attempts to explain sex have since proliferated like libidinous rabbits. They present an unusual spectacle to the observer of science. Most of the time scientists are groping around in a barrel of ignorance trying to find a fact or a theory or to discern a pattern where none had been seen before. But this was a rather different game. The fact—sex—was well known. To explain it—to give sex an advantage—was not sufficient. The proffered explanation had to be better than others. It is like the gazelle running faster than other gazelles rather than running faster than cheetahs. Theories of sex are a dime a dozen, and most are "right" in the sense of making logical sense. But which is most right?[25]

In the pages that follow you will meet three kinds of scientists. The first is a molecular biologist, muttering about enzymes and exonucleolytic degradation. He wants to know what happens to the DNA of which genes are made. His conviction is that sex is all about repairing DNA or some such molecular engineering. He does not understand equations, but he loves long words, usually ones he and his colleagues have invented. The second is a geneticist, all mutations and Mendelism. He will be obsessed with describing what happens to genes during sex. He will demand experiments, such as depriving organisms of sex for many generations to see what happens. Unless you stop him, he will start writing equations and talking of "linkage disequilibria." The third is an ecologist, all parasites and polyploidy. He loves comparative evidence: which species has sex and which does not. He knows a plethora of extraneous facts about the arctic and the tropics. His thinking is a little less rigorous than others, his language a little more colorful. His natural habitat is the graph, his occupation the computer simulation.

Each of these characters champions a type of explanation for sex. The molecular biologist is essentially talking about why sex was invented, which is not necessarily the same question as what sex achieves today, the question the geneticist prefers to address. The ecologist, meanwhile, is asking a slightly different question:

Under what circumstances is sex better than asex? An analogy might be the reasons for the invention of computers. The historian (like the molecular biologist) will insist they were invented to crack the codes used by German submarine commanders. But they are not used for that today. They are used to do repetitive tasks more efficiently and quickly than people can (the geneticist's answer). The ecologist is interested in why computers have replaced telephone operators but not, say, cooks. All three may be "right" on different levels.

THE MASTER-COPY THEORY

The leader of the molecular biologists is Harris Bernstein of the University of Arizona. His argument is that sex was invented to repair genes. The first hint of this was the discovery that mutant fruit flies unable to repair genes are unable to "recombine" them, either. Recombination is the essential procedure in sex, the mixing of genes from the two grandparents of the sperm or egg. Knock out genetic repair, and sex stops, too.

Bernstein noticed that the tools the cell uses for sex are the same as it uses to repair genes. But he has been unable to convince the geneticists or the ecologists that repair is more than the original, long superseded purpose of the machinery sex uses. The geneticists say the machinery of sex did indeed evolve from the machinery of gene repair, but that is not the same thing as saying sex exists today to repair genes. After all, human legs are the descendants of fishes' fins, but they are designed nowadays for walking, not swimming.[26]

A quick digression into molecules is necessary here. DNA, the stuff of genes, is a long, thin molecule that carries information in a simple alphabet of four chemical "bases," like Morse code with two kinds of dots and two kinds of dashes. Call these bases "letters": A, C, G, and T. The beauty of DNA is that each letter is complementary to another, meaning that it prefers to align itself opposite that other letter. Thus A pairs with T and vice versa, C with G and vice versa. This means there is an automatic way of copying DNA: by going along the strand of the molecule, stitching together another from the complementary letters. The sequence AAGTTC becomes, on the complementary strand, TTCAAG; copy that and you get the original sequence back again. Every gene normally consists of a strand of DNA and its complementary copy closely entwined in the famous double helix. Special enzymes move up and down the strands, and where they find a break, repair it by reference to the complementary strand. DNA is continually being damaged by sunlight and chemicals. If it were not for the repair enzymes, it would quite quickly become meaningless gobbledygook.

But what happens when both strands are damaged at the same place? This can be quite common—for example, when the two strands get fused together like a spot of glue on a closed zipper. The repair enzymes have no way of knowing what to repair the DNA to. They need a template of what the gene used to look like. Sex provides it. It introduces a copy of the same gene from another creature (outcrossing) or from another chromosome (recombination) in the same creature. Repair can now refer to a fresh template.

Of course, the fresh template may also be damaged at the same place, but the chances of that are small. A shopkeeper adding up a list of prices makes sure he has it right the first time by simply repeating the task. His reasoning is that he is unlikely to make the same mistake twice.

The repair theory is supported by some good circumstantial evidence. For example, if you expose a creature to damaging ultraviolet light, it generally fares better if it is capable of recombination than if it is not, and it fares better still if it has two chromosomes in its cells. If a mutant strain appears that eschews recombination, it proves to be especially susceptible to damage by ultraviolet light. Moreover, Bernstein can explain details that his rivals cannot—for example, the curious fact that just before dividing its chromosome pairs in two to make an egg, a cell will double the number and then dispose of three-quarters of the proceeds. In the repair theory, this is to find, and convert to a "common currency," the errors that are to be repaired.[27]

Nonetheless, the repair theory remains inadequate to the task it has set itself. It is silent on outcrossing. Indeed, if sex is about getting spare copies of genes, it would be better to get them from relatives rather than seek out unrelated members of the species. Bernstein says outcrossing is a way of masking mutations, but this amounts to no more than a restatement of the reason why inbreeding is a bad thing; and sex is the cause of inbreeding, not the consequence.

Moreover, every argument that the repair people give for recombination is merely an argument for keeping backup copies of genes, and there is a far simpler way of doing that than swapping them at random between chromosomes. It is called "diploidy."[28] An egg or a sperm is "haploid"—it has one copy of each gene. A bacterium or a primitive plant, such as moss, is the same. But most plants and nearly all animals are diploid, meaning they have two copies of every gene, one from each parent. A few creatures, especially plants that are descended from natural hybrids or have been selected by man for large size, are "polyploid." Most hybrid wheat, for example, is "hexaploid"; it has six copies of each gene. In yams, female plants are "octoploid" or hexaploid, males all "tetraploid"— a discrepancy that renders yams sterile. Even some strains of rainbow trout and domestic chicken are "triploid"—plus a single parrot that turned up a few years ago.[29] Ecologists have begun to suspect that polyploidy in plants is a sort of alternative to sex. At high altitudes and high latitudes many plants seem to abandon sex in favor of asexual polyploidy.[30]

But by mentioning ecologists we are getting ahead of ourselves. The point at issue is gene repair. If diploid creatures were to indulge in a little recombination between chromosomes every time their cells divided as the body grew, there would be plenty of opportunity for repair. But they do not. They recombine their genes only at the final peculiar division called meiosis that leads to the formation of an egg or a sperm. Bernstein has an answer for this. He says that there is another, more economical way to repair damage to genes during ordinary cell division, which is to allow the fittest cells to survive. There is no need for repair at that stage because the undamaged cells will soon outgrow the damaged ones. Only when producing germ cells, which go out to face the world alone, need you check for errors.[31]

The verdict on Bernstein: unproven. Certainly the tools of sex seem to be derived from the tools of repair, and certainly recombination achieves some gene repair. But is it the purpose of sex? Probably not.

CAMERAS AND RATCHETS

The geneticists, too, are obsessed with damaged DNA. But whereas the molecular biologists concentrate on the damage that is repaired, the geneticists talk about the damage that cannot be repaired. They call this "mutation."

Scientists used to think of mutations as rare events. But in recent years they have gradually come to realize how many mutations happen. They are accumulated at the rate of about one hundred per genome per generation in mammals. That is, your children will have one hundred differences from you and your spouse in their genes as a result of random copying errors by your enzymes or as a result of mutations in your ovaries or testicles caused by cosmic rays. Of those one hundred, about ninety-nine will not matter: they will be so-called silent or neutral mutations that do not affect the sense of genes. That may not seem many, given that you have seventy-five thousand pairs of genes and that many of the changes will be tiny and harmless or will happen in silent DNA between genes. But it is enough to lead to a steady accumulation of defects and, of course, a steady rate of invention of new ideas.[32]

The received wisdom on mutations is that most of them are bad news and a good proportion kill their owners or inheritors (cancer starts as one or more mutations), but that occasionally among the bad there is a good mutation, a genuine improvement. The sickle cell anemia mutation, for example, can be fatal to those who have two copies of it, but the mutation has actually increased in some parts of Africa because it gives immunity to malaria.

For many years geneticists concentrated on good mutations and viewed sex as a way of distributing them among the population, like the "cross-fertilization" of good ideas in universities and industries. Just as technology needs "sex" to bring in innovations from outside, so an animal or plant that relies on only its own inventions will be slow to innovate. The solution is to beg, borrow, or steal the inventions of other animals and plants, to get hold of their genes in the way that companies copy one another's inventions. Plant breeders who try to combine high yield, short stems, and disease resistance in rice plants are acting like manufacturers with access to many different inventors. Breeders of asexual plants must wait for the inventions to accumulate slowly within the same lineage. One of the reasons the common mushroom has changed very little over the three centuries that it has been in cultivation is that mushrooms are asexual, and so no selective breeding has been possible.[33]

The most obvious reason to borrow genes is to benefit from the ingenuity of others as well as yourself. Sex brings together mutations, constantly rearranging genes into new combinations until fortuitous synergy results. One ancestor of a giraffe, for example, might have invented a longer neck while another invented longer legs. The two together were better than either alone.

But this argument confuses consequence with cause. Its advantages are far too remote; they will appear after a few generations, by which time any asexual competitor will long ago have outpopulated its sexual rivals. Besides, if sex is good at throwing together good combinations of genes, it will be even better at breaking them up. The one thing you can be sure about sexual creatures is that their offspring will be different from them, as many a Caesar, Bourbon, and Plantagenet discovered to their disappointment. Plant breeders much prefer varieties of wheat or corn that are male-sterile and produce seeds without sex because it enables them to be sure their good varieties will breed true.

It is almost the definition of sex that it breaks up combinations of genes. The great cry of the geneticists is that sex reduces "linkage disequilibria." What they mean is that if it were not for recombination, genes that are linked together—such as those for blue eyes and blond hair—would always be linked together, and nobody would ever have blue eyes and brown hair, or blond hair and brown eyes. Thanks to sex, the moment the fabled synergy is found, it is lost again. Sex disobeys that great injunction: "If it ain't broke, don't fix it." Sex increases randomness.[34]

In the late 1980s there was one last revival of interest in theories of "good" mutation. Mark Kirkpatrick and Cheryl Jenkins were interested not in two separate inventions but in the ability to invent the same thing twice. Suppose, for example, that blue eyes double fertility, so that people with blue eyes have twice as many children as people with brown eyes. And suppose that at first everybody has brown eyes. The first mutation in a brown-eyed person to blue eyes will have no effect because blue eyes are a recessive gene, and the dominant brown-eye gene on the person's other chromosome will mask it. Only when the blue-eye genes of two of the descendants of the original mutant person come together will the great benefit of blue eyes be seen. Only sex would allow the people to mate and the genes to meet. This so-called segregation theory of sex is logical and uncontroversial. It is indeed one of the advantageous consequences of sex. Unfortunately, it is far too weak an effect to be the main explanation for sex's prevalence. Mathematical models reveal that it would take five thousand generations to do its good work and asex would long since have won the game.[35]

In recent years the geneticists have turned away from good mutations and begun to think about bad ones. Sex, they suggest, is a way of getting rid of bad mutations. This idea also has its origins in the 1960s, with Hermann Muller, one of the fathers of the Vicar of Bray theory Muller, who spent much of his career at the University of Indiana, published his first scientific paper on genes in 1911, and a veritable flood of ideas and experiments followed in the succeeding decades. In 1964 he had one of his greatest insights; it has come to be known as "Muller's ratchet." A simplified example of it goes like this: There are ten water fleas in a tank, only one of which is entirely free of mutations; the others all have one or several minor defects. On average only five of the water fleas in each generation manage to breed before they are eaten by a fish. The defect-free flea has a one-in-two chance of not breeding. So does the flea with the most defects, of course, but there is a difference: Once the defect-free flea is dead, the only way for it to be re-created is for another mutation to correct the mutation in a flea with a defect—a very unlikely possibility. The one with two defects can be re-created easily by a single mutation in a water flea with one defect anywhere among its genes. In other words, the random loss of certain lines of descent will mean that the average number of defects gradually increases. Just as a ratchet turns easily one way but cannot turn back, so genetic defects inevitably accumulate. The only way to prevent the ratchet from turning is for the perfect flea to have sex and pass its defect-free genes to other fleas before it dies.[16]

Muller's ratchet applies if you use a photocopier to make a copy of a copy of a copy of a document. With each successive copy the quality deteriorates. Only if you guard the unblemished original can you regenerate a clean copy. But suppose the original is stored with the copies in a file and more copies are made when there is only one left in the file. You are just as likely to send out the original as to send out a copy. Once the original is lost, the best copy you can make is less good than it was before. But you can always make a worse copy just by copying the worst copy you have.

Graham Bell of McGill University has disinterred a curious debate that raged among biologists at the turn of the century about whether sex had a rejuvenating effect. What intrigued these early biologists was if and why a population of protozoa kept in a tank with sufficient food but given no chance to have sex inevitably fell into a gradual decline in vigor, size, and rate of (asexual) reproduction. Reanalyzing the experiments, Bell found some clear examples of Muller's ratchet at work. Bad mutations gradually accumulated in the protozoa deprived of sex. The process was accelerated by the habit of this one group of protozoa, the ciliates, of keeping its germ-line genes in one place and keeping copies of them elsewhere for everyday use. The method of reproducing the copies is hasty and inaccurate, so defects accumulate especially fast there. During sex, one of the things the creatures do is throw away their copies and create new ones from the germ-line originals. Bell compares it with a chair maker who copies the last chair he made, errors and all, and returns to his original design only occasionally. Sex therefore does indeed have a rejuvenating effect. It enables these little animals to drop all the accumulated errors of an especially fast asexual ratchet whenever they have sex.[37]

Bell's conclusion was a curious one. If a population is small (less than 10 billion) or the number of genes in the creature is very large, the ratchet has a severe effect on an asexual lineage. This is because it is easier to lose the defect-free class in a smaller population. So those creatures with larger genomes and relatively smaller populations (10 billion is twice as many people as there are on Earth) will be ratcheted into trouble fairly quickly. But those with few genes and vast populations are all right. Bell reckons that being sexual was a prerequisite for being big (and therefore few), or, conversely, sex is unnecessary if you stay small.[38]

Bell calculated the amount of sex—or, rather, of recombination—that is needed to halt the ratchet; for smaller creatures, less sex is necessary. Water fleas need to have sex only once every several generations. Human beings need to have sex in every generation. Moreover, as James Crow at the University of Wisconsin in Madison has suggested, Muller's ratchet may explain why budding is a relatively rare way of reproducing—especially among animals. Most asexual species still go to the trouble of growing their offspring from single cells (eggs). Why? Crow suggests it is because defects that would be fatal in a single cell can be easily smuggled into a bud.[39]

If the ratchet is a problem only for big creatures, why do so many small ones have sex? Besides, to halt the ratchet requires only occasional episodes of sex; it does not require so many animals to abandon asexual reproduction altogether. Aware of these difficulties, in 1982 Alexey Kondrashov of the Research Computer Center in Poschino, near Moscow, came up with a theory that is a sort of reverse Muller's ratchet. He argued that in an asexual population, every time a creature dies because of a mutation it gets rid of that mutation but no more. In a sexual population some of the creatures born have lots of mutations and some have few. If the ones with lots of mutations die, then sex keeps throwing the ratchet into reverse, purging mutations. Since most mutations are harmful, this gives sex a great advantage.[40]

But why purge mutations in this way rather than correct more of them by better proofreading? Kondrashov has an ingenious explanation of why this makes sense. The cost of making proofreading mechanisms perfect gets rapidly higher as you get nearer to perfection; in other words, it is like the law of diminishing returns. Allowing some mistakes through but having sex to purge them out may be cheaper.

Matthew Meselson, a distinguished molecular biologist, has come up with another explanation that expands on Kondrashov's idea. Meselson suggests that "ordinary" mutations that change one letter for another in the genetic code are fairly innocuous because they can be repaired, but insertions—whole chunks of DNA that jump into the middle of genes—cannot be reversed so easily. These "selfish" insertions tend to spread like an infection, but sex defeats them, since sex segregates them into certain individuals whose deaths purge them from the population.[41]

Kondrashov is prepared to stand by an empirical test of his idea. He says that if the rate of deleterious mutations turns out to be more than one per individual per generation, then he is happy; if it proves to be less than one, then his idea is in trouble. The evidence so far is that the deleterious mutation rate teeters on the edge: It is about one per individual per generation in most creatures. But even supposing it is high enough, all that proves is that sex can perhaps play a role in purging mutations. It does not say that is why sex persists.[42]

Meanwhile, there are defects in the theory. It fails to explain how bacteria—of which some species rarely have sex and others not at all—nonetheless suffer from mutation at a low rate and make fewer proofreading mistakes when copying DNA. As one of Kondrashov's critics put it, sex is "a cumbersome strange tool to have evolved for a housekeeping role."[43]

And Kondrashov's theory suffers from the same flaw as all genetic-repair theories and the Vicar of Bray himself: It works too slowly. Pitted against a clone of asexual individuals, a sexual population must inevitably be driven extinct by the clone's greater productivity unless the clone's genetic drawbacks can appear in time. It is a race against time. For how long? Curtis Lively of the University of Indiana has calculated that for every tenfold increase in population size, the advantage of sex is granted six more generations to show its effects or sex will lose the game. If there are a million individuals, sex has forty generations before it goes extinct; if a billion, it has eighty. Yet the genetic repair theories all require thousands of generations to do their work. Kondrashov's is certainly the fastest theory, but it is probably not fast enough.[44]

There is still no purely genetic theory to explain sex that attracts wide support. An increasing number of students of evolution believe that the solution to the great enigma of sex lies in ecology, not genetics.

Chapter 3

THE POWER OF PARASITES

The chessboard is the world; the pieces are the phenomena of the universe; the rules of the game are what we call the laws of Nature. The player on the other side is hidden from us. We know that his play is always fair, just, and patient. But also we know, to our cost, that he never overlooks a mistake or makes the smallest allowance for ignorance.

—Thomas Henry Huxley

Even for microscopic animals, the bdelloid rotifers are peculiar. They live in any kind of fresh water, from puddles in your gutter to hot springs by the Dead Sea and ephemeral ponds on the Antarctic continent. They look like animated commas driven by what appear to be small waterwheels at the front of the body, and when their watery home dries up or freezes, they adopt the shape of an apostrophe and go to sleep. This apostrophe is known as a "tun," and it is astonishingly resistant to abuse. You can boil it for an hour or freeze it to within 1 degree of absolute zero—that is, to −272 degrees Centigrade—for a whole hour. Not only does it fail to disintegrate, it does not even die. Tuns blow about the globe as dust so easily that rotifers are thought to travel regularly between Africa and America. Once thawed out, the tun quickly turns back into a rotifer, paddles its way about the pond with its bow wheels, eating bacteria as it goes, and within a few hours starts producing eggs that hatch into other rotifers. A bdelloid rotifer can fill a medium-sized lake with its progeny in just two months.

But there is another odd thing about bdelloids besides their feats of endurance and fecundity. No male bdelloid rotifer has ever been seen. As far as biologists can tell, every single member of every one of all five hundred species of bdelloid in the world is a female. Sex is simply not in the bdelloid repertoire.

It is possible that bdelloid rotifers mix others' genes with their own by eating their dead comrades and absorbing some of their genes, or something bizarre like that,[1] but recent research by Matthew Meselson and David Welch suggests that they just never do have sex. They have found that the same gene in two different individuals can be up to 30 percent different at points that do not affect its function—a level of difference that implies bdelloids gave up sex between 40 million and 80 million years ago.[2]

There are many other species in the world that never have sex, from dandelions and lizards to bacteria and amoebas, but the bdelloids are the only example of a whole order of animal that entirely lacks the sexual habit. Perhaps as a result the bdelloids all look rather alike, whereas their relatives, the monogonont rotifers, tend to be much more varied; they cover the whole range of shapes of punctuation marks. Nonetheless, the bdelloids are a living rebuke to the conventional wisdom of biology textbooks—that without sex, evolution can barely happen and species cannot adapt to change. The existence of the bdelloid rotifers is, in the words of John Maynard Smith, "an evolutionary scandal."[3]

THE ART OF BEING SLIGHTLY DIFFERENT

Unless a genetic mistake happens, a baby bdelloid rotifer is identical to its mother. A human baby is not identical to its mother. That is the first consequence of sex. Indeed, according to most ecologists, it is the purpose of sex.

In 1966, George Williams exposed the logical flaw at the heart of the textbook explanation of sex. He showed how it required animals to ignore short-term self-interest in order to further the survival and evolution of their species, a form of self-restraint that could have evolved only under very peculiar circumstances. He was very unsure what to put in its place. But he noticed that sex and dispersal often seem to be linked. Thus, grass grows asexual runners to propagate locally but commits its sexually produced seeds to the wind to travel farther. Sexual aphids grow wings; asexual ones do not. The suggestion that immediately follows is that if your young are going to have to travel abroad, then it is better that they vary because abroad may not be like home.[4]

Elaborating on that idea was the main activity throughout

the 1970s of ecologists interested in sex. In 1971, in his first attack on the problem, John Maynard Smith suggested that sex was needed for those cases in which two different creatures migrate into a new habitat in which it helps to combine both their characters.[5] Two years later Williams returned to the fray and suggested that if most of the young are going to die, as most who try their luck as travelers will, then it may be the very fittest ones that will survive. It therefore matters not one bit how many young of average quality a creature has. What counts is having a handful of young that are exceptional. If you want your son to become pope, the best way to achieve this is not to have lots of identical sons but to have lots of different sons in the hope that one is good, clever, and religious enough.[6]

The common analogy for what Williams was describing is a lottery. Breeding asexually is like having lots of lottery tickets all with the same number. To stand a chance of winning the lottery, you need lots of different tickets. Therefore, sex is useful to the individual rather than the species when the offspring are likely to face changed or unusual conditions.

Williams was especially intrigued by creatures such as aphids and monogonont rotifers, which have sex only once every few generations. Aphids multiply during the summer on a rosebush, and monogonont rotifers multiply in a street puddle. But when the summer comes to an end, the last generation of aphids or of monogonont rotifers is entirely sexual: It produces males and females that seek each other out, mate, and produce tough little young that spend the winter or the drought as hardened cysts awaiting the return of better conditions. To Williams this looked like the operation of his lottery. While conditions were favorable and predictable, it paid to reproduce as fast as possible—asexually. When the little world came to an end and the next generation of aphid or rotifer faced the uncertainty of finding a new home or waited for the old one to reappear, then it paid to produce a variety of different young in the hope that one would prove ideal.

Williams contrasted the "aphid-rotifer model" with two others: the strawberry-coral model and the elm-oyster model.

Strawberry plants and the animals that build coral reefs sit in the same place all their lives, but they send out runners or coral branches so that the individual and its clones gradually spread over the surrounding space. However, when they want to send their young much farther away, in search of a new, pristine habitat, the strawberries produce sexual seeds and the corals produce sexual larvae called "planulae." The seeds are carried away by birds; the planulae drift for many days on the ocean currents. To Williams, this looked like a spatial version of the lottery: Those who travel farthest are most likely to encounter different conditions, so it is best that they vary in the hope that one or two of them will suit the place they reach. Elm trees and oysters, which are sexual, produce millions of tiny young that drift on breezes or ocean currents until a few are lucky enough to land in a suitable place and begin a new life. Why do they do this? Because, said Williams, both elms and oysters have saturated their living space already. There are few clearings in an elm forest and few vacancies on an oyster bed. Each vacancy will attract many thousands of applicants in the form of new seeds or larvae. Therefore, it does not matter that your young are good enough to survive. What matters is whether they are the very best. Sex gives variety, so sex makes a few of your offspring exceptional and a few abysmal, whereas asex makes them all average.[7]

THE TANGLED BANK

Williams's proposition has reappeared in many guises over the years, under many names and with many ingenious twists. In general, however, the mathematical models suggest that these lottery models only work if the prize that rewards the right lottery ticket is indeed a huge jackpot. Only if a very few of the dispersers survive and do spectacularly well does sex pay its way. In other cases, it does not.[8]

Because of this limitation, and because most species are not necessarily producing young that will migrate elsewhere, few ecologists wholeheartedly adopted lottery theories. But it was not until Graham Bell in Montreal asked, like the apocryphal king and the goldfish, to see the actual evidence for the pattern the lottery model was designed to explain that the whole edifice tumbled down. Bell set out to catalog species according to their ecology and their sexuality. He was trying to find the correlation between ecological uncertainty and sexuality that Williams and Maynard Smith had more or less assumed existed. So he expected to find that animals and plants were more likely to be sexual at higher latitudes and altitudes (where weather is more variable and conditions harsher); in fresh water rather than the sea (because fresh water varies all the time, flooding, drying up, heating up in summer, freezing in winter, and so on, whereas the sea is predictable); among weeds that live in disturbed habitats; and in small creatures rather than large ones. He found exactly the opposite. Asexual species tend to be small and live at high latitudes and high altitudes, in fresh water or disturbed ground. They live in unsaturated habitats where harsh, unpredictable conditions keep populations from reaching full capacity. Indeed, even the association between sex and hard times in aphids and rotifers turns out to be a myth. Aphids and monogonont rotifers both turn sexual not when winter or drought threaten but when overcrowding affects the food supply. You can make them turn sexual in the laboratory just by letting them get too crowded.

Bell's verdict on the lottery model was scathing: "Accepted, at least as a conceptual foundation, by the best minds which have contemplated the function of sexuality, it seems utterly to fail the test of comparative analysis."[9]

Lottery models predict that sex should be most common where in fact it is rarest—among highly fecund, small creatures in changeable environments. On the contrary, here sex is the exception; but in big, long-lived, slow-breeding creatures in stable environments sex is the rule.

This was a bit unfair toward Williams, whose "elm-oyster model" had at least predicted that fierce competition between saplings for space was the reason elms were sexual. Michael Ghiselin developed this idea further in 1974 and made some telling analogies with economic trends. As Ghiselin put it, "In a saturated economy, it pays to diversify." Ghiselin suggested that most creatures compete with their brothers and sisters, so if everybody is a little different from their brothers and sisters, then more can survive. The fact that your parents thrived doing one thing means that it will probably pay to do something else because the local habitat might well be full already with your parents' friends or relatives doing their thing.[10]

Graham Bell has called this the "tangled bank" theory, after the famous last paragraph of Charles Darwin's *Origin of Species*: "It is interesting to contemplate an entangled bank, clothed with many plants of many kinds, with birds singing on the bushes, with various insects flitting about, and with worms crawling through the damp earth, and to reflect that these elaborately constructed forms, so different from each other and dependent upon each other in so complex a manner, have all been produced by laws acting around us."[11]

Bell used the analogy of a button maker who has no competitors and has already supplied buttons to most of the local market. What does he do? He could either continue selling replacements for buttons or he could diversify the range of his buttons and try to expand the market by encouraging his customers to buy all sorts of different kinds of buttons. Likewise, sexual organisms in saturated environments, rather than churning out more of the same offspring, would be better off varying them a bit in the hope of producing offspring that could avoid the competition by adapting to a new niche. Bell concluded from his exhaustive survey of sex and asex in the animal kingdom that the tangled bank was the most promising of the ecological theories for sex.[12]

The tangled bankers had some circumstantial evidence for their idea, which came from crops of wheat and barley. Mixtures of different varieties generally yield more than a single variety does; plants transplanted to different sites generally do worse than in their home patches, as if genetically suited to their home ground; if allowed to compete with one another in a new site, plants derived from cuttings or tillers generally do worse than plants derived from sexual seed, as if sex provides some sort of variable advantage.[13]

The trouble is, all these results are also predicted by rival theories just as plausibly. Williams wrote: "Fortune will be benevolent indeed if the inference from one theory contradicts that of another."[14] This is an especially acute problem in the debate. One scientist gives the analogy of somebody trying to decide what makes his driveway wet: rain, lawn sprinklers, or flooding from the local river. It is no good turning on the sprinkler and observing that it wets the drive or watching rain fall and seeing that it wets the drive.[15] To conclude anything from such observations would be to fall into the trap that philosophers call "the fallacy of affirming the consequent." Because sprinklers *can* wet the drive does not prove that they *did* wet the drive. Because the tangled bank is consistent with the facts does not prove it is the cause of the facts.

It is hard to find dedicated enthusiasts of tangled banks these days. Their main trouble is a familiar one: If it ain't broke, why does sex need to fix it? An oyster that has grown large enough to breed is a great success, in oyster terms. Most of its siblings are dead. If, as tangled bankers assume, the genes had something to do with that, then why must we automatically assume that the combination of genes that won in this generation will be a flop in the next? There are ways around this difficulty for tangled bankers, but they sound a bit like special pleading. It is easy enough to identify an individual case where sex would have some advantage, but to raise it to a general principle for every habitat of every mammal and bird, for every coniferous tree, a principle that can give a big enough advantage to overcome the fact that asex is twice as fecund as sex—nobody can quite bring himself to do that.

There is a more empirical objection to the tangled bank theory. Tangled banks predict a greater interest in sex in those animals and plants that have many small offspring that then compete with one another than among the plants and animals that have few large young. Superficially, the effort devoted to sex has little to do with how small the offspring are. Blue whales, the biggest animals, have huge young—each may weigh five tons or more. Giant sequoias, the biggest plants, have tiny seeds, so small that the ratio of their weight to the weight of the tree is the same as the ratio of the tree to the planet Earth.[16] Yet both are sexual creatures. By contrast, an amoeba, which splits in half when it breeds, has an enormous "young" as big as "itself." Yet it never has sex.

A student of Graham Bell's named Austin Burt went out and looked at the real world to see if the tangled bank fitted the facts. He looked not at whether mammals have sex but at how much recombination goes on among their genes. He measured this quite easily by counting the number of "crossovers" on a chromosome. These are spots where, quite literally, one chromosome swaps genes with another. What Burt found was that among mammals the amount of recombination bears no relation to the number of young, little relation to body size, and close relation to age at maturity. In other words, long-lived, late-maturing mammals do more genetic mixing regardless of their size or fecundity than short-lived, early maturing mammals. By Burt's measure, man has thirty crossovers, rabbits ten, and mice three. Tangled-bank theories would predict the opposite.[17]

.The tangled bank also conflicted with the evidence from fossils. In the 1970s evolutionary biologists realized that species do not change much. They stay exactly the same for thousands of generations, to be suddenly replaced by other forms of life. The tangled bank is a gradualist idea. If tangled banks were true, then species would gradually drift through the adaptive landscape, changing a little in every generation, instead of remaining true to type for millions of generations. A gradual drifting away of a species from its previous form happens on small islands or in tiny populations precisely because of effects somewhat analogous to Muller's ratchet: the chance extinction of some forms and the chance prosperity of other, mutated forms. In larger populations the process that hinders this is sex itself, for an innovation is donated to the rest of the species and quickly lost in the crowd. In island populations sex cannot do this precisely because the population is so inbred.[18]

It was Williams who first pointed out that a huge false assumption lay, and indeed still lies, at the core of most popular treatments of evolution. The old concept of the ladder of progress still lingers on in the form of a teleology: Evolution is good for species, and so they strive to make it go faster. Yet it is stasis, not change, that is the hallmark of evolution. Sex and gene repair and the sophisticated screening mechanisms of higher animals to ensure that only defect-free eggs and sperm contribute to the next generation—all these are ways of preventing change. The coelacanth, not the human, is the triumph of genetic systems because it has remained faithfully true to type for millions of generations despite endless assaults on the chemicals that carry its heredity. The old "Vicar of Bray" model of sex, in which sex is an aid to faster evolution, implies that organisms would prefer to keep their mutation rate fairly high—since mutation is the source of all variety—and then do a good job of sieving out the bad ones. But, as Williams put it, there is no evidence yet found that any creature ever does anything other than try to keep its mutation rate as low as possible. It strives for a mutation rate of zero. Evolution depends on the fact that it fails.[19]

Tangled banks work mathematically only if there is a sufficient advantage in being odd. The gamble is that what paid off in one generation will not pay off in the next and that the longer the generation, the more this is so—which implies that conditions keep changing.

THE RED QUEEN

Enter, running, the Red Queen. This peculiar monarch became part of biological theory twenty years ago and has been growing ever more important in the years since then. Follow me if you will into a dark labyrinth of stacked shelves in an office at the University of Chicago, past ziggurats of balanced books and three-foot Babels of paper. Squeeze between two filing cabinets and emerge into a Stygian space the size of a broom cupboard, where sits an oldish man in a checked shirt and with a gray beard that is longer than God's but not so long as Charles Darwin's. This is the Red Queen's first prophet, Leigh Van Valen, a single-minded student of evolution.

One day in 1973, before his beard was so gray, Van Valen was searching his capacious mind for a phrase to express a new discovery he had made while studying marine fossils. The discovery was that the probability a family of animals would become extinct does not depend on how long that family has already existed. In other words, species do not get better at surviving (nor do they grow feeble with age, as individuals do). Their chances of extinction are random.

The significance of this discovery had not escaped Van Valen, for it represented a vital truth about evolution that Darwin had not wholly appreciated. The struggle for existence never gets easier. However well a species may adapt to its environment, it can never relax, because its competitors and its enemies are also adapting to their niches. Survival is a zero-sum game. Success only makes one species a more tempting target for a rival species. Van Valen's mind went back to his childhood and lit upon the living chess pieces that Alice encountered beyond the looking glass. The Red Queen is a formidable woman who runs like the wind but never seems to get anywhere:

> "Well, in *our* country," said Alice, still panting a little, "you'd generally get to somewhere else—if you ran very fast for a long time as we've been doing."
>
> "A slow sort of country!" said the Queen. "Now, *here*, you see, it takes all the running *you* can do to keep in the same place. If you want to get to somewhere else, you must run at least twice as fast as that!"[20]

"A new evolutionary law," wrote Van Valen, who sent a manuscript to each of the most prestigious scientific journals, only to see it rejected. Yet his claim was justified. The Red Queen has become a great personage in the biological court. And nowhere has she won a greater reputation than in theories of sex.[21]

Red Queen theories hold that the world is competitive to the death. It does keep changing. But did we not just hear that species are static for many generations and do not change? Yes. The point about the Red Queen is that she runs but stays in the same place. The world keeps coming back to where it started; there is change but not progress.

Sex, according to the Red Queen theory, has nothing to do with adapting to the inanimate world—becoming bigger or better camouflaged or more tolerant of cold or better at flying—but is all about combating the enemy that fights back.

Biologists have persistently overestimated the importance of physical causes of premature death rather than biological ones. In virtually any account of evolution, drought, frost, wind, or starvation looms large as the enemy of life. The great struggle, we are told, is to adapt to these conditions. Marvels of physical adaptation—the camel's hump, the polar bear's fur, the rotifer's boil-resistant tun— are held to be among evolution's greatest achievements. The first ecological theories of sex were all directed at explaining this adaptability to the physical environment. But with the tangled bank, a different theme has begun to be heard, and in the Red Queen's march it is the dominant tune. The things that kill animals or prevent them from reproducing are only rarely physical factors. Far more often other creatures are involved—parasites, predators, and competitors. A water flea that is starving in a crowded pond is the victim not of food shortage but of competition. Predators and parasites probably cause most of the world's deaths, directly or indirectly. When a tree falls in the forest, it has usually been weakened by a fungus. When a herring meets its end, it is usually in the mouth of a bigger fish or a in a net. What killed your ancestors two centuries or more ago? Smallpox, tuberculosis, influenza, pneumonia, plague, scarlet fever, diarrhea. Starvation or accidents may have weakened people, but infection killed them. A few of the wealthier ones died of old age or cancer or heart attacks, but not many.[22]

The "great war" of 1914–18 killed 25 million people in four years. The influenza epidemic that followed killed 25 million in four months.[23] It was merely the latest in a series of devastating plagues to hit the human species after the dawn of civilization. Europe was laid waste by measles after A.D. 165, by smallpox after A.D. 251, by bubonic plague after 1348, by syphilis after 1492, and by tuberculosis after 1800.[24] And those are just the epidemics. Endemic diseases carried away additional vast numbers of people. Just as every plant is perpetually under attack from insects, so every animal is a seething mass of hungry bacteria waiting for an opening. There may be more bacterial than human cells in the object you proudly call "your" body. There may be more bacteria in and on you as you read this than there are human beings in the whole world.

Again and again in recent years evolutionary biologists have found themselves returning to the theme of parasites. As Richard Dawkins put it in a recent paper: "Eavesdrop [over] morning coffee at any major centre of evolutionary theory today, and you will find 'parasite' to be one of the commonest words in the language. Parasites are touted as the prime movers in the evolution of sex, promising a final solution to that problem of problems."[25]

Parasites have a deadlier effect than predators for two reasons. One is that there are more of them. Human beings have no predators except great white sharks and one another, but they have lots of parasites. Even rabbits, which are eaten by stoats, weasels, foxes, buzzards, dogs, and people, are host to far more fleas, lice, ticks, mosquitoes, tapeworms, and uncounted varieties of protozoa, bacteria, fungi, and viruses. The myxomatosis virus has killed far more rabbits than have foxes. The second reason, which is the cause of the first, is that parasites are usually smaller than their hosts, while predators are usually larger. This means that the parasites live shorter lives and pass through more generations in a given time than their hosts. The bacteria in your gut pass through six times as many generations during your lifetime as people have passed through since they were apes.[26] As a consequence, they can multiply faster than their hosts and control or reduce the host population. The predator merely follows the abundance of its prey.

Parasites and their hosts are locked in a close evolutionary embrace. The more successful the parasite's attack (the more hosts it infects or the more resources it gets from each), the more the host's chances of survival will depend on whether it can invent a defense. The better the host defends, the more natural selection will promote the parasites that can overcome the defense. So the advantage will always be swinging from one to the other: The more dire the emergency for one, the better it will fight. This is truly the world of the Red Queen, where you never win, you only gain a temporary respite.

BATTLES OF WIT

It is also the inconstant world of sex. Parasites provide exactly the incentive to change genes every generation that sex seems to demand. The success of the genes that defended you so well in the last generation may be the best of reasons to abandon these same gene combinations in the next. By the time the next generation comes around, the parasites will have surely evolved an answer to the defense that worked best in the last generation. It is a bit like sport. In chess or in football, the tactic that proves most effective is soon the one that people learn to block easily. Every innovation in attack is soon countered by another in defense.

But of course the usual analogy is an arms race. America builds an atom bomb, so Russia does, too. America builds missiles; so must Russia. Tank after tank, helicopter after helicopter, bomber after bomber, submarine after submarine, the two countries run against each other, yet stay in the same place. Weapons that would have been invincible twenty years before are now vulnerable and obsolete. The bigger the lead of one superpower, the harder the other tries to catch up. Neither dares step off the treadmill while it can afford to stay in the race. Only when the economy of Russia collapses does the arms race cease (or pause).[27]

These arms race analogies should not be taken too seriously, but they do lead to some interesting insights. Richard Dawkins and John Krebs raised one argument derived from arms races to the level of a "principle": the "life-dinner principle." A rabbit running from a fox is running for its life, so it has the greater evolutionary incentive to be fast. The fox is merely after its dinner. True enough, but what about a gazelle running from a cheetah? Whereas foxes eat things other than rabbits, cheetahs eat only gazelles. A slow gazelle might never be unlucky enough to meet a cheetah, but a slow cheetah that never catches anything dies. So the downside is greater for the cheetah. As Dawkins and Krebs put it, the specialist will usually win the race.[28]

Parasites are supreme specialists, but arms race analogies are less reliable for them. The flea that lives in the cheetah's ear has what economists call an "identity of interest" with the cheetah: If the cheetah dies, the flea dies. Gary Larson once drew a cartoon of a flea walking through the hairs on a dog's back carrying a placard that read: THE END OF THE DOG IS NEAR. The death of the dog is bad news for the flea, even if the flea hastened it. The question of whether parasites benefit from harming their hosts has vexed parasitologists for many years. When a parasite first encounters a new host (myxomatosis in European rabbits, AIDS in human beings, plague in fourteenth-century Europeans) it usually starts off as extremely virulent and gradually becomes less so. But some diseases remain fatal, while others quickly become almost harmless. The explanation is simple: The more contagious the disease, and the fewer resistant hosts there are around, the easier it will be to find a new host. So contagious diseases in unresistant populations need not worry about killing their hosts, because they have already moved on. But when most potential hosts are already infected or resistant, and the parasite has difficulty moving from host to host, it must take care not to kill its own livelihood. In the same way an industrial boss who pleads with his workers, "Please don't strike or the company will go bust," is likely to be more persuasive if unemployment is high than if the workers already have other job offers. Yet, even where virulence declines, the host is still being hurt by the parasite and is still under pressure to improve its defenses, while the parasite is continually trying to get around those defenses and sequester more resources to itself at the host's expense.[29]

ARTIFICIAL VIRUSES

Startling proof of the fact that parasites and hosts are locked in evolutionary arms races has come from a surprising source: the innards of computers. In the late 1980s evolutionary biologists began to notice a new discipline growing among their more computer-adept colleagues called artificial life. Artificial life is a hubristic name for computer programs that are designed to evolve through the same process of replication, competition, and selection as real life. They are, in a sense, the ultimate proof that life is just a matter of information and that complexity can result from directionless competition, design from randomness.

If life is information and life is riddled with parasites, then information, too, should be vulnerable to parasites. When the history of computers comes to be written, it is possible that the first program to earn the appellation "artificially alive" will be a deceptively simple little two-hundred-line program written in 1983 by Fred Cohen, a graduate student at the California Institute of Technology. The program was a "virus" that would insinuate copies of itself into other programs in the same way a real virus insinuates copies of itself into other hosts. Computer viruses have since become a worldwide problem. It begins to look as if parasites are inevitable in any system of life.[30]

But Cohen's virus and its pesky successors were created by people. It was not until Thomas Ray, a biologist at the University of Delaware, conceived an interest in artificial life that computer parasites first appeared spontaneously. Ray designed a system called Tierra that consisted of competing programs that were constantly being filled by mutation with small errors. Successful programs would thrive at one another's expense.

The effect was astonishing. Within Tierra, programs began to evolve into shorter versions of themselves. Programs that were seventy-nine instructions long began to replace the original eighty-instruction programs. But then suddenly there appeared versions of the program just forty-five instructions long: They borrowed half of the code they needed from longer programs. These were true parasites. Soon a few of the longer programs evolved what Ray called immunity to parasites. One program became impregnable to the attentions of one parasite by concealing part of itself. But the parasites were not beaten. A mutant parasite appeared in the soup that could find the concealed lines.[31]

And so the arms race escalated. Sometimes when he ran the

computer, Ray was confronted with spontaneously appearing hyper-parasites, social hyperparasites, and cheating hyper-hyperparasites—all within an evolving system of (initially) ridiculous simplicity. He had discovered that the notion of a host-parasite arms race is one of the most basic and unavoidable consequences of evolution.[32]

Arms race analogies are flawed, though. In a real arms race, an old weapon rarely regains its advantage. The day of the longbow will not come again. In the contest between a parasite and its host, it is the old weapons, against which the antagonist has forgotten how to defend, that may well be the most effective. So the Red Queen may not stay in the same place so much as end up where she started from, like Sisyphus, the fellow condemned to spend eternity rolling a stone up a hill in Hades only to see it roll down again.

There are three ways for animals to defend their bodies against parasites. One is to grow and divide fast enough to leave them behind. This is well known to plant breeders, for example: The tip of the growing shoot into which the plant is putting all its resources is generally free of parasites. Indeed, one ingenious theory holds that sperm are small specifically so they have no room to carry bacteria with them to infect eggs.[33] A human embryo indulges in a frenzy of cell division soon after it is fertilized, perhaps to leave behind any viruses and bacteria stuck in one of the compartments. The second defense is sex, of which more anon. The third is an immune system, used only by the descendants of reptiles. Plants and many insects and amphibians have an additional method: chemical defense. They produce chemicals that are toxic to their pests. Some species of pests then evolve ways of breaking down the toxins, and so on. An arms race has begun.

Antibiotics are chemicals produced naturally by fungi to kill their rivals: bacteria. But when man began to use antibiotics, he found that, with disappointing speed, the bacteria were evolving the ability to resist the antibiotics. There were two startling things about antibiotic resistance in pathogenic bacteria. One, the genes for resistance seemed to jump from one species to another, from harmless gut bacteria to pathogens, by a form of gene transfer not unlike sex. And two, many of the bugs seemed to have the resistance genes already on their chromosomes; it was just a matter of reinventing the trick of switching them on. The arms race between bacteria and fungi has left many bacteria with the ability to fight antibiotics, an ability they no longer "thought they would need" when inside a human gut.

Because they are so short-lived compared with their hosts, parasites can be quicker to evolve and adapt. In about ten years, the genes of the AIDS virus change as much as human genes change in 10 million years. For bacteria, thirty minutes can be a lifetime. Human beings, whose generations are an eternal thirty years long, are evolutionary tortoises.

PICKING DNA'S LOCKS

Evolutionary tortoises nonetheless do more genetic mixing than evolutionary hares. Austin Burt's discovery of a correlation between generation length and amount of recombination is evidence of the Red Queen at work. The longer your generation time, the more genetic mixing you need to combat your parasites.[34] Bell and Burt also discovered that the mere presence of a rogue parasitic chromosome called a "B-chromosome" is enough to induce extra recombination (more genetic mixing) in a species.[35] Sex seems to be an essential part of combating parasites. But how?

Leaving aside for the moment such things as fleas and mosquitoes, let us concentrate on viruses, bacteria, and fungi, the causes of most diseases. They specialize in breaking into cells—either to eat them, as fungi and bacteria do, or, like viruses, to subvert their genetic machinery for the purpose of making new viruses. Either way, they must get into cells. To do that they employ protein molecules that fit into other molecules on cell surfaces; in the jargon, they "bind." The arms races between parasites and their hosts are all about these binding proteins. Parasites invent new keys; hosts change the locks. There is an obvious group-selectionist argument here for sex: At any one time a sexual species will have lots of different locks; members of an asexual one will all have the same locks. So a parasite with the right key will quickly exterminate the asexual species but not the sexual one. Hence, the well-known fact: By turning our fields over to monocultures of increasingly inbred strains of wheat and maize, we are inviting the very epidemics of disease that can only be fought by the pesticides we are forced to use in ever larger quantities.[36]

The Red Queen's case is both subtler and stronger than that, though. It is that an individual, by having sex, can produce offspring more likely to survive than an individual that produces clones of itself. The advantage of sex can appear in a single generation. This is because whatever lock is common in one generation will produce among the parasites the key that fits it. So you can be sure that it is the very lock not to have a few generations later, for by then the key that fits it will be common. Rarity is at a premium.

Sexual species can call on a sort of library of locks that is unavailable to asexual species. This library is known by two long words that mean roughly the same thing: heterozygosity and polymorphism. They are the things that animals lose when their lineage becomes inbred. What they mean is that in the population at large (polymorphism) and in each individual as well (heterozygosity) there are different versions of the same gene at any one time. The "polymorphic" blue and brown eyes of Westerners are a good example: Many brown-eyed people carry the recessive gene for blue eyes as well; they are heterozygous. Such polymorphisms are almost as puzzling as sex to true Darwinists because they imply that one gene is as good as the other. Surely, if brown eyes were marginally better than blue (or, more to the point, if normal genes were better than sickle-cell-anemia genes), then one would gradually have driven the other extinct. So why on earth are we stuffed full of so many different versions of genes? Why is there so much heterozygosity? In the case of sickle-cell anemia it is because the sickle gene helps to defeat malaria, so the heterozygotes (those with one normal gene and one sickle gene) are better off than those with normal genes where malaria is common, whereas the homozygotes (those with two normal genes or two sickle genes) suffer from malaria and anemia respectively.[37]

This example is so well worn from overuse in biology text-books that it is hard to realize it is not just another anecdote but an example of a common theme. It transpires that many of the most notoriously polymorphic genes, such as the blood groups, the histocompatibility antigens and the like, are the very genes that affect resistance to disease—the genes for locks. Moreover, some of these polymorphisms are astonishingly ancient; they have persisted for geological eons. For example, there are genes that have several versions in mankind, and the equivalent genes in cows also have several versions. But what is bizarre is that the cows have the very same versions of the genes as mankind. This means that you might have a gene that is more like the gene of a certain cow than it is like the equivalent gene in your spouse. This is considerably more astonishing than it would be to discover that the word for, say, "meat" was *viande* in France, *fleisch* in Germany, *viande* again in one uncontacted Stone Age village in New Guinea, and *fleisch* in a neighboring village. Some very powerful force is at work ensuring that most versions of each gene survive and that no version changes very much.[38]

That force is almost certainly disease. As soon as a lock gene becomes rare, the parasite key gene that fits it becomes rare, so that lock gains an advantage. In a case where rarity is at a premium, the advantage is always swinging from one gene to another, and no gene is ever allowed to become extinct. To be sure, there are other mechanisms that can favor polymorphism: anything that gives rare genes a selective advantage over common genes. Predators often give rare genes a selective advantage by overlooking rare forms and picking out common forms. Give a bird in a cage some concealed pieces of food, most of which are painted red but a few painted green; it will quickly get the idea that red things are edible and will initially overlook green things. J. B. S Haldane was the first to realize that parasitism, even more than predation, could help to maintain polymorphism, especially if the parasite's increased success in attacking a new variety of host goes with reduced success against an old variety—which would be the case with keys and locks.[39]

The key and lock metaphor deserves closer scrutiny. In flax, for example, there are twenty-seven versions of five different genes that confer resistance to a rust fungus: twenty-seven versions of five locks. Each lock is fitted by several versions of one key gene in the rust. The virulence of the rust fungus attack is determined by how well its five keys fit the flax's five locks. It is not quite like real keys and locks because there are partial fits: The rust does not have to open every lock before it can infect the flax. But the more locks it opens, the more virulent its effects.[40]

THE SIMILARITY BETWEEN SEX AND VACCINATION

At this point the alert know-it-alls among you will be seething with impatience at my neglect of the immune system. The normal way to fight a disease, you may point out, is not to have sex but to produce antibodies, by vaccination or whatever. The immune system is a fairly recent invention in geological terms. It started in the reptiles perhaps 300 million years ago. Frogs, fish, insects, lobsters, snails, and water fleas do not have immune systems. Even so, there is now an ingenious theory that marries the immune system with sex in an overarching Red Queen hypothesis. Hans Bremermann of the University of California at Berkeley is its author, and he makes a fascinating case for the interdependence of the two. The immune system, he points out, would not work without sex.[41]

The immune system consists of white blood cells that come in about 10 million different types. Each type has a protein lock on it called an "antibody," which corresponds to a key carried by a bacterium called an "antigen." If a key enters that lock, the white cell starts multiplying ferociously in order to produce an army of white cells to gobble up the key-carrying invader, be it a flu virus, a tuberculosis bacterium, or even the cells of a transplanted heart. But the body has a problem. It cannot keep armies of each antibody-lock ready to immobilize all types of keys because there is simply no room for millions of different types, each represented by millions of individual cells. So it keeps only a few copies of each white cell. As soon as one type of white cell meets the antigen that fits its locks, it begins multiplying. Hence the delay between the onset of flu and the immune response that cures it.

Each lock is generated by a sort of random assembly device that tries to maintain as broad a library of kinds of lock as it can, even if some of the keys that fit them have not yet been found in parasites. This is because the parasites are continually changing their keys to try to find ones that fit the host's changing locks. The immune system is therefore prepared. But this randomness means that the host is bound to produce white cells that are designed to attack its own cells among the many types it invents. To get around this, the host's own cells are equipped with a password, which is known as a major histocompatibility antigen. This stops the attack. (Please excuse the mixed metaphor—keys and locks and passwords; it does not get any more mixed.)

To win, then, the parasite must do one of the following: infect somebody else by the time the immune response hits (as flu does), conceal itself inside host cells (as the AIDS virus does), change its own keys frequently (as malaria does), or try to imitate whatever password the host's own cells carry that enable them to escape attention. Bilharzia parasites, for example, grab password molecules from host cells and stick them all over their bodies to camouflage themselves from passing white cells. Trypanosomes, which cause sleeping sickness, keep changing their keys by switching on one gene after another. The AIDS virus is craftiest of all. According to one theory, it seems to keep mutating so that each generation has different keys. Time after time the host has locks that fit the keys and the virus gets suppressed. But eventually, after perhaps ten years, the virus's random mutation hits upon a key that the host does not have a lock for. At that point the virus has won. It has found the gap in the repertoire of the immune system's locks and runs riot. In essence, according to this theory, the AIDS virus evolves until it finds a chink in the body's immune armor.[42]

Other parasites try to mimic the passwords carried by the host. The selective pressure is on all pathogens to mimic the passwords of their hosts. The selective pressure is on all hosts to keep changing the password. This, according to Bremermann, is where sex comes in.

The histocompatibility genes, which determine more than

the passwords but are themselves responsible for susceptibility to disease, are richly polymorphic. There are over one hundred versions of each histocompatibility gene in the average population of mice, and even more in human beings. Every person carries a unique combination, which is why transplants between people other than identical twins are rejected unless special drugs are taken. And without sexual outbreeding, it is impossible to maintain that polymorphism.

Is this conjecture or is there proof? In 1991, Adrian Hill and his colleagues at Oxford University produced the first good evidence that the variability of histocompatibility genes is driven by disease. They found that one kind of histocompatibility gene, HLA-Bw53, is frequent where malaria is common and very rare elsewhere. Moreover, children ill with malaria generally do not have HLA-Bw53. That may be why they are ill.[43] And in an extraordinary discovery made by Wayne Potts of the University of Florida at Gainesville, house mice appear to choose as mates only those house mice that have different histocompatibility genes from their own. They do this by smell. This preference maximizes the variety of genes in mice and makes the young mice more disease-resistant.[44]

WILLIAM HAMILTON AND PARASITE POWER

That sex, polymorphism, and parasites have something to do with one another is an idea with many fathers. With characteristic prescience, J. B. S. Haldane got most of the way there: "I wish to suggest that [heterozygosity] may play a part in disease resistance, a particular race of bacteria or virus being adapted to individuals of a certain range of biochemical constitutions, while the other constitutions are relatively resistant." Haldane wrote that in 1949, four years before the structure of DNA was elucidated.[45] An Indian colleague of Haldane's, Suresh Jayakar, got even closer a few years later.[46] Then the idea lay dormant for many years, until the late 1970s when five people came up with the same notion independently of one another within the space of a few years: John Jaenike of

Rochester, Graham Bell of Montreal, Hans Bremermann of Berkeley, John Tooby of Harvard, and Bill Hamilton of Oxford.[47]

But it was Hamilton who pursued the connection between sex and disease most doggedly and became most associated with it. In appearance, Hamilton was an almost implausibly perfect example of the absentminded professor as he stalked through the streets of Oxford, deep in thought, his spectacles attached umbilically to a string around his neck, his eyes fixed on the ground in front of him. His unassuming manner and relaxed style of writing and storytelling were deceptive. Hamilton had a habit of being at the right place in biology at the right time. In the 1960s he molded the theory of kin selection—the idea that much of animal cooperation and altruism is explained by the success of genes that cause animals to look after close relatives because they share many of the same genes. Then in 1967 he stumbled on the bizarre internecine warfare of the genes that we shall meet in chapter 4. By the 1980s he was anticipating most of his colleagues in pronouncing reciprocity as the key to human cooperation. Again and again in this book we will find we are treading in Hamilton's footsteps.[48]

With the help of two colleagues from the University of Michigan, Hamilton built a computer model of sex and disease, a slice of artificial life. It began with an imaginary population of two hundred creatures. They happened to be rather like humans—each began breeding at fourteen, continued until thirty-five or so, and had one offspring every year. But the computer then made some of them sexual—meaning two parents had to produce and rear each child— and some of them asexual. Death was random. As expected, the sexual race quickly became extinct every time they ran the computer. In a game between sex and asex, asex always won, other things being equal.[49]

Next, they introduced several species of parasites, two hundred of each, whose power depended on "virulence genes" matched by "resistance genes" in the hosts. The least resistant hosts and the least virulent parasites were killed in each generation. Now the asexual race no longer had an automatic advantage. Sex often won the game, mostly if there were lots of genes that determined resistance and virulence in each creature.

What kept happening in the model, as expected, was that resistance genes that worked got more common, then virulence genes that undid those resistance genes got more common in turn, so those resistance genes grew rare again, followed by the virulence genes. As Hamilton put it, "Antiparasite adaptations are in constant obsolescence." But instead of the unfavored genes being driven to extinction, as happened to the asexual species, once rare, they stopped getting rarer; they could therefore be brought back. "The essence of sex in our theory," wrote Hamilton, "is that it stores genes that are currently bad but have promise for reuse. It continually tries them in combination, waiting for the time when the focus of disadvantage has moved elsewhere." There is no permanent ideal of disease resistance, merely the shifting sands of impermanent obsolescence.[50]

When it runs the simulations, Hamilton's computer screen fills with a red transparent cube inside which two lines, one green and one blue, chase each other like fireworks on a slow-exposure photograph. What is happening is that the parasite is pursuing the host through genetic "space," or, to put it more precisely, each axis of the cube represents different versions of the same gene, and the host and the parasite keep changing their gene combinations. About half the time the host eventually ends up in one corner of the cube, having run out of variety in its genes, and stays there. Mutation mistakes are especially good at preventing it from doing that, but even without them it will do so spontaneously. What happens is entirely unpredictable even though the starting conditions are ruthlessly "deterministic"—there is no element of chance. Sometimes the two lines pursue each other on exactly the same steady course around the edge of the cube, gradually changing one gene for fifty generations, then another, and so on. Sometimes strange waves and cycles appear. Sometimes there is pure chaos: The two lines just fill the cube with colored spaghetti. It is strangely alive.[51]

Of course the model is hardly the real world; it no more clinches the argument than building a model of a battleship proves that a real battleship will float. But it helps identify the conditions under which the Red Queen is running forever: A hugely simplified version of a human being and a grotesquely simplified version of a parasite will continually change their genes in cyclical and random ways, never settling, always running, but never going anywhere, eventually coming back to where they started—as long as they both have sex.[52]

SEX AT ALTITUDE

Hamilton's disease theory makes many of the same predictions as Alexey Kondrashov's mutation theory, which we met in the last chapter. To return to the analogy of the lawn sprinkler and the rainstorm, both can explain how the driveway got wet. But which is correct? In recent years ecological evidence has begun to tip the scales Hamilton's way. In certain habitats, mutation is common and diseases rare—mountaintops, for example, where there is much more ultraviolet light of the type that damages genes and causes mutations. So if Kondrashov is right, sex should be more common on mountaintops. It is not. Alpine flowers are often among the most asexual of flowers. In some groups of flowers, the ones that live near the tops of mountains are asexual, while those that live lower down are sexual. In five species of *Townsendia*, the alpine daisy, the asexuals are all found at higher altitudes than the sexuals. In *Townsendia condensata*, which lives only at very high altitudes, only one sexual population has ever been found, and that was the one nearest sea level.[53]

There are all sorts of explanations of this that have little to do with parasites, of course: The higher you go, the colder it gets, and the less you can rely on insects to pollinate a sexual flower. But if Kondrashov were right, such factors should be overwhelmed by the need to fight mutation. And the altitude effect is mirrored by a latitude effect. In the words of one textbook: "There are ticks and lice, bugs and flies, moths, beetles, grasshoppers, millipedes, and more, in all of which males disappear as one moves from the tropics toward the poles."[54]

Another trend that fits the parasite theory is that most asexual plants are short-lived annuals. Long-lived trees face a particular problem because their parasites have time to adapt to their genetic defenses—to evolve. For example, among Douglas firs infested by scale insects (which are amorphous blobs of insectness that barely even look like animals), the older trees are more heavily infested than the younger ones. By transplanting scale insects from one tree to another, two scientists were able to show that this is an effect of better-adapted insects, not weaker old trees. Such trees would do their offspring no favors by having identical young, on whom the well-adapted insects would immediately descend. Instead, the trees are sexual and have different young.[55]

Disease might almost put a sort of limit on longevity: There is little point in living much longer than it takes your parasites to adapt to you. How yew trees, bristlecone pines, and giant sequoias get away with living for thousands of years is not clear, but what is clear is that, by virtue of chemicals in their bark and wood, they are remarkably resistant to decay. In the Sierra Nevada mountains of California lie the trunks of fallen sequoias, partly covered by the roots of huge pine trees that are hundreds of years old, yet the wood of the sequoia stumps is hard and true.[56]

In the same vein it is tempting to speculate that the peculiar synchronized flowering of bamboo might have something to do with sex and disease. Some bamboos flower only once every 121 years, and they do so at exactly the same moment all over the world, then die. This gives their young all sorts of advantages: They do not have living parents to compete with, and the parasites are wiped out when the bamboo parent plants die. (Their predators have problems, too; flowering causes a crisis for pandas.)[57]

Moreover, it is a curious fact that parasites themselves are often sexual, despite the enormous inconvenience this causes. A bilharzia worm inside a human vein cannot travel abroad to seek a mate, but if it encounters a genetically different worm, infected on a separate occasion, they have sex. To compete with their sexual hosts, parasites, too, need sex.

SEXLESS SNAILS

But these are all hints from natural history, not careful scientific experiments. There is also a small amount of more direct evidence in favor of the parasite theory of sex. By far the most thorough study of the Red Queen was done in New Zealand by a soft-spoken American biologist named Curtis Lively who became intrigued by the evolution of sex when told to write an essay on the subject as a student. He soon abandoned his other research, determined to solve the problem of sex. He went to New Zealand and examined water snails from streams and lakes and found that in many populations there are no males and the females give birth as virgins, but in other populations the females mate with males and produce sexual offspring. So he was able to sample the snails, count the males, and get a rough measure of the predominance of sex. His prediction was that if the Vicar of Bray was right and snails needed sex to adjust to changes, he would find more males in streams than in lakes because streams are changeable habitats; if the tangled bank was right and competition between snails was the cause of sex, he would find more males in lakes than in streams because lakes are stable, crowded habitats; if the Red Queen was right, he would find more males where there were more parasites.[58]

There were more males in lakes. About 12 percent of snails in the average lake are male, compared to 2 percent in the average stream. So the Vicar of Bray is ruled out. But there are also more parasites in lakes, so the Red Queen is not ruled out. Indeed, the closer he looked, the more promising the Red Queen seemed to be. There were no highly sexual populations without parasites.[59]

But Lively could not rule out the tangled bank, so he returned to New Zealand and repeated his survey, this time intent on finding out whether the snails and their parasites were genetically adapted to each other. He took parasites from one lake and tried to infect snails from another lake on the other side of the Southern Alps. In every case the parasites were better at infecting snails from their own lake. At first this sounds like bad news for the Red Queen, but Lively realized it was not. It is a very host cen-

tered view to expect greater resistance in the home lake. The parasite is constantly trying to outwit the snail's defenses, so it is likely
to be only one molecular step behind the snail in changing its keys
to suit the snail's locks. Snails from another lake have altogether
different locks. But since the parasite in question, a little creature
called *Microphallus*, actually castrates the snail, it grants enormous
relative success to the snails with new locks. Lively is now doing
the crucial experiment in the laboratory—to see whether the presence of parasites actually prevents an asexual snail from displacing
a sexual one.[60]

The case of the New Zealand snails has done much to satisfy critics of the Red Queen, but they have been even more
impressed by another of Lively's studies—of a little fish in Mexico
called the topminnow. The topminnow sometimes hybridizes with a
similar fish to produce a triploid hybrid (that is, a fish that stores
its genes in triplicate, like a bureaucrat). The hybrid fish are incapable of sexual reproduction, but each female will as a virgin produce clones of herself as long as she receives sperm from a normal
fish. Lively and Robert Vrijenhoek of Rutgers University in New
Jersey caught topminnows in each of three different pools and
counted the number of cysts caused by black spot disease, a form
of worm infection. The bigger the fish, the more black spots. But
in the first pool, Log pool, the hybrids had far more spots than the
sexual topminnows, especially when large. In the second pool, San-
dal pool, where two different asexual clones coexisted, those from
the more common clone were the more parasitized; the rarer clones
and the sexual topminnows were largely immune. This was what
Lively had predicted, reasoning that the worms would adjust their
keys to the most common locks in the pond, which would be those
of the most common clone. Why? Because a worm would always
have a greater chance of encountering the most common lock than
any other lock. The rare clone would be safe, as would the sexual
topminnows, each of which had a different lock.

But even more intriguing was the third pool, Heart pool.
This pool had dried up in a drought in 1976 and had been recolo-
nized two years later by just a few topminnows. By 1983 all the

topminnows there were highly inbred, and the sexual ones were more susceptible to black spots than the clones in the same pool. Soon more than 95 percent of the topminnows in Heart pool were asexual clones. This, too, fits the Red Queen theory, for sex is no good if there is no genetic variety: It's no good changing the locks if there is only one type of lock available. Lively and Vrijenhoek introduced some more sexual female topminnows into the pool as a source of new kinds of lock. Within two years the sexual topminnows had become virtually immune to black spot, which had now switched to attacking the hybrid clones. More than 80 percent of the topminnows in the pool were sexual again. So all it took for sex to overcome its twofold disadvantage was a little bit of genetic variety.[61]

The topminnow study beautifully illustrates the way in which sex enables hosts to impale their parasites on the horns of a dilemma. As John Tooby has pointed out, parasites simply cannot keep their options open. They must always "choose." In competition with one another they must be continually chasing the most common kind of host and so poisoning their own well by encouraging the less common type of host. The better their keys fit the locks of the host, the quicker the host is induced to change its locks.[62]

Sex keeps the parasite guessing. In Chile, where introduced European bramble plants became a pest, rust fungus was introduced to control them. It worked against an asexual species of bramble and failed against a sexual species. And when mixtures of different varieties of barley or wheat do better than pure stands of one variety (as they do), roughly two-thirds of the advantage can be accounted for by the fact that mildew spreads less easily through the mixture than through a pure stand.[63]

THE SEARCH FOR INSTABILITY

The history of the Red Queen explanation of sex is an excellent example of how science works by synthesizing different approaches to a problem. Hamilton and others did not pluck the idea of parasites and sex from thin air. They are the beneficiaries of three separate lines of research that have only now converged. The first was the discovery that parasites can control populations and cause them to go in cycles. This was hinted at by Alfred Lotka and Vito Volterra in the 1920s and fleshed out by Robert May and Roy Anderson in London in the 1970s. The second was the discovery by J. B. S. Haldane and others in the 1940s of abundant polymorphism, the curious phenomenon that for almost every gene there seemed to be several different versions, and something was keeping one from driving out all the others. The third was the discovery by Walter Bodmer and other medical scientists of how defense against parasites works—the notion of genes for resistance providing a sort of lock-and-key system. Hamilton put all three lines of inquiry together and said: Parasites are in a constant battle with hosts, a battle that is fought by switching from one resistance gene to another; hence the battery of different versions of genes. None of this would work without sex.[64]

In all three fields the breakthrough was to abandon notions of stability. Lotka and Volterra were interested in knowing whether parasites could stably control populations of hosts; Haldane was interested in what kept polymorphisms stable for so long. Hamilton was different. "Where others seem to want stability I always hope to find, for the benefit of my idea of sex, as much change and motion . . . as I can get."[65]

The main weakness of the theory remains the fact that it requires some kind of cycles of susceptibility and resistance; the advantage should always be swinging back and forth like a pendulum, though not necessarily with such regularity.[66] There are some examples of regular cycles in nature: Lemmings and other rodents often grow abundant every three years and rare in between. Grouse on Scottish moors go through regular cycles of abundance and scarcity, with about four years between peaks, and this is caused by a parasitic worm. But chaotic surges, such as locust plagues, or much more steady growth or decline, such as in human beings, are more normal. It remains possible that versions of the genes for resistance to disease do indeed show cycles of abundance and scarcity. But nobody has looked.[67]

THE RIDDLE OF THE ROTIFER

Having explained why sex exists, I must now return to the case of the bdelloid rotifers, the tiny freshwater creatures that never have sex at all—a fact that John Maynard Smith called a "scandal." For the Red Queen theory to be right, the bdelloids must in some manner be immune from disease; they must have an alternative antiparasite mechanism to sex. That way they could be exceptions that prove the rule rather than embarrass it.

As it happens, the rotifer scandal may be on the verge of a solution. But in the best traditions of the science of sex, it could still go either way. Two new theories to explain the sexlessness of bdelloid rotifers point to two different explanations.

The first is Matthew Meselson's. He thinks that genetic insertions—jumping genes that insert copies of themselves into parts of the genome where they do not belong—are for some reason not a problem for rotifers. They do not need sex to purge them from their genes. It's a Kondrashov-like explanation, though with a touch of Hamilton. (Meselson calls insertions a form of venereal genetic infection.)[68] The second is a more conventional Hamiltonian idea. Richard Ladle of Oxford University noticed that there are groups of animals capable of drying out altogether without dying—losing about 90 percent of their water content. This requires remarkable biochemical skill. And none of them have sex. They are tardigrades, nematodes, and bdelloid rotifers. Some rotifers, remember, dry themselves out into little "tuns" and blow around the world in dust. This is something sexual monogonont rotifers cannot do (although their eggs can). Ladle thinks that drying yourself out may be an effective antiparasite strategy, a way of purging the parasites from your body. He cannot yet explain exactly why the parasites mind being dried out more than their hosts do; viruses are little more than molecular particles, in any case, and so could surely survive a good drying. But he seems to be on to something. Those nematode or tardigrade species that do not dry out are sexual. Those that can dry out are all female.[69]

The Red Queen has by no means conquered all her rivals. Pockets of resistance remain. Genetic repair diehards hold out in places like Arizona, Wisconsin, and Texas. Kondrashov's banner still attracts fresh followers. A few lonely tangled bankers snipe from their laboratories. John Maynard Smith pointedly calls himself a pluralist still. Graham Bell says he has abandoned the "monolithic confidence" (in the tangled bank) that infused his book *The Masterpiece of Nature*, but has not become an undoubting Red Queener. George Williams still hankers after his notion that sex is a historical accident that we are stuck with. Joe Felsenstein maintains that the whole argument was misconceived, like a discussion of why goldfish do not add to the weight of the water when added to a bowl. Austin Burt takes the surprising view that the Red Queen and the Kondrashov mutation theory are merely detailed vindications of Weismann's original idea that sex supports the variation needed to speed up evolution—that we have come full circle. Even Bill Hamilton concedes that the pure Red Queen probably needs some variation in space as well as time to make her work. Hamilton and Kondrashov met for the first time in Ohio in July 1992 and agreed convivially to differ until more evidence was in. But scientists always say that: Advocates never concede defeat. I believe that a century hence biologists will look back and declare that the Vicar of Bray fell down a tangled bank and was slain by the Red Queen.[70]

Sex is about disease. It is used to combat the threat from parasites. Organisms need sex to keep their genes one step ahead of their parasites. Men are not redundant after all; they are woman's insurance policy against her children being wiped out by influenza and smallpox (if that is a consolation). Women add sperm to their eggs because if they did not, the resulting babies would be identically vulnerable to the first parasite that picked their genetic locks.

Yet before men begin to celebrate their new role, before the fireside drum-beating sessions incorporate songs about pathogens, let them tremble before a new threat to the purpose of their existence. Let them consider the fungus. Many fungi are sexual, but they do not have males. They have tens of thousands of different sexes, all physically identical, all capable of mating on equal terms, but all incapable of mating with themselves.[71] Even among animals there are many, such as the earthworm, that are hermaphrodites. To be sexual does not necessarily imply the need for sexes, let alone for just two sexes, let alone for two sexes as different as men and women. Indeed, at first sight, the most foolish system of all is two sexes because it means that fully 50 percent of the people you meet are incompatible as breeding partners. If we were hermaphrodites, everybody would be a potential partner. If we had ten thousand sexes, as does the average toadstool, 99 percent of those we meet would be potential partners. If we had three sexes, two-thirds would be available. It turns out that the Red Queen's solution to the problem of why people are sexual is only the beginning of a long story.

Chapter 4

GENETIC MUTINY AND GENDER

The turtle lives

'twixt plated decks

Which practically conceal its sex.

I think it clever of the turtle

In such a fix to be so fertile.

—Ogden Nash

In the Middle Ages, the archetypal British village owned one common field for grazing cattle. Every villager shared the common and was allowed to graze as many cattle on it as he wanted. The result was that the common was often overgrazed until it could support only a few cattle. Had each villager been encouraged to exercise a little restraint, the common could have supported far more cattle than it did.

This "tragedy of the commons"[1] has been repeated again and again throughout the history of human affairs. Every sea fishery that has ever been exploited is soon overfished and its fishermen driven into penury. Whales, forests, and aquifers have been treated in the same way. The tragedy of the commons is, for economists, a matter of ownership. The lack of a single ownership of the commons or the fishery means that everybody shares equally in the cost of overgrazing or overfishing. But the individual who grazes one too many cows or the fisherman who catches one too many netfuls still gets the whole of the reward of that cow or netful. So he reaps the benefits privately and shares the costs publicly. It is a one-way ticket to riches for the individual and a one-way ticket to poverty for the village. Individually rational behavior leads to a collectively irrational outcome. The free-rider wins at the expense of the good citizen.

Exactly the same problem plagues the world of the genes. It is, oddly, the reason that boys are different from girls.

WHY ARE PEOPLE NOT HERMAPHRODITES?

None of the theories discussed so far explains why there are two separate genders.[2] Why is every creature not a hermaphrodite, mixing its genes with those of others, but avoiding the cost of maleness by being a female, too? For that matter, why are there two genders at all, even in hermaphrodites? Why not just give each other parcels of genes, as equals? "Why sex?" makes no sense without "why sexes?" As it happens, there is an answer. This chapter is about perhaps the strangest of all the Red Queen theories, the one that goes under the unprepossessing name of "intragenomic conflict." Translated, it is about harmony and selfishness, about conflicts of interest between genes inside bodies, about free-rider genes and outlaw genes. And it claims that many of the features of a sexual creature arose as reactions to this conflict, not to be of use to the individual. It "gives an unstable, interactive, and historical character to the evolutionary process."[3]

The thirty thousand pairs of genes that make and run the average human body find themselves in much the same position as seventy-five thousand human beings inhabiting a small town. Just as human society is an uneasy coexistence of free enterprise and social cooperation, so is the activity of genes within a body. Without cooperation, the town would not be a community. Everybody would lie and cheat and steal his way to wealth at the expense of everybody else, and all social activities—commerce, government, education, sport—would grind to a mistrustful halt. Without cooperation between the genes, the body they inhabit could not be used to transmit those genes to future generations because it would never get built.

A generation ago, most biologists would have found that paragraph baffling. Genes are not conscious and do not choose to cooperate; they are inanimate molecules switched on and off by chemical messages. What causes them to work in the right order and create a human body is some mysterious biochemical program, not a democratic decision. But in the last few years the revolution begun by Williams, Hamilton, and others has caused more and more biologists to think of genes as analogous to active and cunning individuals. Not that genes are conscious or driven by future goals—no serious biologist believes that—but the extraordinary teleological fact is that evolution works by natural selection, and natural selection means the enhanced survival of genes that enhance their own survival. Therefore, a gene is by definition the descendant of a gene that was good at getting into future generations. A gene that does things that enhance its own survival may be said, teleologically, to be doing them *because* they enhance its survival. Cooperating to build a body is as effective a survival "strategy" for genes as cooperating to run a town is a successful social strategy for human beings.

But society is not all cooperation; a measure of competitive free enterprise is inevitable. A gigantic experiment called communism in a laboratory called Russia proved that. The simple, beautiful suggestion that society should be organized on the principle "from each according to his ability, to each according to his need" proved disastrously unrealistic because each did not see why he should share the fruits of his labors with a system that gave him no reward for working harder. Enforced cooperation of the Communist kind is as vulnerable to the selfish ambitions of the individual as a free-for-all would be. Likewise, if a gene has the effect of enhancing the survival of the body it inhabits but prevents that body from breeding or is never itself transmitted through breeding, then that gene will by definition become extinct and its effect will disappear.

Finding the right balance between cooperation and competition has been the goal and bane of Western politics for centuries. Adam Smith recognized that the economic needs of the individual are better met by unleashing the ambitions of all individuals than by planning to meet those needs in advance. But even Adam Smith could not claim that free markets produce Utopia. Even the most libertarian politician today believes in the need to regulate, oversee, and tax the efforts of ambitious individuals so as to ensure that they do not satisfy their ambitions entirely at the expense of others. In the words of Egbert Leigh, a biologist at the Smithsonian Tropical Research Institute, "Human intelligence has yet to design a society where free competition among the members works for the good of the whole."[4] The society of genes faces exactly the same problem. Each gene is descended from a gene that unwittingly jostled to get into the next generation by whatever means was in its power. Cooperation between them is marked, but so is competition. And it is that competition that led to the invention of gender.

As life emerged from the primeval soup several billion years ago, the molecules that caused themselves to be replicated at the expense of others became more numerous. Then some of those molecules discovered the virtues of cooperation and specialization, so they began to assemble in groups called chromosomes to run machines called cells that could replicate these chromosomes efficiently. In just the same way little groups of agriculturalists joined with blacksmiths and carpenters to form cooperative units called villages. The chromosomes then discovered that several kinds of cells could merge to form a supercell, just as villages began to group together as tribes. This was the invention of the modern cell from a team of different bacteria. The cells then grouped together to make animals and plants and fungi, great big conglomerates of conglomerates of genes, just as tribes merged into countries and countries into empires.[5]

None of this would have been possible for society without laws to enforce the social interest over the individual, selfish drive; it was the same with genes. A gene has only one criterion by which posterity judges it: whether it becomes an ancestor of other genes. To a large extent it must achieve that at the expense of other genes, just as a man acquires wealth largely by persuading others to part with it (legally or illegally). If the gene is on its own, all other genes are its enemies—every man for himself. If the gene is part of a coalition, then the coalition shares the same interest in defeating a rival coalition, just as employees of Hertz share the same interest in its thriving at the expense of Avis.

This broadly describes the world of viruses and bacteria. They are disposable vehicles for simple teams of genes, each team highly competitive with other teams but with largely harmonious relations among team members. For reasons that will soon become apparent, this harmony breaks down when bacteria merge to become cells and cells merge to become organisms. It has to be reasserted by laws and bureaucracies.

And even at the bacterial level it does not entirely hold true. Consider the case of a new, supercharged mutant gene that appears in a bacterium. It is superior to all other genes of its type, but its fate is determined largely by the quality of its team. It is like a brilliant engineer finding himself employed by a doomed, small firm or a brilliant athlete stuck on a second-rate team. Just as the engineer or the athlete seeks a transfer, so we might expect that bacterial genes would have invented a way to transfer themselves from one bacterium to another.

They have. It is called "conjugation," and it is widely agreed to be a form of sex itself. Two bacteria simply connect to each other by a narrow pipe and shunt some copies of genes across. Unlike sex, it has nothing to do with reproduction, and it is a relatively rare event. But in every other respect it is sex. It is genetic trade.

Donal Hickey of the University of Ottawa and Michael Rose of the University of California at Irvine were the first to suggest in the early 1980s that bacterial "sex" was invented not for the bacteria but for the genes—not for the team but for the players.[6] It was a case of a gene achieving its selfish end at the expense of its teammates, abandoning them for a better team. Their theory is not a full explanation of why sex is so common throughout the animal and plant kingdoms; it is not a rival to the theories discussed heretofore. But it does suggest how the whole process got itself started. It suggests an origin for sex.

From the point of view of an individual gene, then, sex is a way to spread laterally as well as vertically. If a gene were able to make its owner-vehicle have sex, therefore, it would have done something to its own advantage (more properly, it would be more likely to leave descendants if it could), even if it were to the disadvantage of the individual. Just as the rabies virus makes the dog want to bite anything, thus subverting the dog to its own purpose of spreading to another dog, so a gene might make its owner have sex just to get into another lineage.

Hickey and Rose are especially intrigued by genes called transposons, or jumping genes, that seem to be able to cut themselves out of chromosomes and stitch themselves back into other chromosomes. In 1980 two teams of scientists simultaneously came to the conclusion that the transposons seemed to be examples of "selfish" or parasitic DNA, which spreads copies of itself at the expense of other genes. Instead of looking for some reason that transposons exist for the benefit of the individual, as scientists had done before, they simply saw it as bad for the individual and good for the transposons.[7] Muggers and outlaws do not exist for the benefit of society but to its detriment and for the benefit of themselves. Perhaps transposons were, in Richard Dawkins's words, "outlaw genes."[8] Hickey then noticed that transposons were much more common among outbreeding sexual creatures than among inbreeding or asexual ones. He ran some mathematical models which showed that parasitic genes would do well even if they had a bad effect on the individual they inhabited. He even found some cases of parasitic genes of yeast that spread quickly in sexual species and slowly in asexual ones. Such genes were on "plasmids," or separate little loops of DNA, and it turns out that in bacteria such plasmids actually provoke the very act of conjugation by which they spread. They are like rabies viruses making dogs bite one another. The line between a rogue gene and an infectious virus is a blurred one.[9]

NOBODY IS DESCENDED FROM ABEL

Despite this little rebellion, life is fairly harmonious in the bacterial team. Even in a more complicated organism such as an amoeba, formed by an agglomeration of ancestral bacteria sometime in the distant past,[10] there is little difference between the interests of the team and the individual members. But in more complicated creatures the opportunities for genes to thrive at the expense of their fellows are greater.

The genes of animals and plants turn out to be full of half-suppressed mutinies against the social harmony. In some female flour beetles there exists a gene called Medea that kills those off-·spring that do *not* inherit it.[11] It is as if the gene booby-traps all the female's young and defuses only those that it itself inhabits. Whole selfish chromosomes called B chromosomes exist that do nothing but ensure their transmission to the next generation by invading every egg the insect makes.[12] Another insect, a scale insect, has an even more bizarre genetic parasite. When its eggs are fertilized, sometimes more than one sperm penetrates the egg. If this happens, one of the sperm fuses with the egg's nucleus in the normal way; the spare sperm hang around and begin dividing as the egg divides. When the creature matures, the parasitic sperm cells eat out its gonads and replace them with themselves. So the insect produces sperm or eggs that are barely related to itself, an astonishing piece of genetic cuckoldry.[13]

The greatest opportunity for selfish genes comes during sex. Most animals and plants are diploid: Their genes come in pairs. But diploidy is an uneasy partnership between two sets of genes, and when partnerships end, things often get acrimonious. The partnerships end with sex. During meiosis, the central genetic procedure of sex, the paired genes are separated to make haploid sperm and eggs. Suddenly each gene has an opportunity to be selfish at its partner's expense. If it can monopolize the eggs or sperm, it thrives and its partner does not.[14]

This opportunity has been explored in recent years by a group of young biologists, prominent among them Steve Frank of the University of California at Irvine, and Laurence Hurst, Andrew Pomiankowski, David Haig, and Alan Grafen at Oxford University. Their logic goes like this: When a woman conceives, her embryo gets only half of her genes. They are the lucky ones; the unlucky other half languish in obscurity in the hope of another toss of the coin when she next breeds. For, to recapitulate, you have twenty-three pairs of chromosomes, twenty-three from your father and twenty-three from your mother. When you make an egg or a sperm, you pick one from each pair to give a total of twenty-three chromosomes. You could give all the ones you inherited from your mother or all the ones from your father, or more likely a mixture of the two. Now a selfish gene that loaded the dice so that it stood a better than fifty-fifty chance of getting into the embryo might do rather well. Suppose it simply killed off its opposite number, the one that came from the other grandparent of the embryo.

Such a gene exists. On chromosome two of a certain kind of fruit fly there is a gene called "segregation distorter," which simply kills all sperm containing the other copy of chromosome two. The fly therefore produces half as much sperm as normal. But all of the sperm contains the segregation distorter gene, which has thereby ensured a monopoly of the fly's offspring.[15]

Call such a gene Cain. Now it so happens that Cain is Abel's virtually identical twin, so he cannot kill his brother without killing himself. This is because the weapon he uses against Abel is merely a destructive enzyme released into the cell—a chemical weapon, as it were. His only hope is to attach to himself a device that protects him—a gas mask (though it in fact consists of a gene that repels the destructive enzyme). The "mask of Cain" protects him from the gas he uses against Abel. Cain becomes an ancestor, and Abel does not. Thus a gene for chromosomal fratricide will spread as surely as a murderer will inherit the Earth. Segregation distorters and other fratricidal genes go under the general name of "meiotic drive" because they drive the process of meiosis, the division of the partnerships, into a biased outcome.[16]

Meiotic-drive genes are known in flies and mice and a few other creatures, but they are rare. Why? For the same reason that murder is rare. The interest of the other genes has been reasserted through laws. Genes, like people, have other things to do than kill each other. Those genes that shared Abel's chromosome and died with him would have survived had they invented some technique to foil Cain. Or, to put it another way, genes that foil meiotic drivers will spread as surely as meiotic drivers will spread. A Red Queen race is the result.

David Haig and Alan Grafen believe that such a response is indeed common and that it consists of a sort of genetic scrambling, the swapping of chunks of chromosomes. If a chunk of chromosome lying next to Abel suddenly swapped places with the chunk lying next to Cain, then the mask of Cain would be unceremoniously removed from Cain's chromosome and plonked onto Abel's. The result: Cain would commit suicide, and Abel would live happily ever after.[17]

This swapping is called "crossing over." It happens between virtually all pairs of chromosomes in most species of animal and plant. It achieves nothing except a more thorough mixing of the genes—which is what most people thought its purpose was before Haig and Grafen suggested otherwise. But Haig and Grafen are implying that crossing over need not serve any such function; it is merely a piece of intracellular law enforcement. In a perfect world policemen would not exist because people would never commit murder. Policemen were not invented because they adorn society but because they prevent the disruption of society. So, according to the Haig-Grafen theory, crossing over polices the division of chromosomes to keep it fair.

This is not, by its nature, the sort of theory that lends itself to easy confirmation. As Haig remarks, in a dry Australian manner, crossing over is like an elephant repellent. You know it's working because you don't see any elephants.[18]

Cain genes survive in mice and flies by hugging their masks close to them so that they are not likely to be parted by crossing over. But there is one pair of chromosomes that is especially plagued by Cain genes, the "sex chromosomes," because these peculiar chromosomes do not engage in crossing over. In people and many other animals, gender is determined by genetic lottery. If you receive a pair of X chromosomes from your parents, you become a female; if you receive an X and a Y, you become a male (unless you are a bird, spider, or butterfly, in which case it is the other way around). Because Y chromosomes contain the genes for determining maleness, they are not compatible with Xs and do not cross over with them. Consequently, a Cain gene on an X chromosome can safely kill the Y chromosome and not risk suicide. It biases the sex ratio of the next generation in favor of females, but that is a cost borne by the whole population equally, whereas the benefit of monopolizing the offspring is received by the Cain gene itself—just as in the case of free-riders causing the tragedy of the commons.[19]

IN PRAISE OF UNILATERAL DISARMAMENT

By and large, however, the common interest of the genes prevails over the ambitions of the outlaws. As Egbert Leigh has put it, "a parliament of genes" asserts its will. Yet the reader may be getting restless. "This little tour of the cellular bureaucracy," he says, "fun though it was, has brought us no closer to the question asked at the beginning of the chapter—why there are two genders."[20]

Have patience. The road we have chosen—to seek conflicts between sets of genes—leads to the answer. For gender itself may prove to be a piece of cellular bureaucracy. A male is defined as the gender that produces sperm or pollen: small, mobile, multitudinous gametes. A female produces few, large, immobile gametes called eggs. But size is not the only difference between male and female gametes. A much more significant difference is that there are a few genes that come only from the mother. In 1981 two scientists at Harvard whose perspicacity we will reencounter throughout the book, Leda Cosmides and John Tooby, pieced together the history of an even more ambitious genetic rebellion against this parliament of genes, one that forced the evolution of animals and plants into strange new directions and resulted in the invention of two genders.[21]

So far I have treated all genes as similar in their pattern of inheritance. But this is not quite accurate. When a sperm fertilizes an egg, it donates just one thing to that egg: a bagful of genes called a nucleus. The rest of it stays outside the egg. A few of the father's genes are left behind because they are not in the nucleus at all; they are in little structures called "organelles." There are two main kinds of organelles, mitochondria, which use oxygen to extract energy from food, and chloroplasts (in plants), which use sunlight to make food from air and water. These organelles are almost certainly the descendants of bacteria that lived inside cells and were "domesticated" because their biochemical skills were of use to the host cells. Being descendants of free-living bacteria, they came with their own genes, and they still have many of these genes. Human mitochondria, for example, have thirty-seven genes of their own. To ask, "Why are there two genders?" is to ask, "Why are organelle genes inherited through the maternal line?"[22] Why not just let the sperm's organelles into the egg, too? Evolution seems to have gone to extraordinary lengths to keep the father's organelles out. In plants a narrow constriction prevents the father's organelles from passing into the pollen tube. In animals the sperm is given a sort of strip search as it enters the egg to remove all the organelles. Why should this be?

The answer lies in the exception to this rule: an alga called *Chlamydomonas* that has two genders called plus and minus rather than male and female. In this species the two parents' chloroplasts engage in a war of attrition that destroys 95 percent of them. The 5 percent remaining are those of the plus parent, which by force of sheer numbers overwhelm the minus ones.[23] This war impoverishes the whole cell. The nuclear genes take the same dim view of it as the prince takes in *Romeo and Juliet* of the war between two of his subjects:

> Rebellious subjects, enemies to peace,
> Profaners of this neighbour-stained steel, —
> Will they not hear? What, ho! you men, you beasts,
> That quench the fire of your pernicious rage
> With purple fountains issuing from your veins,
> On pain of torture, from those bloody hands
> Throw your mistemper'd weapons to the ground,
> And hear the sentence of your moved prince.
> Three civil brawls, bred of an airy word,
> By thee, Old Capulet, and Montague,
> Have thrice disturb'd the quiet of our streets.
> . . . If ever you disturb our streets again,
> Your lives shall pay the forfeit of the peace.

As the prince soon discovers, even this severe sentence is insufficient to suppress the quarrel. Had he followed the example of the nuclear genes, he would have killed all the Montagues. The nuclear genes of both father and mother between them arrange that the organelles of the male are slaughtered. It is an advantage (to the male nucleus, not to the male organelles) to be of the type that allows its organelles to be killed, so that a viable offspring results. So owners of docile, suicidal organelles (in the minus gender) would proliferate. Soon any deviation from a ratio of fifty-fifty killers and victims would benefit the rarer type and cause the ratio to correct itself. Two genders have been invented: killer, which provides the organelles, and victim, which does not.

Laurence Hurst of Oxford uses these arguments to predict that two genders are a consequence of sex by fusion. That is, where sex consists of the fusing of two cells, as in *Chlamydomonas* and most animals and plants, you find two genders. Where it consists of "conjugation"—the formation of a pipe between the two cells and the transfer of a nucleus of genes down the pipe—and there is no fusion of cells, then there is no conflict and no need for killer and victim genders. Sure enough, in those species with sex by conjugation, such as ciliated protozoa and mushrooms, there are many different genders. In those species with sex by fusion, there are almost invariably two genders. In one especially satisfying case there is a "hypotrich" ciliate that can have sex in either fashion. When it has fusion sex, it behaves as if it had two genders. When it has conjugation sex, there are many genders.

In 1991, just as he was putting the finishing touches on this tidy story, Hurst came across a case that seemed to contradict it: a form of slime-mold that has thirteen genders and fusion sex. But he delved deeper and discovered that the thirteen genders were arranged in a hierarchy. Gender thirteen always contributes the organelles, whomever it mates with. Gender twelve contributes them only if it mates with gender eleven and downward. And so on. This works just as well as having two genders but is a great deal more complicated.[24]

SAFE SEX TIPS FOR SPERM

Along with most of the animal and plant kingdoms, we practice fusion sex and we have two genders. But it is a much modified form of fusion sex. Males do not submit their organelles to be slaughtered; they leave them behind at the border. The sperm carries just a nucleus cargo, a mitochondrial engine, and a flagellum propeller. The sperm-making cells go to great lengths to strip off the rest of the cytoplasm before the sperm is complete and redigest it at some expense. Even the propeller and engine are jettisoned when the sperm meets the egg; only the nucleus travels farther.

Hurst explains this by raising once again the matter of disease.[25] Organelles are not the only genetic rebels inside cells; bacteria and viruses are there as well. And exactly the same logic applies to them as to organelles. When cells fuse, the rival bacteria in each engage in a struggle to the death. If a bacterium living happily inside an egg suddenly finds its patch invaded by a rival carried by a sperm, it will have to compete, and that might well mean abandoning its latency and manifesting itself as disease. There is ample evidence that diseases are reawakened by other "rival" infections. For example, the virus that causes AIDS, known as HIV, infects human brain cells but lies dormant there. If, however, cytomegalovirus, an entirely different kind of virus, infects a brain cell already infected with HIV, then the effect is to reawaken the HIV virus, which proliferates rapidly. This is one of the reasons HIV seems more likely to go on to cause AIDS if the infected person gets a second, complicating infection. Also, one of the features of AIDS is that all sorts of normally innocuous bacteria and viruses, such as Pneumocystis, or cytomegalovirus or herpes, which live calmly inside many of our bodies, can suddenly become virulent and aggressive during the progression of AIDS. This is partly because AIDS is a disease of the immune system, and immune surveillance of these diseases is therefore lifted, but it also makes evolutionary sense. If your host is going to die, you had better multiply as fast as possible. So-called opportunist infections therefore hit you when you are down. Incidentally, one scientist has suggested that the cross-reactivity of the immune system (infection with one strain causes immune resistance to another strain of the same species of parasite) might be the parasite's way of slamming the door on rival members of its species once it is inside.[26]

If it pays a parasite to go for broke when a rival appears, then it pays a host to prevent cross-infection with two strains of parasite. And nowhere is the risk of cross-infection greater than during sex. A sperm fusing with an egg risks bringing its cargo of bacteria and viruses as well; their arrival would awaken the egg's own parasites and cause a battle for possession that would leave the egg sick or dead. To avoid this, therefore, the sperm tries to avoid bringing into the egg material that might harbor bacteria or viruses. It passes just the nucleus into the egg. Safe sex indeed.

Proof of this theory will be hard to come by, but suggestive support comes from *Paramecium*, a protozoan that mates by conjugation—passing spare nuclei through a narrow tube. The procedure is hygienic in the sense that only the nuclei travel through the tube. Two paramecia stay linked for only two minutes or so; any longer and cytoplasm would also pass through the tube. The tube is too narrow even for the nucleus, which only just squeezes through. And it may be no accident that *Paramecium* and its relatives are the only creatures that possess such tiny nuclei, which are used as stores of genes ("coding vaults" they have been called) and from which larger, working copies are made for everyday use.[27]

DECISION TIME

Gender, then, was invented as a means of resolving the conflict between the cytoplasmic genes of the two parents. Rather than let such conflict destroy the offspring, a sensible agreement was reached: All the cytoplasmic genes would come from the mother, none from the father. Since this made the father's gametes smaller, they could specialize in being more numerous and mobile, the better to find eggs. Gender is a bureaucratic solution to an antisocial habit.

This explains why there are two genders, one with small gametes, the other with large ones. But it does not explain why every creature cannot have both genders on board. Why are people not hermaphrodites? Were I a plant, the question might not arise: Most plants are hermaphrodites. There is a general pattern for mobile creatures to be "dioecious" (with separate genders) and sessile creatures, such as plants and barnacles, to be hermaphroditic. This makes a sort of ecological sense. Given that pollen is lighter than seed, a flower that produces only seed can have only local offspring. One that also produces pollen can generate plants that spread far and wide. A law of diminishing returns applies to seed but not to pollen.

But it does not explain why animals took a different route. The answer lies in those muttering organelles left behind at the gate when the sperm entered the egg. In a male any gene in an organelle is in a cul-de-sac because it will be left behind by the sperm. All of the organelles in your body and all of the genes in them came from your mother; none came from your father. This is bad news for the genes, whose life's work, remember, is to pass into the next generation. Every man is a dead end for organelle genes. Not surprisingly, there is a "temptation" for such genes to invent solutions to their difficulty (that is, those that do solve the problem spread at the expense of those that do not). The most attractive solution for an organelle gene in a hermaphrodite is to divert all of the owner's resources into female and away from male reproduction.

This is not pure fantasy. Hermaphrodites are in a state of constant battle against rebellious organelle genes trying to destroy their male parts. Male-killer genes have been found in more than 140 species of plant. They grow flowers, but the male anthers are stunted or withered: Seed but no pollen is produced. Invariably the cause of this sterility is a gene that lies inside an organelle, not a nuclear gene. By killing the anthers, the rebellious gene diverts more of the plant's resources into female seed, through which it can be inherited. The nucleus has no such bias toward females; indeed, if the rebels are achieving their aims in many members of the species, the nucleus would benefit greatly from being the only plant on the block capable of producing pollen. So wherever they appear, male-sterility genes are soon blocked by nuclear fertility restorers. In maize, for example, there are two male-sterility organelle genes, each suppressed by a separate nuclear restorer. In tobacco there are no less than eight such pairs of genes. By hybridizing different strains of maize, plant breeders can release the male-sterile genes from nuclear suppression because the suppressor from one parent no longer recognizes the rebel from the other. They wish to do this because a field of male-sterile maize cannot fertilize itself. By planting a different, male-fertile strain among it, the breeders can collect hybrid seed. And hybrid seed, benefiting from the mysterious boost known as hybrid vigor, out-yields both its parents. Male-sterile/female-fertile strains of sunflower, sorghum, cabbage, tomato, maize, and other crops are a mainstay of farmers all over the world.[28]

It is easy to spot when male-sterile genes are at work. The plants have two types: hermaphrodite and female. Such populations of plants are known as gynodioecious; androdioecious plants, with males and hermaphrodites only, are almost unknown. In wild thyme, for example, about half the plants are usually female, the rest hermaphrodites. The only way to explain the fact that they have stopped halfway along the one-way street is to posit a continuing battle between the organelles' male-killer genes and nuclear fertility restorer genes. Under certain conditions the battle will reach a stalemate; any further advance by one side gives the other an advantage and the ability to force it back. The more common male-killers get, the more restorer genes will be favored, and vice-versa.[29]

The same logic does not apply to animals, many of which are not hermaphrodites. It pays an organelle gene to kill males only if by doing so some energy or resource is diverted to the sisters of the killed males; hence, male-killing is rarer. In hermaphroditic plants, if the male function dies, the female function of the plant grows more vigorously or produces more seed. But a male-killer gene in, say, a mouse, by killing the males in a brood, does not benefit those mice's sisters at all. Killing males because they are evolutionary culs-de-sac for organelles would be pure spite.[30]

Consequently, the battle is resolved rather differently in

animals. Imagine a population of happy hermaphroditic mice. There arrives in its midst a mutation, which happens to kill male gonads (testes). It spreads because females that have the gene do rather well: They have twice as many babies because they put no effort into making sperm. Soon the population consists of hermaphrodites and females, the latter possessing the male-killing gene. It is possible for the species to escape back to hermaphroditism by suppressing the male-killer gene, as many plants have obviously done, but it is just as likely that something else will happen before a mutation that causes the suppression can appear and take effect.

Maleness is a rather rare commodity at this stage. The few remaining hermaphroditic mice are at a premium because only they can produce the sperm that the all-female mice still need. The rarer they get, the better they do. No longer does it pay to have the male-killing mutation. Rather, the reverse. What would really pay the nuclear genes would be a female-killer gene so that one of the hermaphrodites could give up its female function altogether and concentrate on selling sperm to the rest. But if such a female-killing gene appeared, then the remaining hermaphrodites, which lack both the female-killer and the male-killer genes, are no longer at a premium. They are competing with pure males and pure females. Most of the sperm on offer comes complete with female-killer genes, and most of the eggs available to fertilize come complete with male-killer genes, so their offspring are constantly forced to specialize. The genders are separated.[31]

The answer to the question "Would you not avoid paying the cost of maleness by being a hermaphrodite?" is simple: Yes, but there is no way to get there from here. We are stuck with two genders.

THE CASE OF THE IMMACULATE TURKEYS

By separating their genders, animals ended the first mutiny of the organelles. But it was a temporary victory. The organelle genes renewed their mutiny, this time with the "aim" of driving all males into extinction and leaving the species all-female. This might seem to be a suicidal ambition because a male-less sexual species would become extinct in one generation, taking all of its genes with it, but there are two reasons this does not faze the organelles. First, they can and do convert the species into a parthenogenetic species, able to give virgin birth without sperm—in effect, they try to abolish sex—and second, they behave like cod fishermen or whale hunters or the grazers of commons. They seek short-term competitive advantage even when it leads to long-term suicide. A rational whale hunter does not spare the last pair of whales so that they can breed; he kills them before his rival does and banks the proceeds. Likewise, an organelle does not spare the last male lest the species become extinct, for it faces extinction anyway if it is in a male.

Consider a ladybird beetle's brood. If the male eggs die, the female eggs in the brood eat them and get a free meal as a result. Not surprisingly, there are male-killing genes at work in ladybirds, flies, butterflies, wasps, and bugs—about thirty species of insects have been studied so far—if and only if the young in a brood are in competition with one another. Those male-killing genes are not in organelles, however, but in bacteria that live inside the insects' cells. Those bacteria, like the organelles, are excluded from sperm but not from eggs.[32]

In animals such genes are called sex-ratio distorters. In at least twelve species of small parasitic wasps called *Trichogramma*, a bacterial infection makes the female produce only female young even from unfertilized eggs. Since all wasps have a peculiar system of sex determination in which unfertilized eggs become male, this does not condemn the race to extinction and helps the bacterium get into the next generation via the cytoplasm of the egg. The whole species becomes parthenogenetic for as many generations as the bacterium is there. Treat the wasps with an antibiotic and, lo and behold, two genders reappear among the offspring. Penicillin cures virgin birth.[33]

In the 1950s scientists at an agricultural research center in Beltsville, Maryland, noticed that some turkey eggs began to develop without being fertilized. Despite heroic efforts by the scientists, these virgin-born turkeys rarely progressed beyond the stage of simple embryos. But the scientists did notice that vaccinating the fowl against fowl pox with a live virus increased the proportion of eggs likely to begin developing without sperm, from 1–2 percent to 3–16 percent. By selective breeding and the use of three live viruses they were able to produce a strain of Pozo Gray turkeys nearly half of whose eggs would begin to develop without sperm.[34]

If turkeys, why not people? Laurence Hurst has pursued an obscure hint of a gender-altering parasite among human beings. In a small French scientific journal there appeared in 1946 an astonishing story. A woman came to the attention of a doctor in Nancy when she was having her second child; her first, a daughter, had died in infancy. She expressed no surprise on learning that the second child was also a daughter. In her family, she said, no sons were ever born.

Her tale was this: She was the ninth daughter of a sixth daughter. Her mother had no brothers, nor did she. Her eight sisters had thirty-seven daughters and no sons. Her five aunts had eighteen daughters and no sons. In all, seventy-two women had been born in two generations of her family and not one man.[35]

That such a thing should happen by chance is possible but amazingly unlikely: less than one chance in a thousand billion billion. The two French scientists who described the case, R. Lienhart and H. Vermelin, also ruled out selective spontaneous abortion of males on the grounds that there were no signs of it. Indeed, many of the women were unusually fecund. One had twelve daughters, two had nine, and one had eight. Instead, the scientists conjectured that the woman and her relatives contained some kind of cytoplasmic gene that feminized every embryo it infected, regardless of the sex chromosomes present. (There is no evidence, incidentally, that virgin birth was involved. The woman's eldest sister was a celibate nun and childless.)

The case of Madame B, as she was described, is tantalizing in the extreme. Did her daughters and nieces have only daughters? Did her first cousins? Is there still, in Nancy, an ever-growing dynasty of women, so that the city's sex ratio will soon be unbalanced? Was the explanation proffered by the French doctors the right one? If so, what was the gene and wherein did it live? It might have been in a parasite or in an organelle. How did it work? We may never know.

THE ALPHABETICAL BATTLE OF THE LEMMINGS

With the exception of some female inhabitants of the city of Nancy, the gender of a human being is determined by his or her sex chromosomes. When you were conceived, your mother's egg was chased by two kinds of your father's sperm, one containing an X chromosome and one containing a Y chromosome. Whichever got there first decided your gender. Among mammals, birds, most other animals, and many plants, this is the usual way of going about things: Gender is determined genetically, by sex chromosomes. Those with an X and a Y are male, those with two Xs are female.

But even the invention of sex chromosomes and their success in largely suppressing the rebellion of cytoplasmic genes did not succeed in making life harmonious in the society of genes. The sex chromosomes themselves began to have an interest in the gender of their owners' children. In man, for instance, the genes that control gender are on the Y chromosome. Half of a man's sperm are X carriers and half are Y carriers. To father a daughter, the man must fertilize his mate with an X carrier. In doing so he passes none of the Y's genes to her. From the Y's point of view, his daughter is unrelated to him. Therefore, a Y gene that causes the death of all the man's X-bearing sperm and ensures its own monopoly of the man's children will thrive at the expense of all other kinds of Y genes. That all those children are sons and the species will therefore go extinct matters not in the least to the Y; he has no foresight.

This phenomenon of the "driving Y" was first predicted by Bill Hamilton in 1967.[36] He saw it as a powerful danger that was liable to drive species extinct suddenly and silently. He wondered what prevented it from happening, if anything did. One solution was to gag the Y chromosome, removing all but its gender-determining role. Indeed, Y chromosomes are kept in a kind of house arrest most of the time: Only a few of their genes are expressed, and the rest are entirely silent. In many species gender is determined not by the Y chromosome but by the ratio of the number of X chromosomes to the number of ordinary chromosomes. One X fails to masculinize a bird, two succeed; and in most birds, the Y chromosome has withered away altogether.

The Red Queen is at work. Far from settling down to a fair and reasonable way of determining gender, nature has to face an infinite series of rebellions. It suppresses one only to find it has opened the way to another. For this reason gender determination is a mechanism full of, in the words of Cosmides and Tooby, "meaningless complexity manifesting unreliability, aberrations, and (from the individual's point of view) waste."[37]

But if the Y chromosome can drive, so can the X. The lemming is a fat arctic mouse famous among cartoonists for apocryphally throwing itself off cliffs in hordes. It is famous among biologists for its tendency to explode in numbers and then collapse again when overcrowding has destroyed its food supply. But it is notable for another reason: It has a peculiar way of determining the gender of its babies. It has three sex chromosomes, W, X, and Y. XY is a male; XX, WX, and WY are all females. YY cannot survive at all. What has happened is that a mutant form of driving X chromosome, W, has appeared that overrules the masculinizing power of the Y. The result is an excess of females. Since this puts males at a premium, you might expect that males would soon evolve the ability to produce more Y-bearing than X-bearing sperm, but they have not done so. Why? At first biologists thought it had something to do with population explosions during which an excess of daughters is a good idea, but recently they have determined that this is unnecessary. The female-biased sex ratio is stable for genetic, not ecological, reasons.[38]

A male that produces only Y sperm can mate with an XX female and produce all sons (XY) or with a WX female and produce half sons and half daughters or with a WY female. In the last case he has only WY daughters because YY sons die. The net result, therefore, is that if he mates with one of each, he will have as many daughters as sons, and all his daughters will be WY females, who can have only daughters. So, far from restoring the sex ratio to equality by producing only Y sperm, he has kept it unbalanced toward females. The case of the lemming demonstrates that even the invention of sex chromosomes did not prevent mutinous chromosomes from altering the sex ratio.[39]

LOTTERY OR CHOICE?

Not all animals have sex chromosomes. Indeed, it is hard to see why so many do. They make gender a pure lottery, governed by an arbitrary convention with the sole advantage of (usually) keeping the sex ratio at fifty-fifty. If the first sperm to reach your mother's egg carried a Y chromosome, you are a male; if it carried an X chromosome, you are a female. There are at least three different and better ways to determine your gender.

The first, for sedentary creatures, is to choose the gender appropriate to your sexual opportunities. For example, be a different gender from your neighbor because he or she will probably turn out to be your mate. A slipper limpet, which delights in the Latin name *Crepidula fornicata*, begins life as a male and becomes a female when it ceases peregrinating and settles on a rock; another male lands on it, and gradually it, too, turns female; a third male lands, and so on, until there is a tower of ten or more slipper limpets, the bottom ones being female, the top ones male. A similar method of gender determination is employed by certain reef fish. The shoal consists of lots of females and a single large male. When he dies, the largest female simply changes gender. The blue-headed wrasse changes gender from female to male when it reaches a certain size.[40]

This sex change makes good sense from the fish's point of view because there is a basic difference between the risks and rewards of being male or female. A large female fish can lay only a few more eggs than a small one, but a large male fish, by fighting for and winning a harem of females, can have a great many more offspring than a small male. Conversely, a small male does worse than a small female because he fails to win a mate at all. Therefore, among polygamists the following strategy often appears: If small, be female; if large, be male.[41]

There is a lot to be said for such stratagems. It is profitable to be a female while growing up and get some breeding done, and then change sex and hit the jackpot as a polygamist male once you are big enough to command a harem. Indeed, the surprise is that more mammals and birds do not adopt this system. Half-grown male deer spend years in a state of celibacy awaiting the chance to breed, while their sisters produce a fawn a year.

A second way of determining gender is to leave it to the environment. In some fish, shrimp, and reptiles, gender is determined by the temperature at which the egg is incubated. Among turtles, warm eggs hatch into females; among alligators, warm eggs hatch into males; among crocodiles, warm and cool eggs hatch into females, intermediate ones into males. (Reptiles are the most adventurous sex determiners of all. Many lizards and snakes use genetic means, but whereas XY iguanas become male and XX female, XY snakes become female and XX male.) Atlantic silverside fish are even more unusual. Those in the North Atlantic determine their gender by genes as we do; those farther south use the temperature of the water to set the gender of the embryo.[42]

This environmental method seems a peculiar way of going about it. It means that unusually warm conditions can lead to too many male alligators and too few females. It leads to "intersexes," animals that are neither one thing nor the other.[43] Indeed, no biologist has a watertight explanation for why alligators, crocodiles, and turtles employ this technique. The best one is that it is all size related. The warm eggs hatch as larger babies than the cool ones. If being large is more of an advantage to males than females (true of crocodiles, in which males compete for females) or vice versa (true of turtles, in which large females lay more eggs than small ones, whereas small males are just as capable of fertilizing females as large ones), then it would pay to make warm eggs hatch as the gender that most benefits from being large.[44] A clearer example of the same phenomenon is the case of a nematode worm that lives inside an insect larva. Its size is set by the size of the insect; once it has eaten all of its home and host, it grows no more. But whereas a big female worm can lay more eggs, a big male worm cannot fertilize more females. So big worms tend to become female and small ones male.[45]

A third way of determining gender is for the mother to choose the sex of each child. One way of achieving this is peculiar to monogonont rotifers, bees, and wasps. Their eggs become female only if fertilized. Unfertilized eggs hatch into males (which means that males are haploid and have only one set of genes to the females' two). Again, this makes some sort of sense. It means that a female can found a dynasty even if she never meets a male. Since most wasps are parasites that live inside other insects, this may help a single female who happens on an insect host to start a colony without waiting for a male to arrive. But haplodiploidy is vulnerable to certain kinds of genetic mutiny. For example, in a wasp called *Nasonia*, there is a rare supernumerary chromosome called PSR, inherited through the male line, that causes any female egg in which it finds itself to become a male by the simple expedient of getting rid of all the father's chromosomes except itself. Reduced to just the haploid maternal complement of chromosomes, the egg develops into a male. PSR is found where females predominate and has the advantage that it is in the rare, and therefore sought-after, gender.[46]

This, briefly, is the theory of sex allocation: Animals choose the appropriate gender for their circumstances unless forced to rely on the genetic lottery of sex chromosomes. But in recent years biologists have begun to realize that the genetic lottery of sex chromosomes is not incompatible with sex allocation. If they could distinguish between X and Y sperm, even birds and mammals could bias the sex ratios of their offpsring, and they would be selected to do so in exactly the same way as crocodiles and nematodes—to produce more of the gender that most benefits from being bigger when the offspring are likely to be big.[47]

PRIMOGENITURE AND PRIMATOLOGY

In the course of the neo-Darwinian revolution of the 1960s and 1970s, Britain and America each produced a grand old revolutionary whose intellectual dominance remains secure to this day: John Maynard Smith and George Williams, respectively. But each country also produced a brilliant young Turk whose precocious intellect exploded on the world of biology like a flare. Britain's prodigy was Bill Hamilton, whom we have already met. America's was Robert Trivers, who as a Harvard student in the early 1970s conceived a whole raft of new ideas that proved far ahead of his time. Trivers is a legend in biology, as he is the first ingenuously to confirm. Unconventional to the point of eccentricity, he divides his time between watching lizards in Jamaica and thinking in a redwood grove near Santa Cruz, California. One of his most provocative ideas, conceived jointly with fellow student Dan Willard in 1973, may hold the key to understanding one of the most potent and yet simple questions a human being ever asks: "Is it a boy or a girl?"[48]

If you include Barbara and Jenna Bush, daughters of the forty-third president of the United States, it is a curious statistical fact that all the presidents have between them had ninety sons and only sixty-three daughters. A sex ratio of 60 percent male in such a large sample is markedly different from the population at large, though how it came about nobody can guess—probably by pure chance. Yet presidents are not alone. Royalty, aristocrats, and even well-off American settlers have all consistently produced slightly more sons than daughters. So do well-fed opossums, hamsters, coypus, and high-ranking spider monkeys. The Trivers-Willard theory links these diverse facts.[49]

Trivers and Willard realized that the same general principle of sex allocation, which determines the gender of nematodes and fish, applies even to those creatures that cannot change sex but that take care of their young. They predicted that animals would be found to have some systematic control over the sex ratio of their own young. Think of it as a competition to have the most grandchildren. If males are polygamous, a successful son can give you far more grandchildren than a successful daughter, and an unsuccessful son will do far worse than an unsuccessful daughter because he will fail to win any mates at all. A son is a high-risk, high-reward reproductive option compared with a daughter. A mother in good condition gives her offspring a good start in life, increasing the chances of her sons' winning harems as they mature. A mother in poor condition is likely to produce a feeble son who will fail to mate at all, whereas her daughters can join harems and reproduce even when not in top condition. So you should have sons if you have reason to think they will do well and daughters if you have reason to think they will do poorly—relative to others in the population.[50]

Therefore, said Trivers and Willard, especially in polygamous animals, parents in good condition probably have male-biased litters of young; parents in poor condition probably have female-biased litters. Initially this was scoffed at as farfetched conjecture, but gradually it has received grudging respect and empirical support.

Consider the case of the Venezuelan opossum, a marsupial that looks like a large rat and lives in burrows. Steven Austad and Mel Sunquist of Harvard were intent on disproving the Trivers-Willard theory. They trapped and marked forty virgin female opossums in their burrows in Venezuela. Then they fed 125 grams of sardines to each of twenty opossums every two days by leaving the sardines outside the burrows, no doubt to the delight and astonishment of the opossums. Every month thereafter they trapped the animals again, opened their pouches, and sexed their babies. Among the 256 young belonging to the mothers who had not been fed sardines, the ratio of males to females was exactly one to one. Among the 270 from mothers who had been fed sardines, the sex ratio was nearly 1.4 to 1. Well-fed opossums are significantly more likely to have sons than poorly fed ones.[51]

The reason? The well-fed opossums had bigger babies; bigger males were much more likely to win a harem of females in later life than smaller males. Bigger females were not much more likely to have more babies than small females. Hence, the mother opossums were investing in the gender most likely to reward them with many grandchildren.

Opossums are not alone. Hamsters reared in the laboratory can be made to have female-biased litters by keeping them hungry during adolescence or pregnancy. Among coypus (large aquatic rodents), females in good condition give birth to male-biased litters; those in poor condition give birth to female-biased litters. In white-tailed deer, older mothers or yearlings in poor condition have female fawns more often than by chance alone. So do rats kept in conditions of stress. But in many ungulates (hoofed animals), stress or poor habitat has the opposite effect, inducing a male-biased sex ratio.[52]

Some of these effects can be easily explained by rival theories. Because males are often bigger than females, male embryos generally grow faster and are more of a strain on the mother. Therefore, it pays a hungry hamster or a weak deer to miscarry a male-biased litter and retain a female-biased one. Moreover, proving biased sex ratios at birth is not easy, and there have been so many negative results that some scientists maintain the positive ones are merely statistical flukes. (If you toss a coin long enough, sooner or later you will get twenty heads in a row.) But neither explanation can address the opossum study and others like it. By the late 1980s many biologists were convinced that Trivers and Willard were right at least some of the time.[53]

The most intriguing results, however, were those that concerned social status. Tim Clutton-Brock of Cambridge University studied red deer on the island of Rhum off the Scottish coast. He found that the mother's condition had little effect on the gender of her calves, but her rank within the social group did have an effect. Dominant females were slightly more likely to have sons than daughters.[54]

Clutton-Brock's results alerted primatologists, who had long suspected biased sex ratios in various species of monkey. In the Peruvian spider monkeys studied by Meg Symington, there was a clear association between rank and gender of offspring. Of twenty-one offspring born to lowest-ranked females, twenty-one were female; of eight born to highest-ranked females, six were male; those in the middle ranks had an equal sex ratio.[55]

But an even greater surprise was in store when other monkeys revealed their gender preferences. Among baboons, howler monkeys, rhesus macaques, and bonnet macaques, the opposite preference prevailed: high-ranking females gave birth to female offspring, and low-ranking females give birth to male offspring. In the eighty births to twenty female Kenyan baboons studied by Jeanne Altmann of the University of Chicago, the effect was so pronounced that high-ranking females were twice as likely to have daughters as low-ranking ones. Subsequent studies have come to less clear conclusions, and a few scientists believe that the monkey results are explained by chance. But one intriguing hint suggests otherwise.[56]

Symington's spider monkeys preferred sons when dominant, whereas the other monkeys preferred daughters. This may be no accident. In most monkeys (including howlers, baboons, and macaques) males leave the troop of their birth and join another at puberty—so-called male-exogamy. In spider monkeys the reverse applies: Females leave home. If a monkey leaves the troop it is born into, it has no chance to inherit its mother's rank. Therefore, high-ranking females will have young of whatever gender stays at home in order to pass on the high rank to them. Low-ranking females will have young of whatever gender leaves the troop in order not to saddle the young with low rank. Thus high-ranking howlers, baboons, and macaques have daughters; high-ranking spider monkeys have sons.[57]

This is a highly modified Trivers-Willard effect, known in the trade as a local-resource competition model.[58] High rank leads to a sex bias in favor of the gender that does not leave at puberty. Could it possibly apply to human beings?

DOMINANT WOMEN HAVE SONS?

Mankind is an ape. Of the five species of ape, three are social, and in two of those, chimpanzees and gorillas, it is the females that leave the home troop. In the chimpanzees of Gombe Stream in Tanzania studied by Jane Goodall, young males born to senior females tend to rise to the top faster than males born to junior females. Therefore, female apes of high social status "should"—according to the Trivers-Willard logic—have male young and those of low social status "should" have female young.[59] Now men are not excessively polygamous, so the rewards of large size to men is not great: big men do not necessarily win more wives, and big boys do not necessarily become big men. But humans are a highly social species whose society is nearly always stratified in some way. One of the prime, indeed, ubiquitous perquisites of high social status in human males, as in male chimpanzees, is high reproductive success. Wherever you look, from tribal aborigines to Victorian Englishmen, high-status males have had—and mostly still do have—more children than low-status ones. And the social status of males is very much inherited, or rather passed on from parent to child, whereas females generally leave home when they marry. I am not implying that the tendency for the female to travel to the male's home when she marries is instinctive, natural, inevitable, or even desirable, but I am noting that it has been general. Cultures in which the opposite happens are rare. So human society, like ape society but unlike most monkey society, is a female-exogamous patriarchy, and sons inherit their father's (or mother's) status more than daughters inherit their parents' status. Therefore, says Trivers-Willard, it would pay dominant fathers and high-ranking mothers, or both, to have sons and subordinates to have daughters. Do they?

The short answer is that nobody knows. American presidents, European aristocrats, various royals, and a few other elites have been suspected of having male-biased progeny at birth. In racist societies, subject races seem to be slightly more likely to have daughters than sons. But the subject is too fraught with potential complicating factors for any such statistics to be reliable. For example, merely by ceasing to breed once they have a boy—which those interested in dynastic succession might do—people would have male-biased sex ratios at birth. However, there certainly are no studies showing reliably unbiased sex ratios. And there is one tantalizing study from New Zealand that hints at what might be found if anthropologists and sociologists cared to look into the matter.[60] As early as 1966, Valerie Grant, a psychiatrist at the University of Auckland in New Zealand, noticed an apparent tendency for women who subsequently gave birth to boys to be more emotionally independent and dominating than those who gave birth to girls. She tested the personalities of eighty-five women in the first trimester of pregnancy using a standard test designed to distinguish "dominant" from "subordinate" personalities—whatever that may mean. Those who later gave birth to daughters averaged 1.35 on the dominance scale (from 0 to 6). Those who later gave birth to sons averaged 2.26, a highly significant difference. The interesting thing about Grant's work is that she began before the Trivers-Willard theory was published, in the 1960s. "I arrived at the idea quite independently of any study in any of the areas in which such a notion might reasonably arise," she told me, "For me the idea arose out of an unwillingness to burden women with the responsibility for the 'wrong' sex child."[61] Her work remains the only hint that maternal social rank affects the gender of children in the way that the Trivers-Willard-Symington theory would predict. If it proves to be more than a chance result, it immediately leads to the question of how people are unconsciously achieving something that they have been consciously striving to achieve for generations unnumbered.

SELLING GENDER

Almost no subject is more steeped in myth and lore than the business of choosing the gender of children. Aristotle and the Talmud both recommended placing the bed on a north-south axis for those wanting boys. Anaxagoras's belief that lying on the right side during sex would produce a boy was so influential that centuries later some French aristocrats had their left testicles amputated. At least posterity had its revenge on Anaxagoras, a Greek philosopher and client of Pericles. He was killed by a stone dropped by a crow, no doubt a retrospective reincarnation of some future French marquis who cut off his left testicle and had six girls in a row.[62]

It is a subject that has always drawn charlatans like blow-flies to a carcass. The old wives' tales that have answered the pleas of fathers for centuries are mostly ineffective. The Japanese Sex Selection Society promotes the use of calcium to increase the chances of having a son—with little effect. A book published in 1991 by two French gynecologists claimed precisely the opposite: that a diet rich in potassium and sodium but poor in calcium and magnesium gives a woman an 80 percent chance of conceiving a son if consumed for six weeks before fertilization. A company offering Americans "gender kits" for $50 was driven into bankruptcy after the regulators claimed it was deceiving the consumer.[63]

The more modern and scientific methods are somewhat more reliable. They all rely on trying to separate in the laboratory Y-bearing (male) sperm from X-bearing (female) sperm based on the fact that the latter possess 3.5 percent more DNA. The widely licensed technique invented by an American scientist, Ronald Ericsson, claims a 70 percent success rate from forcing the sperm to swim through albumen, which supposedly slows down the heavier X-bearing sperm more than it does the Y-bearing sperm, thus separating them. By contrast, Larry Johnson of the United States Department of Agriculture has developed a technique that works efficiently (about 70 percent male offspring and 90 percent female.) It dyes the sperm DNA with a fluorescent dye and then allows the sperm to swim in Indian file past a detector. According to the brightness of the sperm's fluorescence, the detector sorts them into two channels. The Y-bearing sperm, having smaller amounts of DNA, are slightly less brightly fluorescent. The detectors can sort sperm at 100,000 a second. Early concerns that the dyes might cause genetic damage have been largely allayed by animal experiments and this technique is now being used in the United States, mostly by people who wish to "balance the family"—have a girl after a string of boys, or vice versa.

Curiously, if humans were birds, it would be much easier to alter the chances of having young of one gender or the other because in birds the mother determines the gender of the embryo, not the father. Female birds have X and Y chromosomes (or sometimes just one X), while male birds have two Xs. So a female bird can simply release an egg of the desired gender and let any sperm fertilize it. Birds do make use of this facility. Bald eagles and some other hawks often give birth to females first and males second. This enables the female to get a head start on the male in the nest, which enables it to grow larger (and female hawks are always larger than males). Red-cockaded woodpeckers raise twice as many sons as daughters and use spare sons as nannies for subsequent broods. Among zebra finches, as Nancy Burley of the University of California at Santa Cruz discovered, "attractive" males mated with "unattractive" females usually have more sons than daughters, and vice versa. Attractiveness in this species can be altered by the simple expedient of putting red (attractive) or green (unattractive) bands on the male's legs, and black (attractive) or light blue (unattractive) on the female's legs. This makes them more or less desirable to other zebra finches as mates.[65]

But we are not birds. The only way to be certain of rearing a boy is to kill a girl child at birth and start again, or to use amniocentesis to identify the gender of the fetus and then abort it if it's a girl. These repugnant practices are undoubtedly on offer in various parts of the world. The Chinese, deprived of the chance to have more than one child, killed more than 250,000 girls after birth between 1979 and 1984.[66] In some age groups in China, there are 122 boys for every 100 girls. In one recent study of clinics in Bombay, of 8,000 abortions, 7,997 were of female fetuses.[67]

It is possible that selective spontaneous abortion also explains much of the animal data. In the case of the coypu, studied by Morris Gosling of the University of East Anglia, females in good condition miscarry whole litters if they are too female-biased, and they start again. Magnus Nordborg of Stanford University, who has studied the implications of sex-selective infanticide in China, believes that such biased miscarriage could explain the baboon data. But it seems a wasteful way to proceed.[68]

There are many well-established natural factors that bias the sex ratio of human offspring, proving that it is at least possible. The most famous is the returning-soldier effect. During and immediately after major wars, more sons are born than usual in the belligerent countries as if to replace the men that died. (This would make little sense; the men born after wars will mate with their contemporaries, not with those widowed by the war). Older fathers are more likely to have girls, but older mothers are more likely to have boys. Women with infectious hepatitis or schizophrenia have slightly more daughters than sons. So do women who smoke or drink. So did women who gave birth after the thick London smog of 1952. So do the wives of test pilots, abalone divers, clergymen, and anesthetists. In parts of Australia that depend on rainfall for drinking water, there is a clear drop in the proportion of sons born 320 days after a heavy storm fills the dams and churns up the mud. Women with multiple sclerosis have more sons, as do women who consume small amounts of arsenic.[69]

Finding the logic in this plethora of statistics is beyond most scientists at this stage. William James of the Medical Research Council in London has for some years been elaborating a hypothesis that hormones can influence the relative success of X and Y sperm. There is a good deal of circumstantial evidence that high levels of the hormone gonadotrophin in the mother can increase the proportion of daughters and that testosterone in the father can increase the proportion of sons.[70]

Indeed, Valerie Grant's theory suggests a hormonal explanation for the returning-soldier effect: that during wars women adopt more dominant roles, which affects their hormone levels and their tendency to have sons. Hormones and social status are closely related in many species; and so, as we have seen, are social status and sex ratio of offspring. How the hormones work, nobody knows, but it is possible that they change the consistency of the mucus in the cervix or even that they alter the acidity of the vagina. Putting baking soda in the vagina of a rabbit was proved to affect the sex ratio of its babies as early as 1932.[71]

Moreover, a hormone theory would tackle one of the most persistent objections to the Trivers-Willard theory: that there seems to be no genetic control of the sex ratio. The failure of animal breeders to produce a strain that can bias the gender of its offspring is glaring. It is not for want of trying. As Richard Dawkins put it: "Cattle breeders have had no trouble in breeding for high milk yield, high beef production, large size, small size, hornlessness, resistance to various diseases, and fearlessness in fighting bulls. It would obviously be of immense interest to the dairy industry if cattle could be bred with a bias toward producing heifer calves rather than bull calves. All attempts to do this have singularly failed."[72]

The poultry industry is even more desperate to learn how to breed chickens that lay eggs that hatch into chicks of only one gender. At present it employs teams of highly trained Koreans, who guard a close secret that enables them to sex day-old chicks at great speed (though a computer program may soon match them[73]). They travel all over the world plying their peculiar trade. It is hard to believe that nature is simply unable to do what the Korean experts can do so easily.

Yet this objection is easily answered once the hormonal theory is taken into account. Munching enchiladas in sight of the Pacific Ocean one day, Robert Trivers explained to me why the failure to breed sex-biased animals is entirely understandable. Suppose you find a cow that produces only heifer calves. With whom do you mate those heifers to perpetuate the strain? With ordinary bulls—diluting the genes in half at once.

Another way of putting it is that the very fact that one segment of the population is having sons makes it rewarding for the other segment to have daughters. Every animal is the child of one male and one female. So if dominant animals are having sons, then it will pay subordinate ones to have daughters. The sex ratio of the population as a whole will always revert to 1:1, however biased it becomes in one part of the population, because if it strays from that, it will pay somebody to have more of the rare gender. This insight occurred first to Sir Ronald Fisher, a Cambridge mathematician and biologist, in the 1920s, and Trivers believes it lies at the heart of why the ability to manipulate the sex ratio is never in the genes.

Besides, if social rank is a principal determinant of sex ratio, it would be crazy to put it in the genes, for social rank is almost by definition something that cannot be in the genes. Breeding for high social rank is a futile exercise in Red Queen running. Rank is relative. "You can't breed for subordinate cows," said Trivers as he munched. "You just create a new hierarchy and reset the thermostat. If all your cows are more subordinate, then the least subordinate will be the most dominant and have appropriate levels of hormones." Instead, rank determines hormones, which determine sex ratio of offspring.[74]

REASON'S CONVERGENT CONCLUSION

Trivers and Willard predict that evolution will build in an unconscious mechanism for altering the sex ratio of an individual's progeny. But we like to think we are rational, conscious decision makers, and a reasoning person can arrive at the same conclusions as evolution. Some of the strongest data to support Trivers and Willard comes not from animals but from the human cultural rediscovery of the same logic.

Many cultures bias their legacies, parental care, sustenance, and favoritism toward sons at the expense of daughters. Until recently this was seen as just another example of irrational sexism or the cruel fact that sons have more economic value than daughters. But by explicitly using the logic of Trivers-Willard, anthropologists have now begun to notice that male favoritism is far from universal and that female favoritism occurs exactly where you would most expect it.

Contrary to popular belief a preference for boys over girls is not universal. Indeed, there is a close relationship between social status and the degree to which sons are preferred. Laura Betzig of the University of Michigan noticed that, in feudal times, lords favored their sons, but peasants were more likely to leave possessions to daughters. While their feudal superiors killed or neglected daughters or banished them to convents, peasants left them more possessions. Sexism was more a feature of elites than of the unchronicled masses.[75]

As Sarah Blaffer Hrdy of the University of California at

Davis has concluded, wherever you look in the historical record, the elites favored sons more than other classes: farmers in eighteenth-century Germany, castes in nineteenth-century India, genealogies in medieval Portugal, wills in modern Canada, and pastoralists in modern Africa. This favoritism took the form of inheritance of land and wealth, but it also took the form of simple care. In India even today, girls are often given less milk and less medical attention than boys.[76]

Lower down the social scale, daughters are preferred even today. A poor son is often forced to remain single, but a poor daughter can marry a rich man. In modern Kenya, Mukogodo people are more likely to take daughters than sons to clinics for treatment when they are sick, and therefore more daughters than sons survive to the age of four. This is rational of the Mukogodo parents because their daughters can marry into the harems of rich Samburu and Maasai men and thrive, whereas their sons inherit Mukogodo poverty. In the calculus of Trivers-Willard, daughters are better grandchildren-production devices than sons.[77]

Of course, this assumes that societies are stratified. As Mildred Dickemann of California State University has postulated, the channeling of resources to sons represents the best investment rich people can make when society is class-ridden. The clearest patterns come from Dickemann's own studies of traditional Indian marriage practices. She found that extreme habits of female infanticide, which the British tried and failed to stamp out, coincided with relatively high social rank in the distinctly stratified society of nineteenth-century India. High-caste Indians killed daughters more than low-caste ones. One clan of wealthy Sikhs used to kill all daughters and live off their wives' dowries.[78]

There are rival theories to explain these patterns, of which the strongest is that economic, not reproductive, currency determines a sexual preference. Boys can earn a living and marry without a dowry. But this fails to explain the correlation with rank. It predicts, instead, that lower social classes would favor sons, not higher ones, for they can least afford daughters. If instead grandchildren production was the currency that mattered, Indian marriage practices make more sense. Throughout India it has always been the case that women more than men can "marry up," into a higher social and economic caste, so daughters of poor people are more likely to do well than sons. In Dickemann's analysis, dowries are merely a distorted echo of the Trivers-Willard effect in a female-exogamous species: Sons inherit the status necessary for successful breeding; daughters have to buy it. If you have no wealth to pass on, use what you have to buy your daughter a good husband.[79]

Trivers and Willard predict that male favoritism in one part of society will be balanced by female favoritism elsewhere if only because it takes one of each to have a baby—the Fisher logic again. In rodents the division seems to be based on maternal condition. In primates it seems to be based on social rank. But baboons and spider monkeys take for granted the fact that their societies are strictly stratified. Human beings do not. What happens in a modern, relatively egalitarian society?

In that uniform middle-class Eden known as California, Hrdy and her colleague Debra Judge have so far been unable to detect any wealth-related sex bias in the wills people leave when they die. Perhaps the old elite habit of preferring boys to girls has at last been vanquished by the rhetoric of equality.[80]

But there is another, more sinister consequence of modern egalitarianism. In some societies the boy-preferring habit seems to have spread from elites to the society at large. China and India are the best examples of this. In China a one-child policy may have led to the deaths of 17 percent of girls. In one Indian hospital 96 percent of women who were told they were carrying daughters aborted them, while nearly 100 percent of women carrying sons carried them to term.[81] This implies that a cheap technology allowing people to choose the gender of their children would indeed unbalance the population sex ratio.

Choosing the gender of your baby is an individual decision of no consequence to anybody else. Why, then, is the idea inherently unpopular? It is a tragedy of the commons—a collective harm that results from the rational pursuit of self-interest by individuals. One person choosing to have only sons does nobody else any harm, but if everybody does it, everybody suffers. The dire predictions range from a male-dominated society in which rape, lawlessness, and a general frontier mentality would hold sway to further increases in male domination of positions of power and influence. At the very least, sexual frustration would be the lot of many men.

Laws are passed to enforce the collective interest at the expense of the individual, just as crossing over was invented to foil outlaw genes. If gender selection were cheap, a fifty-fifty sex ratio would be imposed by parliaments of people as surely as equitable meiosis was imposed by the parliament of the genes.

Chapter 5

THE PEACOCK'S TALE

Tut, You saw her fair, none else being by,

Herself poised with herself in either eye.

But in that crystal scales let there be weigh'd

Your lady's love against some other maid

That I will show you shining at this feast,

And she shall scant show well that now seems best.

—William Shakespeare, *Romeo and Juliet*

The Australian brush turkey builds the best compost heaps in the world. Each male constructs a layered mound of two tons of leaves, twigs, earth, and sand. The mound is just the right size and shape to heat up to the perfect temperature to cook an egg slowly into a chick. Female brush turkeys visit the males' mounds, lay eggs in them, and depart. When the eggs hatch, the young struggle slowly to the surface of the mound, emerging ready to fend for themselves.

To paraphrase Samuel Butler ("a hen is just an egg's way of making another egg"), if the eggs are just the female's way of making another brush turkey, then the mound is just the male's way of making another brush turkey. The mound is almost as precisely a product of his genes as the egg is of hers. Unlike the female, though, the male has a residual uncertainty. How does he know that he is the father of the eggs in the mound? The answer, discovered recently by Australian scientists, is that he does not know and, in fact, is often not the father. So why does he build vast mounds to raise other males' offspring when the whole point of sexual reproduction is for his genes to find a way into the next generation? It turns out that the female is not allowed to lay an egg in the mound until she has agreed to mate with the male; that is his price for the use of the mound. Her price is that he must then accept an egg. It is a fair bargain.

But this puts the mound in an entirely different light. From the male's point of view the mound is not, after all, his way of making young brush turkeys. It is his way of attracting female brush turkeys to mate with him. Sure enough, the females select the best mounds, and therefore the best mound makers, when deciding where to lay their eggs. The males sometimes usurp one another's mounds, so the best mound owner may actually be the best mound stealer.

Even if a mediocre mound would do, a female is wise to pick the best so that her sons inherit the mound-building, mound-stealing, and female-attracting qualities of their father. The male brush turkey's mound is both his contribution to child rearing and a solid expression of his courtship.[1]

The story of the brush turkey's mound is a story from the theory of sexual selection, an intricate and surprising collection of insights about the evolution of seduction in animals, which is the subject of this chapter. And, as will become clear in later chapters, much of human nature can be explained by sexual selection.

IS LOVE RATIONAL?

It is sometimes hard even for biologists to remember that sex is merely a genetic joint venture. The process of choosing somebody to have sex with, which used to be known as falling in love, is mysterious, cerebral, and highly selective. We do not regard any and all members of the opposite sex as adequate partners for genetic joint venture. We consciously decide whether to consider people, we fall in love despite ourselves, we entirely fail to fall in love with people who fall in love with us. It is a mightily complicated business.

It is also nonrandom. The urge to have sex is in us because we are all descended from people who had an urge to have sex with each other; those that felt no urge left behind no descendants. A woman who has sex with a man (or vice versa) is running the risk of ending up with a set of genes to partner hers in the next generation. Little wonder that she is prepared to pick those partner genes carefully. Even the most promiscuous woman does not have sex indiscriminately with anyone who comes along.

The goal for every female animal is to find a mate with sufficient genetic quality to make a good husband, a good father, or a good sire. The goal for every male animal is often to find as many wives as possible and sometimes to find good mothers and dams, only rarely to find good wives. In 1972, Robert Trivers noticed the reason for this asymmetry, which runs right through the animal kingdom; the rare exceptions to his rule prove why it generally holds. The sex that invests most in rearing the young—by carrying a fetus for nine months in its belly, for example—is the sex that makes the least profit from an extra mating. The sex that invests the least has time to spare to seek other mates. Therefore, broadly speaking, males invest less and seek quantity of mates, while females invest more and seek quality of mates.[2]

The result is that males compete for the attention of females, which means that males have a greater opportunity to leave large numbers of offspring than females and a greater risk of not breeding at all. Males act as a kind of genetic sieve: Only the best males get to breed, and the constant reproductive extinction of bad males constantly purges bad genes from the population.[3] From time to time it has been suggested that this is the "purpose" of males, but that commits the fallacy of assuming evolution designs what is best for the species.

The sieve works better in some species than in others. Elephant seals are so severely sieved that in each generation a handful of males father all the offspring. Male albatrosses are so faithful to their single wives that virtually every male that reaches the right age will breed. Nonetheless, it is fair to state that in the matter of choosing mates, males are usually after quantity and females after quality. In the case of a bird such as a peacock, males will go through their ritual courtship display for any passing female; females will mate with only one male, usually the one with the most elaborately decorated tail. Indeed, according to sexual selection theory, it is the female's fault that the male has such a ridiculous tail at all. Males evolved long tails to charm females. Females evolved the ability to be charmed to be sure of picking the best males.

This chapter is about a kind of Red Queen contest, one that resulted in the invention of beauty. In human beings, when all

practical criteria for choosing a mate—wealth, health, compatibility, fertility—are ignored, what is left is the apparently arbitrary criterion of beauty. It is much the same in other animals. In species
where the females get nothing useful from their mates, they seem
to choose on aesthetic criteria alone.

ORNAMENTS AND CHOOSINESS

To put it in human terms, we are asking of animals (as we later will
of human beings): Are they marrying for money, for breeding, or
for beauty? Sexual selection theory suggests that much of the
behavior and some of the appearance of an animal is adapted not to
help it survive but to help it acquire the best or the most mates.
Sometimes these two—survival and acquiring a mate—are conflict-
ing goals. The idea goes back to Charles Darwin, though his thinking on the matter was uncharacteristically fuzzy. He first touched
on the subject in *On the Origin of Species* but later wrote an entire
book about it: *The Descent of Man and Selection in Relation to Sex.*[4]

Darwin's aim was to suggest that the reason human races
differed from one another was that for many generations the
women in each race had preferred to mate with men who looked,
say, black or white. In other words, at a loss to explain the useful-
ness of black or white skin, he suspected instead that black women
preferred black men and white women preferred white men—and
posited this as cause rather than effect. Just as pigeon fanciers
could develop breeds by allowing only their favorite strains to
reproduce, so animals could do the same to one another through
selective mate choice.

His racial theory was almost certainly a red herring,[5] but
the notion of selective mate choice was not. Darwin wondered if
selective "breeding" by females was the reason that so many male
birds and other animals were gaudy, colorful, and ornamented.
Gaudy males seemed a peculiar result of natural selection since it
was hard to imagine that gaudiness helped the animal to survive. In
fact, it would seem to be quite the reverse: Gaudy males should be
more conspicuous to their enemies.

Taking the example of the peacock, with its great tail decked with iridescent eyes, Darwin suggested that peacocks have long tails (they are not actually tails but elongated rump feathers that cover the tail) because peahens will mate only with peacocks that have long tails. After all, he observed, peacocks seem to use their tail when courting females. Ever since then the peacock has been the crest, mascot, emblem, and quarry of sexual selection.

Why should peahens like long tails? Darwin could only reply: Because I say so. Peahens prefer long trains, he said, because of an innate aesthetic sense—which is no answer at all. And peahens choose peacocks for their tails rather than vice versa because, sperm being active and eggs passive, that is usually the way of the world: Males seduce, females are seduced.

Of all Darwin's ideas, female choice proved the least persuasive. Naturalists were quite happy to accept the notion that male weapons, such as antlers, could have arisen to help males in the battle for females, but they instinctively recoiled at the frivolous idea that a peacock's tail should be there to seduce peahens. They wanted, rightly, to know why females would find long tails sexy and what possible value they could bring the hens. For a century after he proposed it, Darwin's theory of female choice was ignored while biologists tied themselves in furious knots to come up with other explanations. The preference of Darwin's contemporary, Alfred Russel Wallace, was initially that no ornaments, not even the peacock's tail, required any explanation other than that they served some useful purpose of camouflage. Later he thought they were the simple expression of surplus male vigor. Julian Huxley, who dominated the discussion of the matter for many years, much preferred to believe that almost all ornaments and ritual displays were for intimidating other males. Others believed that the ornaments were aids to females for telling species apart, so that they chose a mate of the right species.[6] The naturalist Hugh Cott was so impressed by the bright colors of poisonous insects that he suggested all bright colors and gaudy accessories were about warning predators of dangers. Some are. In the Amazon rain forest the butterflies are color-coded: yellow and black means distasteful, blue and green means too quick to catch.[7] In the 1980s a new version of this theory was adapted to birds, suggesting that colorful birds are the fastest fliers and are flaunting the fact to hawks and other predators: I'm fast, so don't even think of trying to chase me. When a scientist put stuffed male and female pied flycatchers out on perches in a wood, it was the dull females that were attacked first by hawks, not the colorful males.[8] Any theory, it seemed, was preferred to the idea of female preference for male beauty.

Yet it is impossible to watch peacocks displaying and not come away believing that the tail has something to do with the seduction of peahens. After all, that was how Darwin got the idea in the first place; he knew that the gaudiest plumes of male birds were used in courting females and not in other activities. When two peacocks fight or when one runs away from a predator, the tail is kept carefully folded away.[9]

TO WIN OR TO WOO

It took more than this to establish the fact of female choice. There were plenty of diehards who followed Huxley in thinking courtship was all a matter of competition between males. "Where female choice has been described, it plays an ancillary, and probably less significant, role than competition between males," wrote British biologist Tim Halliday as late as 1983.[10] Just as a female red deer accepts her harem master, who has fought for the harem, so perhaps a peahen accepts that she will mate with the champion male.

In one sense the distinction does not matter much. Peahens that all pick the same cock and red deer hinds that indifferently submit to the same harem master both end up "choosing" one male from among many. In any case, the peahens' "choice" may be no more voluntary or conscious than the hinds'. The peahens have merely been seduced rather than won. They may have been seduced by the display of the best male without ever having given the matter a conscious thought—let alone realized that what they were doing was "choosing." Think of human analogies. Two caricatured cavemen who fought to the death so that the winner could sling the loser's wife over his shoulder and take her away are at one extreme; Cyrano de Bergerac, who hoped to seduce Roxanne with words alone, is at the other. But in between there are thousands of permutations. A man can "win" a woman by competing with other men, or he can woo her, or both.

The two techniques—wooing and winning—are equally likely to sieve out the "best" male. The difference is that whereas the first technique will select dandies, the second will select bruisers. Thus, bull elephant seals and red deer stags are big, armed, and dangerous. Peacocks and nightingales are aesthetic show-offs.

By the mid-1980s evidence had begun to accumulate that, in many species, females had a large say in the matter of their mating partner. Where males gather on communal display arenas, a male's success owes more to his ability to dance and strut than to his ability to fight other males.[11]

It took a series of ingenious Scandinavians to establish that female birds really do pay attention to male plumes when choosing a mate. Anders Møller, a Danish scientist whose experiments are famously clever and thorough, found that male swallows with artificially lenghtened tails acquired mates more quickly, reared more young, and had more adulterous affairs than males of normal length.[12] Jakob Hoglund proved that male great snipe, which display by flashing their white tail feathers at passing females, could be made to lure more females by the simple expedient of having white typing-correction fluid painted onto their tails.[13] The best experiment of all was by Malte Andersson, who studied the widow bird of Africa. Widow birds have thick black tails many times the lengths of their bodies, which they flaunt while flying above the grass. Andersson caught thirty-six of these males, cut their tails, and either spliced on a longer set of tail feathers or left them shortened. Those with elongated tails won more mates than those with shortened tails or tails of unchanged length.[14] Tail-lengthening experiments in other species that have unusually long tails have similarly boosted male success.[15]

So females choose. Definitive evidence that the female preference itself is heritable has so far been hard to come by, but it would be odd if it were not. A suggestive hint comes from Trinidad where small fish called guppies vary in color according to the stretch of water they inhabit. Two American scientists proved that in those types of guppies in which the males are brightest orange in color, the females show the strongest preference for orange males.[16]

This female preference for male ornaments can actually be a threat to the survival of the males. The scarlet-tufted malachite sunbird is an iridescent green bird that lives high on the slopes of Mount Kenya where it feeds on the nectar of flowers and on insects that it catches on the wing. The male has two long tail streamers, and females prefer the males with the longest streamers. By lengthening the tail streamers of some males, shortening those of others, adding weight to those of a third group, and merely adding rings of similar weight to the legs of a fourth, two scientists were able to prove that female-preferred tail streamers are a burden to their bearers. The ones with lengthened or weighted tails were worse at catching insects; the ones with shortened tails were better; the ones with only rings on their legs were as good as normal.[17]

Females choose; their choosiness is inherited; they prefer exaggerated ornaments; exaggerated ornaments are a burden to males. That much is now uncontroversial. Thus far Darwin was right.

DESPOTIC FASHIONS

The question Darwin failed to answer was *why*. Why on earth should females prefer gaudiness in males? Even if the "preference" was entirely unconscious and was merely an instinctive response to the superior seduction technique of gaudy males, it was the evolution of the female preference, not the male trait, that was hard to explain.

Sometime during the 1970s it began to dawn on people that a perfectly good answer to the question had been available since 1930. Sir Ronald Fisher had suggested then that females need no better reason for preferring long tails than that other females also prefer long tails. At first such logic sounds suspiciously circular, but that is its beauty. Once most females are choosing to mate with some males rather than others and are using tail length as the criterion—a big *once*, granted, but we'll return to that—then any female who bucks the trend and chooses a short-tailed male will have short-tailed sons. (This presumes that the sons inherit their father's short tail.) All the other females are looking for long-tailed males, so those short-tailed sons will not have much success. At this point, choosing long-tailed males need be no more than an arbitrary fashion; it is still despotic. Each peahen is on a treadmill and dare not jump off lest she condemn her sons to celibacy. The result is that the females' arbitrary preferences have saddled the males of their species with ever more grotesque encumbrances. Even when those encumbrances themselves threaten the life of the male, the process can continue—as long as the threat to his life is smaller than the enhancement of his breeding success. In Fisher's words: "The two characteristics affected by such a process, namely plumage development in the male and sexual preference in the female, must thus advance together, and so long as the process is unchecked by severe counter-selection, will advance with ever-increasing speed."[18]

Polygamy, incidentally, is not essential to the argument. Darwin noticed that some monogamous birds have very colorful males: mallards, for example, and blackbirds. He suggested that it would still benefit males to be seductive and so win the first females that are ready to breed, if not the most, and his conjecture has largely been borne out by recent studies. Early-nesting females rear more young than late-nesting ones, and the most vigorous songster or gaudiest dandy tends to catch the early female. In those monogamous species in which both males and females are colorful (such as parrots, puffins, and peewits) there seems to be a sort of mutual sexual selection at work: Males follow a fashion for picking gaudy females and vice versa.[19]

Notice, though, that in the monogamous case the male is choosing as well as seducing. A male tern will present his intended with fish, both to feed her and to prove that he can fish well enough to feed her babies. If he is choosing the earliest female to arrive and she is choosing the best fisherman, they are both employing eminently sensible criteria. It is bizarre even to suggest that choice plays no part in their mating. From terns to peafowl, there is a kind of continuum of different criteria. A hen pheasant, for example, who will get no help from a cock in rearing her young, happily chooses to ignore a nearby cock who is unmated to join the harem of a cock who already has several wives. He runs a sort of protection racket within his territory, guarding his females while they feed in exchange for sexual monopoly over them. The best protector is more use to her than a faithful house-husband. A peahen, on the other hand, does not even get such protection. The peacock provides her with nothing but sperm.[20]

Yet there is a paradox here. In the tern's case, choosing a poor male is a disastrous decision that will leave her chicks liable to starve. In the hen pheasant's case, choosing the less effective harem defender will apparently leave her inconvenienced. In the peahen's case, picking the poorest male will leave her hardly affected at all. She gets nothing practical from her mate, so it seems there is nothing to be lost. You would expect, therefore, that the choice would be made most carefully by the tern and least carefully by the peahen.

Appearances suggest the exact opposite. Peahens survey several males and take their time over their decision, allowing each to parade his tail to best advantage. What is more, most of the peahens choose the same male. Terns mate with little fuss. Females are the most choosy where the least seems to be at stake.[21]

RUNNING OUT OF GENES

Least at stake? One very important thing is at stake in the peafowl case: a bunch of genes. Genes are the only thing a peahen gets from a peacock, whereas a female tern gets tangible help from the male as well. A tern must demonstrate only paternal proficiency; a peacock must demonstrate that he has the best genes on offer.

Peacocks are among the few birds that run a kind of market

in seduction techniques, called a "lek," after the Swedish word for play. Some grouse, several birds of paradise and manakins, plus a number of antelope, deer, bats, fish, moths, butterflies, and other insects also indulge in lekking. A lek is a place where males gather in the breeding season, mark out little territories that are clustered together, and parade their wares for visiting females. The characteristic of the lek is that one or a few males, usually those that display near its center, achieve most of the matings. But the central position of a successful male is not the cause of his success so much as the consequence: Other males gather around him.

The sage grouse of the American West has been the best studied of lekking birds. It is an extraordinary experience to drive out to the middle of Wyoming before dawn, stop the car on a featureless plain that looks like every other one, and see it come alive with dancing grouse. Each knows his place; each runs through his routine of inflating the air sacs in his breast and strutting forward, bouncing the fleshy sacs through his feathers for all the world like a dancer at the Folies Bergère. The females wander through this market, and after several days of contemplating the goods on offer, they mate with one of the males. That they are choosing, not being forced to choose, seems obvious: The male does not mount the female until she squats in front of him. Minutes later his job is done, and her long and lonely parenthood is beginning. She has received only one thing from her mate—genes—and it looks as if she has tried hard to get the best there were to be had.

Yet the problem of greatest choosiness in the species where choice least matters reappears. A single sage grouse cock may perform half of all the matings at one lek; it is not unknown for this top male to mate thirty or more times in a morning.[22] The result is that in the first generation the genetic cream is skimmed from the surface of the population, in the second the cream of the cream, in the third the cream of the cream of the cream, and so on. As any dairy farmer can attest, this is a procedure that quickly becomes pointless. There is just not enough separability in cream to keep taking the thickest layer. It is the same for sage grouse. If 10 percent of the males father the next generation, pretty soon all the females and all the males will be genetically identical, and there will be no point in selecting one male over another because they are all the same. This is known as the "lek paradox," and it is the hurdle that all modern theories of sexual selection attempt to leap. How they do so is the subject of the rest of this chapter.

MONTAGUES AND CAPULETS

It is time to introduce the great dichotomy. Sexual selection theory is split into two warring factions. There is no accepted name for each party; most people call them "Fisher" and "Good-genes." Helena Cronin, who has written a masterful history of the sexual selection debate,[23] prefers "good-taste" and "good-sense." They are sometimes also known as the "sexy-son" versus the "healthy-offspring" theories.

The Fisher (sexy-son, good-taste) advocates are those who insist that the reason peahens prefer beautiful males is that they seek heritable beauty itself to pass on to their sons, so that those sons may in turn attract females. The Good-geners (healthy-offspring, good-sense) are those who believe that peahens prefer beautiful males because beauty is a sign of good genetic qualities—disease resistance, vigor, strength—and that the females seek to pass these qualities on to their offspring.

Not all biologists admit to being members of one school or the other. Some insist there can be a reconciliation; others would like to form a third party and cry with Mercutio, "A plague on both your houses." But nonetheless the distinction is as real as the enduring feud between Capulets and Montagues in *Romeo and Juliet*. This is biological civil war.

The Fisherians derive their ideas mostly from Sir Ronald Fisher's great insight about despotic fashion, and they follow Darwin in thinking the female's preference for gaudiness is arbitrary and without purpose. Their position is that females choose males according to the gaudiness of their colors, the length of their plumes, the virtuosity of their songs, or whatever, because the species is ruled by an arbitrary fashion for preferring beauty that none dares buck. The Good-gene people follow Alfred Russel Wallace (though they do not know it) in arguing that arbitrary and foolish as it may seem for a female to choose a male because his tail is long or his song loud, there is method in her madness. The tail or the song tells each female exactly how good the genes are of each male. The fact that he can sing loudly or grow and look after a long tail proves that he can father healthy and vigorous daughters and sons just as surely as the fishing ability of a tern tells his mate that he can feed a growing family. Ornaments and displays are designed to reveal the quality of genes.

The split between Fisher and Good-genes began to emerge in the 1970s once the fact of female choice had been established to the satisfaction of most. Those of a theoretical or mathematical bent—the pale, eccentric types umbilically attached to their computers—became Fisherians. Field biologists and naturalists—bearded, besweatered, and booted—gradually found themselves Good-geners.[24]

IS CHOOSING CHEAP?

The first round went to the Fisherians. Fisher's intuition was fed into mathematical models and emerged intact. In the early 1980s three scientists programmed their computers to play an imaginary game of females choosing long-tailed males and bearing sons that had the long tails and daughters that shared the preference of their mothers. The longer the male's tail, the greater his mating success but the smaller his chances of surviving to mate at all. The scientists' key discovery was that there exists a "line of equilibrium" on which the game can stop at any point. On that line the handicap to a female's sons of having a long tail is exactly balanced by the advantage those sons have in attracting a mate.[25]

In other words, the choosier the females, the brighter and more elaborate the male ornaments will be, which is exactly what you find in nature. Sage grouse are elaborately ornamented, and only a few males get chosen; terns are unornamented, and most males win mates.

The models also showed that the process could run away from the line of equilibrium with Fisher's "ever-increasing speed" but only if females vary in their (heritable) preference and if the male's ornament is not much of an encumbrance to him. These are fairly unlikely conditions except early in the process when a new preference and a new trait have just emerged.

But the mathematicians discovered more. It mattered greatly if the process of choosing was costly to females. If in deciding which male to mate with a female wastes time that could be more profitably spent incubating eggs or she exposes herself to the risk of being caught by an eagle, then the line no longer stands. As soon as the species reaches it, and the advantages of long tails are balanced by their disadvantages, there is no net benefit to being choosy, so the costs of choice will drive females into indifference. This looked to be fatal to the whole Fisher idea, and there was brief interest in another version of it (which is known as the "sexy-son" theory) that suggested sexy husbands made bad fathers—a clear cost to being a choosy female.[26]

Luckily, another mathematical insight came to the rescue. The genes that cause the elaborate ornament or long tail to appear are subject to random mutation. The more elaborate the ornament, the more likely that a random mutation will make the ornament less elaborate, not more. Why? A mutation is a wrench thrown into the genetic works. Throwing a wrench into a simple device, such as a bucket, may not alter its function much, but throwing a wrench into a more complicated device, such as a bicycle, will almost certainly make it less good as a bicycle. Thus, any change in a gene will tend to make the ornament smaller, less symmetrical, or less colorful. This "mutational bias" is sufficient, according to the mathematicians, to make it worth the female's while to choose an ornamented male because it means that any defect in the ornament might otherwise be inherited by the sons; by choosing the most elaborate ornament she is choosing the male with the fewest mutations. The mutational bias is also sufficient, perhaps, to defeat the central conundrum that we set the theories earlier—the fact that if the best genetic cream of the cream is taken off each generation, there will soon be no separability left in the cream. Mutational bias keeps turning some of the cream back into milk.[27]

The result of a decade of mathematical games, then, has been to prove that the Fisherians are not wrong. Arbitrary ornaments can grow elaborate for no other reason than that females discriminate between males and end up following arbitrary fashions; and the more they discriminate, the more elaborate the ornaments become. What Fisher said in 1930 was right, but it left a lot of naturalists unconvinced for two reasons. First, Fisher assumed part of what he set out to prove: That females are already choosy is crucial to the theory. Fisher himself had an answer for this, which was that initially females chose long-tailed males for more utilitarian reasons—for example, that it indicated their superior size or vigor. This is not a foolish idea; after all, even the most monogamous species, in which every male wins a female (such as terns), are choosy. But it is an idea borrowed from the enemy camp. And the Good-geners can reply: "If you admit that our idea works initially, why rule it out later on?"

The second reason is more mundane. Proving that Fisher's runaway selection could happen and the ornament get bigger with ever-increasing speed does not prove that it does happen. Computers are not the real world. Nothing could satisfy the naturalists but an experiment, one demonstrating that the sexiness of sons drove the evolution of an ornament.

Such an experiment has never been devised, but those, like me, with a bias toward the Fisherians find several lines of argument fairly persuasive. Look around the world and what do you see? You see that the ornaments we are discussing are nothing if not arbitrary. Peacocks have eyes in their train; sage grouse have inflatable air sacs and pointed tails; nightingales have melodies of great variety and no particular pattern; birds of paradise grow bizarre feathers like pennants; bower birds collect blue objects. It is a cacophony of caprice and color. Surely if sexually selected ornaments told a tale of their owner's vigor, they would not be so utterly random.

One other piece of evidence seems to weigh in the balance on the side of Fisher—the phenomenon of copying. If you watch a lek carefully, you see that the females often do not make up their own minds individually; they follow one another. Sage grouse hens are more likely to mate with a cock who has just mated with another hen. In black grouse, which also lek, the cocks tend to mate several times in a row if at all. A stuffed female black grouse (known in this species as a greyhen) placed in a male's territory tends to draw other females to that territory—though not necessarily causing them to mate.[28] In guppy fish, females that have been allowed to see two males, one of which is already courting a female, subsequently prefer that male to the other even if the female that was being courted is no longer present.[29]

Such copying is just what you would expect if Fisher was right because it is fashion-following for its own sake. It hardly matters whether the male chosen is the "best" male; what counts is that he is the most fashionable, as his sons will be. If the Good-geners are right, females should not be so influenced by each other's views. There is even a hint that peahens try to prevent one another from copying, which would also make sense to a follower of Fisher.[30] If the goal is to have the sexiest son in the next generation, then one way of doing that is to mate with the sexiest male; a second way is to prevent other females from mating with the sexiest male.

ORNAMENTAL HANDICAPS

If females choose males for the sexiness of their future sons, why shouldn't they go for other genetic qualities, too? The Good-geners think that beauty has a purpose. Peahens choose genetically superior males in order to have sons and daughters who are equipped to survive as well as equipped to attract mates.

The Good-geners can marshal as much experimental support as the Fisherians. Fruit flies given a free choice of mate produce young that prove tougher in competition with the young of those not allowed to choose.[31] Female sage grouse, black grouse, great snipe, fallow deer, and widow birds all seem to prefer the males on their leks that display most vigorously.[32] If a stuffed grey-hen is put on the boundary between two blackcocks' dancing grounds, the two males fight over the right to monopolistic necrophilia. The winner is usually the male who is most attractive to females, and he is also more likely to survive the next six months than the other male. This seems to imply that attracting females is not the only thing he is good at; he is also good at surviving.[33] The brighter red a male house finch is, the more popular he is with the females; but he is also a better father—he provides more food for the babies—and will live longer because he is genetically more dis-ease-resistant. By choosing the reddest male on offer, females are therefore getting superior survival genes as well as attractiveness genes.[34]

It is hardly surprising to find that the males best at seduc-tion tend to be the best at other things as well; it does not prove that females are seeking good genes for their offspring. They might be avoiding feeble males lest they catch a virus from them. Nor do such observations damage the idea that the most important thing a sexy male can pass on to his sons is his sexiness—the Fisher idea. They merely suggest that he can also pass on other attributes.

Consider, though, the case of Archbold's bowerbird, which lives in New Guinea. As in other bowerbirds, the male builds an elaborate bower of twigs and ferns and therein tries to seduce females. The female inspects the bower and mates with the male if she likes the workmanship and the decorations, which are usually objects of one unusual color. What is peculiar about Archbold's bowerbird is that the best decorations consist of feathers from one particular kind of bird of paradise, known as the King of Saxony. These feathers, which are several times longer than the original owner's body and stem from just above his eye, are like a car's antenna sporting dozens of square blue pennants. Because they are molted once a year, do not grow until the bird of paradise is four years old, and are much in demand among local tribesmen, the plumes must be very hard for the bowerbird to acquire. Once acquired they must be guarded against other jealous male bower-birds anxious to steal them for their own bowers. So, in the words of Jared Diamond, a female bowerbird who finds a male that has decorated his bower with King of Saxony plumes knows "that she has located a dominant male who is terrific at finding or stealing rare objects and defeating would-be thieves."[35]

So much for the bowerbird. What about the bird of paradise itself, the rightful owner of the plumes? The fact that he survived long enough to grow plumes, grew longer ones than any other male nearby, and kept them in good condition would be an equally reliable indicator of his genetic quality. But it reminds us of the thing that most puzzled Darwin and got the whole debate started: If the point of the plumes is to indicate his quality, might not the plumes themselves affect his quality? After all, every tribesman in New Guinea is out to get him, and every hawk will find him easier to spot. He may have indicated that he is good at surviving, but his chances of survival are now lower for having the plumes. They are a handicap. How can a system of females choosing males that are good at surviving encumber those males with handicaps to survival?

It is a good question with a paradoxical answer, for which we owe a debt of thanks to Amotz Zahavi, a mercurial Israeli scientist. He saw in 1975 that the more a peacock's tail or a bird of paradise's plumes handicapped the male, the more honest the signal was that he sent the female. She could be assured by the very fact of his survival that the long-tailed male in front of her had been through a trial and passed. He had survived despite being handicapped. The more costly the handicap, the better it was as a signal of his genetic quality; therefore, peacocks' tails would evolve faster if they were handicaps than if they were not. This is the reverse of Fisher's prediction that peacocks' tails should gradually cease evolving once they become severe handicaps.[36]

It is an appealing—and familiar—thought. When a Maasai warrior killed a fierce beast to prove himself to a potential mate, he was running the risk of being killed but was also showing that he had the necessary courage to defend a herd of cattle. Zahavi's "handicap" was only a version of such initiation rituals, yet it was attacked from all sides, and the consensus was that he was wrong.

The most telling argument against it was that the sons would inherit the handicap as well as the good genes, so they would be encumbered to the same degree as they were endowed. They would be no better off than if they were unencumbered and unsexy.[37]

In recent years, however, Zahavi has been vindicated. Mathematical models proved that he might be right and his critics wrong.[38] His vindicators have added to his theory two subtleties that lend it special relevance to the Good-gene theory of sexual selection. The first is that handicaps might (perhaps must) not only affect survival and reflect quality but also do so in a graduated way; the weaker the male, the harder it would be to produce or maintain a tail of a given length. And indeed, experiments on swallows have shown that birds promoted above their station, by being given longer tail streamers than they grew naturally, could not the next time grow as long a tail as before; carrying the extra handicap had taken its toll.[39] The second is that the handicapping ornament might be designed so as best to reveal deficiency. After all, life would be a lot easier for swans if they were not white, as anybody who has tried swimming in a lake in a wedding dress would know. Swans do not become white until they are a few years old and ready to breed; perhaps being whiter than white proves to a skeptical swan that its suitor can spare the time from feeding to clean his plumage.

The vindication of Zahavi played a critical role in reigniting the debate between Fishererians and Good-geners. Until that happened, Good-gene theories could work only if the ornaments they resulted in were not encumbrances to the males. Thus, a male might advertise the quality of his genes, but to do so at a high cost to himself would be counterproductive unless there were a sexy-son effect.

LOUSY MALES

The handicap theory now comes face-to-face with the central conundrum of sexual selection. This is the lek paradox: that peahens are constantly skimming off the cream of the genetic cream by choosing only the very few best males to mate with, and as a result, within a very few generations, no variety is left to choose from. The good-gene assertion that mutations are likely to make ornaments and displays less effective provides a partial answer, but it is not a persuasive one. After all, it argues only for not choosing the worst rather than for choosing the best.

Only the Red Queen can solve our dilemma. What sexual selection theory seems to have concluded is that females are constantly running (by being so selective) but are staying in the same place (having no variety to select from). When we find that, we should be on the lookout for some ever-changing enemy, some arms-race rival. It is here that we meet Bill Hamilton again. We last encountered him when discussing the idea that sex itself is an essential part of the battle against disease. If the main purpose of sex is to grant your descendants immunity from parasites, then it follows directly that it makes sense to seek a mate with parasite-resistance genes. AIDS has reminded us all too forcibly of the value of choosing a healthy sexual partner, but similar logic applies to all diseases and parasites. In 1982, Hamilton and a colleague, Marlene Zuk (now at the University of California at Riverside), suggested that parasites might hold the key to the lek paradox and to gaudy colors and peacocks' tails, for parasites and their hosts are continually changing their genetic locks and keys to outwit each other. The more common a particular strain of host is in one generation, the more common the strain of parasite is that can overcome its defenses in the next. And vice versa: Whatever strain of host is most resistant to the prevalent strain of parasite will itself be the prevalent strain of host in the next generation. Thus, the most disease resistant male might often turn out to be the descendant of the least resistant one in a previous generation. The lek paradox is thus solved at a stroke. By choosing the healthiest male in each generation, females will be picking a different set of genes each time and never run out of genetic variety to select from.[40]

The Hamilton-Zuk parasite theory was bold enough, but the two scientists did not stop there. They looked up the data for 109 species of bird and found that the most brightly colored species were also the ones most troubled by blood parasites. That claim has been challenged and much debated, but it seems to hold up. Zuk found the same in a survey of 526 tropical birds, and others found it to be true of birds of paradise and some species of freshwater fish[41]—the more parasites, the showier the species. Even among human beings, the more polygamous a society, the greater its parasite burden, though it is not clear if this means anything.[42] And these might be no more than suggestive coincidences; correlation does not imply cause. Three kinds of evidence are needed to turn their conjecture into a fact: first, that there are regular genetic cycles in hosts and parasites; second, that ornaments are especially good at demonstrating freedom from parasites; third, that females choose the most resistant males for that reason rather than the males just happening to be the most resistant.

The evidence has been pouring in since Hamilton and Zuk first published their theory. Some of it supports them, some does not. None quite meets all the criteria set forth above. Just as the theory predicts that the more flamboyant *species* should be the ones most troubled by parasites, so it predicts that *within a species* the more flamboyant a male's ornament, the lower his parasite burden. This proves to be true in diverse cases; it is also true that females generally favor males with fewer parasites. This holds for sage grouse, bowerbirds, frogs, guppies, even crickets.[43] In swallows, females prefer males with longer tails; those males have fewer lice, and their offspring inherit louse resistance even when reared by foster swallow parents.[44] Something similar is suspected in pheasants and jungle fowl (the wild species to which domestic chickens belong).[45] Yet these are deeply unshocking results. It would have been far more surprising to find females being seduced by sick, scrawny males than to find them succumbing to the charms of the healthiest. After all, they might be avoiding a sick male for no better reason than that they do not wish to catch his bug.[46]

Experiments done on sage grouse have begun to satisfy some of the skeptics. Mark Boyce and his colleagues at the University of Wyoming found that male grouse sick with malaria do poorly, and so do males covered with lice. They noticed, too, that the

lice were easy to notice because they left spots on the males' inflat-
ed air sacs. By painting such spots on a healthy male's sac, Boyce
and his colleagues were able to reduce his mating success.[47] If they
could go on to show cycles from one resistance gene to another
mediated by female choice, they would have given the Good-gene
theory a significant boost.

THE SYMMETRY OF BEAUTY

In 1991, Anders Møller and Andrew Pomiankowski stumbled on a
possible way of settling the civil war between Fisher and Good-
genes: symmetry. It is a well-known developmental accident that
animals' bodies are more symmetrical if they were in good condition when growing up, and they are less symmetrical if they were
stressed while growing. For example, scorpionflies develop more
symmetrically when fathered by well-fed fathers that could afford
to feed their wives. The reason for this is simply the old wrench-in-
the-works argument: Making something symmetrical is not easy. If
things go wrong, the chances are it will come out asymmetrical.[48]

   Most body parts, such as wings and beaks, should therefore
be most symmetrical when they are just the right size and be the
least symmetrical when stress has left them too small or too large.
If Good-geners are right, ornaments should be the most symmetrical when they are the largest because large ornaments indicate the
best genes and the least stress. If Fisherians are right, you would
expect no relationship between ornament size and symmetry; if
anything, the largest ornaments should be the least symmetrical
because they reflect nothing about the owner other than that he can
grow the largest ornament.

   Møller noticed that, among the swallows he studied, the
longest tails of the males were also the most symmetrical. This was
quite unlike the pattern of other feathers, such as wings, which
obeyed the usual rule: The most symmetrical were the ones closest
to the average length. In other words, whereas most feathers show a
U-shaped curve of asymmetry against length, tail streamers show a

steady upward progression. Since the swallows with the longest tails are the most successful in securing mates, it follows that the most symmetrical tails are also doing better. So Møller cut or elongated the tail feathers of certain males and at the same time enhanced or reduced the symmetry of the tails. Those with longer tails got mates sooner and reared more offspring, but within each class of length, those with enhanced symmetry did better than those with reduced symmetry.[49]

Møller interprets this as unambiguous evidence in favor of Good-genes, for it shows that a condition-dependent trait—symmetry—is sexually selected. He joined forces with Pomiankowski to begin to separate those ornaments that show a correlation between symmetry and size from those that do not—in effect, to separate Good-genes from Fisher. Their initial conclusion was that animals with single ornaments—such as a swallow with a long tail—are Good-geners and show increasing symmetry with increasing size, whereas animals with multiple ornaments—such as a pheasant with its long tail, red facial roses, and colorful feather patterns—are mostly Fisherian, showing no relationship between size and symmetry. Since then, Pomiankowski has returned to the subject from a different angle, arguing that Fisher and many ornaments are likely to predominate when the cost to females of choosing is cheap; Good-genes will predominate when the cost of choosing is high. Again we reach the same conclusion: Peacocks are Fisherian; swallows are Good-geners.[50]

HONEST JUNGLE FOWL

So far I have considered the evolution of male ornaments mainly from the female's point of view because it is her preferences that drive that evolution. But in a species such as a peafowl, where female choice of mate rules, the male is not entirely a passive spectator of his evolutionary fate. He is both an ardent suitor and an eager salesman. He has a product to sell—his genes, perhaps—and information to impart about that product, but he does not simply hand the information over and await the peahen's decision. He is out to persuade her, to seduce her. And just as she is descended from females who made a careful choice, so he is descended from males who made a hard sell.

The analogy of the sales pitch is revealing, for advertisers do not promote their product merely by providing information about it. They fib, exaggerate, and try to associate it with pleasurable images. They sell ice cream using sexy pictures, airplane tickets using couples walking hand in hand on beaches, instant coffee using romance, and cigarettes using cowboys.

When a man wants to seduce a woman, he does not send her a copy of his bank statement but a pearl necklace. He does not send her his doctor's report but lets slip that he runs twenty miles a week and never gets colds. He does not tell her what degree he got but instead dazzles her with wit. He does not display testaments to how thoughtful he is but sends her roses on her birthday. Each gesture has a message: I'm rich, I'm fit, I'm clever, I'm nice. But the information is packaged to be more seductive and more effective, just as the message "Buy my ice cream" catches the eye when it is accompanied by a picture of two good-looking people seducing each other.

In courtship, as in the world of advertising, there is a discrepancy of interests between the buyer and the seller. The female needs to know the truth about the male: his health, wealth, and genes. The male wants to exaggerate the information. The female wants the truth; the male wants to lie. The very word seduction implies trickery and manipulation.[51]

Seduction therefore becomes a classic Red Queen contest, although this time the two protagonists are male and female, not host and disease. Zahavi's handicap theory, as explored by Hamilton and Zuk, predicted that honesty would eventually prevail and males who cheat would be revealed. This is because the handicap is the female's criterion of choice for the very reason that it reveals the male's state of health.

The red jungle fowl is the ancestor of the domestic chicken. Like a farmyard rooster, the cock is equipped with a good many

ornaments that his mate does not share: long, curved tail feathers, a
bright ruff around the neck, a red comb on the crown of his head,
and a loud dawn call, to name the most obvious. Marlene Zuk
wanted to find out which of these mattered to female jungle fowl,
so she presented sexually receptive hens with two tethered males
and examined which they chose. In some of the trials one of the
cocks was reared with a roundworm infection in his gut, which
affected his plumage, beak, and leg length very little but showed
clearly in his comb and eye color, both of which were less colorful
than in healthy males. Zuk found that hens preferred cocks with
good combs and eyes but paid less attention to plumage. She failed
to make hens go for males with fake red elastic combs on their
heads, however; they found them too bizarre. Nonetheless, it was
clear that hens paid most attention to the most health informative
feature of a cock.[52]

Zuk knew that poultry farmers, too, observe the comb and
wattles of a cockerel to judge his health. What intrigued her was
the idea that the wattles were more "honest" about the state of a
cockerel than his feathers. Many birds, especially in the pheasant
family, grow fleshy structures about their faces to emphasize during display: Turkeys grow long wattles over their beaks, pheasants
have fleshy red "roses" on their faces, sage grouse bare their air
sacs, and tragopans have expandable electric blue bibs beneath their
chins.

A cockerel's comb is red because of the carotenoid pigments
in it. A male guppy fish is rendered orange by carotenoids also, and
a housefinch's and a flamingo's red plumage also depends on
carotenoids. The peculiar thing about carotenoids is that birds and
fish cannot synthesize them within their own tissues; they extract
them from their food—from fruit, shellfish, or other plants and
invertebrates. But their ability to extract carotenoids from their
food and deliver it to their tissues is greatly affected by certain
parasites. A cockerel affected by the bacterial disease coccidiosis,
for example, accumulates less carotenoid in his comb than a healthy
cockerel—even when both animals have been fed equal quantities
of carotenoid. Nobody knows exactly why the parasites have this

specific biochemical effect, but it seems to be unavoidable and is therefore extremely useful to the female: The brightness of carotenoid-filled tissues is a visible sign of the levels of parasite infection. It is not surprising that red and orange are common colors in fleshy ornaments used in display, such as the combs, wattles, and lappets of pheasants and grouse.[53]

The size and brightness of such combs may be *affected* by parasites, but they are *effected* by hormones. The higher the level of testosterone in the blood of a cockerel, the bigger and brighter his comb and wattles will be. The problem for the cockerel is that the higher his level of testosterone, the greater his parasite infestation. The hormone itself seems to lower his resistance to parasites.[54] Once again nobody knows why, but cortisol, the "stress" hormone that is released into the bloodstream during times of emotional crisis, also has a marked effect on the immune system. A long study of cortisol levels in children in the West Indies revealed that the children are much more likely to catch an infection shortly after their cortisol levels have been high because of family tension or other stress.[55] Cortisol and testosterone are both steroid hormones, and they have a remarkably similar molecular structure. Of the five biochemical steps needed to make cholesterol into either cortisol or testosterone, only the last two steps are different.[56] There seems to be something about steroid hormones that unavoidably depresses immune defense. This immune effect of testosterone is the reason that men are more susceptible to infectious diseases than women, a trend that occurs throughout the animal kingdom. Eunuchs live longer than other men, and male creatures generally suffer from higher mortality and strain. In a small Australian creature called the marsupial mouse, all the males contract fatal diseases during the frantic breeding season and die. It is as if male animals have a finite sum of energy that they can spend on testosterone or immunity to disease, but not both at the same time.[57]

The implication for sexual selection is that it does not pay to lie. Having sex-hormone levels that are too high increases the size of your ornaments but makes you more vulnerable to parasites, which are revealed in the state of those ornaments. It is possible that it works in the other direction: The immune system suppresses the production of testosterone. In Zuk's words, "Males are thus necessarily more vulnerable to disease as they acquire the accoutrements of maleness."[58]

The best proof of these conjectures comes from a study of roach, which are small fish with reddish fins, in the Lake of Biel in Switzerland. Male roach grow little tubercules all over their bodies during the breeding season, which seem to stimulate females during courtship as the fish rub against each other. The more parasites a male has, the fewer tubercules he grows. It is possible for a zoologist to judge, just from a male's tubercules, whether he is infested with a roundworm or a flatworm. The implication follows: If a zoologist can deduce which parasite is present, a female roach probably can as well. This pattern results from different kinds of sex hormones; one can be raised in concentration only at the expense of leaving the roach vulnerable to one kind of parasite; the other can be raised only at the expense of lowering defenses against another kind of parasite.[59]

If cockerels' wattles and roach tubercules are honest signals, so presumably are songs. A nightingale that can sing loud and long must be in vigorous health, and one that has a large repertoire of different melodies must be experienced or ingenious, or both. An energetic display such as the pas de deux of a pair of male manakins may also be an honest signal. A bird that merely shows its feathers, such as a peacock or a bird of paradise, might be a cheat whose strength has been sapped by bad habits since he grew the plumes. After all, peacock feathers still shine brightly when their owner is dead and stuffed. Perhaps it is no surprise, then, that most male birds do not molt just before the breeding season but adopt their spring plumage the autumn before. They have to keep it tidy all winter. The very fact that a male has looked after his plumes for six months tells a female something about his enduring vigor. Bill Hamilton points out that white fluffy feathers around a bird's rear end, which are common in grouse of various kinds, must be especially hard to keep clean if the bird has diarrhea.[60]

Zahavi certainly believed that honesty was a prerequisite of handicaps, and vice versa. To be honest, he thought, an ornament must be costly; otherwise it could be used to cheat. A deer cannot grow large antlers without consuming five times its normal daily intake of calcium; a pupfish cannot be iridescent blue unless it is genuinely in good condition, a fact that will be tested by other male fish in fights. On the assumption that anybody who refuses to play the game and use an honest signal must have something to hide, males are likely to find themselves dragged into honest displays. Therefore, display ornaments are examples of "truth in advertising."[61]

All this is very logical, but in about 1990 it started to make one group of biologists uneasy. They had an instinctive aversion to the idea that sexual advertising is about the truth because they knew that television advertising is not about passing on information; it is about manipulating the viewer. In the same way, they argued, all animal communication is about manipulating the receiver.

The first and most eloquent (manipulative?) champions of this view were two Oxford biologists, Richard Dawkins and John Krebs. According to them, a nightingale does not sing to inform potential mates about himself; he sings to seduce them. If that means lying about his true prowess, so be it.[62] Perhaps an ice cream advertisement is honest in a simplistic sense because it gives the name of the brand, but it is not honest in implying that sex is sure to follow after every spoonful. Such a crude lie can surely be perceived by that genius of the animal kingdom, humans. But it is not. Advertising works. Brand names are better known if they are advertised with sexy or alluring pictures, and better-known brands sell better. Why does it work? Because the price the consumer would have to pay in ignoring the subliminal message is just too high. It is better to be fooled into buying the second-best ice cream than go to the bother of educating yourself to resist the salesmanship.

Any peahens reading this might begin to recognize their dilemma. For they, too, may be fooled by the male's display into buying the second-best male. Remember, the lek paradox argues that there is little to choose between males on a lek anyway because they were all fathered by the same few males in the previous generation. So two theories—truth in advertising and dishonest manipulation—seem to come to opposite conclusions. Truth in advertising concludes that females will discover a cheating seducer; dishonest manipulation concludes that males will seduce females against their better judgment.

WHY DO YOUNG WOMEN HAVE NARROW WAISTS?

Marian Dawkins and Tim Guilford of Oxford have recently suggested a resolution to this conundrum. As long as detecting the dishonesty in the signal is costly to the female, it might not be worth her while to do so. In other words, if she has to risk her life seeking out and comparing many males to ensure that she has chosen the best one, then the marginal advantage she gains by picking the best one is outweighed by the risk she has run. It is better to let herself be seduced by a good one than to have the best become the enemy of the good. After all, if she cannot easily distinguish the truthful from the dishonest badge of quality, then other females will not, either, and so her sons will not be punished for any dishonesty they inherit from their father.[63]

A startling example of this sort of logic comes from a controversial theory about human beings that was developed a few years ago by Bobbi Low and her colleagues at the University of Michigan. Low was looking to explain why young women have fat on their breasts and buttocks more than on other parts of their bodies. The reason this requires explaining is that young women are different from other human beings in this respect. Older women, young girls, and men of all ages gain fat on their torsos and limbs much more evenly. If a woman of twenty or so gains weight, it largely takes the form of fat on the breasts and buttocks; her waist can remain remarkably narrow.

So much is undisputed fact. What follows is entirely conjecture, and it was a conjecture that caused Low a good deal of sometimes vicious (and mostly foolish) criticism when she published the idea in 1987.

Twenty-year-old women are in their breeding prime; there-fore, the unusual pattern of fat distribution might be expected to be connected with getting a mate or bearing children. Standard explanations concern the bearing of children; for example, fat is inconvenient if it competes for space about the waist with a fetus. Low's explanation concerns the attraction of mates and takes the form of a Red Queen race between males and females. A man look-ing for a wife is likely to be descended from men who found two things attractive (among many others): big breasts, for feeding his children, and wide hips, for bearing them. Death during infancy due to a mother's milk shortage would have been common before modern affluence—and still is in some parts of the world. Death of the mother and infant from a birth canal that was too narrow must also have been common. Birth complications are peculiarly frequent in humans for the obvious reason that the head size of a baby at birth has been increasing quickly in the past 5 million years. The only way birth canals kept pace (before Julius Caesar's mother was cut open) was through the selective death of narrow-hipped women.

Grant, then, that early men may have preferred women with relatively wide hips and large breasts. That still does not explain the gaining of fat on breasts and hips; fat breasts do not produce more milk than lean ones, and fat hips are no farther apart than lean ones of the same bone structure. Low thinks women who gained fat in those places may have deceived men into thinking they had milkful breasts and wide hip bones. Men fell for it—because the cost of distinguishing fat from heavy breasts or of dis-tinguishing fat from wide hips was just too great, and the opportu-nity to do so was lacking. Men have counterattacked, evolutionarily speaking, by "demanding" small waists as proof of the fact that there is little subcutaneous fat, but women have easily overcome this by keeping waists slim even while gaining fat elsewhere.[64]

Low's theory might not be right, as she is the first to admit, but it is no less logical or farfetched than any of its rivals, and for our purposes it serves to demonstrate that a Red Queen race between a dishonest advertiser (in this case, unusually, a female) and a receiver who demands honesty may not always be won by the honesty-demanding gender. It is essential, if Low is right, that fat be cheaper to gain than mammary tissue, just as it is essential, for Dawkins and Guilford, that cheating be cheaper than telling the truth.[65]

CHUCKING FROGS

The male's goal is seduction: He is trying to manipulate the female into falling for his charms, to get inside her head and steer her mind his way. The evolutionary pressure is on him to perfect displays that make her well disposed toward him and sexually aroused so that he can be certain of mating. Male scorpions lull females into the mood for sex at great risk to their lives. One false step in the seduction, and the female's mood changes so that she looks upon the male as a meal.

The evolutionary pressure on a female—assuming she benefits from choosing the best male—is to invent resistance to all but the most charming displays. To say this is merely to rephrase the whole argument of female choice with a greater emphasis on the *how* than the *why*. But such rephrasings can be illuminating, and this one has proved exceptionally so. Michael Ryan of the University of Texas rephrased the question a few years ago, and he did so partly because he studies frogs. It is easy to measure female preferences in frogs because the male sits in one spot and calls, and the female moves toward the sound of the male she likes the most. Ryan replaced the males with loudspeakers and offered each female different recordings of males to test her preference.

The male tungara frog attracts a female by making a long whine followed by a "chuck" noise. All of its close relatives except one make the whine but not the chuck. But at least one of the chuckless relatives turns out to prefer calls with chucks to those without. This was rather like discovering that a New Guinea tribesman found women in white wedding dresses more attractive than women dressed in tribal gear. It seems to indicate that the preference for the chuck just happens to exist in the fact that the female's ear (to be precise, the basilar papillae of the inner ear) is tuned to the chuck's frequency; the male has, in evolutionary terms, discovered and exploited this. In Ryan's mind this deals a blow to the whole house of female-choice theory. That theory, whether in Fisher's sexy-son form or the Good-genes form, predicts that the male's ornament and the female's preference for such an ornament will evolve together. Ryan's result seems to suggest that the preference existed fully formed before the male ever had the ornament. Peahens preferred eyed tails a million years ago when peacocks still looked like big chickens.[66]

Lest the tungara frog be thought a fluke, a colleague of Ryan's, Alexandra Basolo, has found exactly the same thing in a fish called the platyfish. Females prefer males who have had long sword-shaped extensions stuck onto their tales. Males of a different species called the swordtail have such swords on their tails, yet none of the platyfish's other relatives have swords, and it stretches belief to argue that they all got rid of the sword rather than that the swordtail acquired it. The preference for sworded tails was there, latent, in platyfish before there were swords.[67]

In one sense what Ryan is saying is unremarkable. That male displays should be suited to the sensory systems of females is only to be expected. Monkeys and apes are the only mammals with good color vision. Therefore, it is not surprising that they are the only mammals decorated with bright colors such as blue and pink. Likewise, it is hardly remarkable that snakes, which are deaf, do not sing to each other. (They hiss to scare hearing creatures.) Indeed, one could list a whole panoply of "peacocks' tails" for each of the five senses and more: the peacock's tail for vision, the nightingale's song for hearing, the scent of the musk deer for smell;[68] the pheromones of the moth for taste; the "morphological exuberance" of some insect "penises" for touch;[69] even the elaborate electrical courtship signals of some electric fish[70] for a sixth sense. Each species chooses to exploit the senses that its females are best at detecting. This is, in a sense, to return to Darwin's original idea: that females have aesthetic senses, for whatever reason, and that those senses shape male ornaments.[71]

Moreover, you would expect the males to pick the method of display that is least dangerous or costly. Those that did so would last longer and leave more descendants than those that did not. As every bird-watcher knows, the beauty of a bird's song is inversely correlated with the colorfulness of its plumage. The operatic male nightingales, warblers, and larks are brown and usually almost indistinguishable from their females. Birds of paradise and pheasants (in which the males are gorgeous, the females dull) are monotonous, simple songsters given to uninspired squawks. Intriguingly, the same pattern holds among the bowerbirds of New Guinea and Australia: The duller the bird, the more elaborate and decorated its bower. What this suggests is that nightingales and bowerbirds have transferred their color to their songs and bowers. There are clear advantages to doing so. A songster can switch his ornament off when danger threatens. A bower builder can leave his behind.[72]

More direct evidence of this pattern comes from fish. John Endler of the University of California at Santa Barbara studies the courtship of guppies and is especially interested in the colors adopted by male guppies. Fish have magnificent color vision; whereas we use three different types of color-detecting cells in the eye (red, blue, and green), fish have four, and birds have up to seven. Compared to the way birds see the world, our lives are monochrome. But fish also have a very different experience from us because their world filters out light of different colors in all sorts of variable ways. The deeper they live, the less red light penetrates compared with blue. The browner the water, the less blue light penetrates. The greener the water, the less red or blue light penetrates. And so on. Endler's guppies live in South American rivers; when courting, they are usually in clear water where orange, red, and blue are the colors that show up best. Their enemies, however, are fish that live in water where yellow light penetrates best. Not surprisingly, male guppies are never yellow.

The males use two kinds of color, one red-orange, which is produced by a carotenoid pigment that the guppy must acquire from its food, and the other blue-green, which is caused by guanine crystals in the skin that are laid down when the guppy reaches maturity. Female guppies that live in tea-colored water, where red-orange is more easily seen, are more sensitive to red-orange light than to blue, which makes sense. The brains of such guppies are tuned to exactly the wavelength of the red-orange carotenoid pigment the male uses in display—and perhaps vice versa.[73]

OF MOZART AND GRACKLE SONG

Down the corridor from Ryan at the University of Texas is Mark Kirkpatrick, who is prepared to upset even more apple carts. Kirkpatrick is acknowledged as one of those who understands sexual selection theory most thoroughly; indeed, he was one of those who made Fisher's idea mathematically respectable in the early 1980s. But he now refuses to accept that we must choose between Fisher and Zahavi. He does so partly because of what Ryan has discovered.

This does not mean Kirkpatrick rejects female choice, as Julian Huxley did. Whereas Huxley thought males did the choosing by fighting among themselves, Kirkpatrick prefers to believe that in many species the females do choose, but their preferences do not evolve. They merely saddle the males with their own idiosyncratic tastes.

Both Good-genes and Fisher theories are obsessed with trying to find a reason for exuberant display that benefits the male. Kirkpatrick looks at it from the female's point of view. Suppose, he says, that peahens' preferences have indeed saddled peacocks with their tails. Why must we explain these female preferences only in terms of the effects on their sons and daughters? Might the peahens not have perfectly good direct reasons for choosing as they do? Might their preferences not be determined by something else entirely? He thinks "other evolutionary forces acting on the preferences will overwhelm the Good-genes factor and often establish female preferences for traits that decrease male survival."[74]

Two recent experiments support the idea that females simply have idiosyncratic tastes that have not evolved. Male grackles—blackish birds of medium size—sing only one kind of song. Female grackles prefer to mate with males that sing more than one kind of song. William Searcy of the University of Pittsburgh discovered why. He made use of the fact that a female grackle will go up to singing loudspeakers and adopt a soliciting posture as if waiting to be mated. Her tendency to do so declines, however, as she gets bored with the song. Only if the loudspeaker starts singing a new song will her soliciting start afresh. Such "habituation" is just a property of the way brains work; our senses, and those of grackles, notice novelty and change, not steady states. The female preference did not evolve; it just is that way.[75]

Perhaps the most startling discovery in sexual selection theory was Nancy Burley's work on zebra finches in the early 1980s. She was studying how these small Australian finches choose their mates, and to make it easier she kept them in aviaries and marked each one with a colored ring on its leg. After a while she noticed something odd: The males with red rings seemed to be preferred by the females. Further experiments proved that the rings were drastically affecting the "attractiveness" of both males and females. Males with red rings were attractive; those with green rings unattractive; females with black or pink rings were preferred; those with light blue rings disliked. It was not just rings. Little paper hats glued to the birds' heads also altered their attractiveness. Female zebra finches have a rather simple rule for assessing potential mates: The more red he has on his body (or the less green, which comes to the same thing given that red and green are seen as opposites by the brain), the more attractive he is.[76]

If females have an existing aesthetic preference, it is only logical that males will evolve to exploit that preference. For example, it is possible that the "eyes" on a peacock's tail are seductive to peahens because they resemble huge versions of real eyes. Real eyes are visually arresting—perhaps even hypnotic—to many kinds of animals, and the sudden appearance of many huge staring eyes may induce a state of mild hypnosis in the peahen, which allows the peacock to lunge at her.[77] This would be consistent with the common discovery that "supernormal stimuli" are often more effective than normal ones. For example, many birds prefer a ridiculous giant egg in their nest to a normal one; a goose will prefer to try to sit on an egg the size of a soccer ball than one of normal size. It is as if their brains have a program that says "like eggs," and the bigger the egg, the more it likes them. So perhaps the bigger the eye-spot, the more attractive or startling it is for a peahen, and the male has simply exploited this by evolving lots of giant eyes without any evolutionary change in the female's preference.[78]

HANDICAPPED ADVERTISERS

Andrew Pomiankowski of London accepts much of what Ryan and Kirkpatrick say but parts company with them on the matter of female choice. He says that what they are considering is merely a constraint that channels the male's trait into the preferred direction of the female's sensory bias. But that does not mean the exaggeration happens without the female's preference changing. It is almost impossible to see how females could avoid the Fisher effect as the male's ornament gets more exaggerated generation by generation. The female who is most discriminating picks the sexiest male and so has the sexiest sons; the one who has the sexiest sons has the most granddaughters. So females get more and more discriminating and more and more difficult to seduce or hypnotize. "The crucial question," wrote Pomiankowksi, "is not whether sensory exploitation has been involved but why females have allowed themselves to be exploited." Besides, it is an impoverished view of selection to believe that a frog's ear can be tuned for detecting predators but not tuned simultaneously and differently to choosing males.[79]

Thus, it is possible to argue with Ryan and Kirkpatrick that male courtship extravagances reflect the innate tastes of females without abandoning the idea that those tastes are of use to the females in that they select the best genes for the next generation. A peacock's tail is, simultaneously, a testament to naturally selected female preferences for eyelike objects, a runaway product of despotic fashion among peahens, and a handicap that reveals its possessor's condition. Such tolerant pluralism is not to everybody's taste, but Pomiankowski insists it does not stem from misguided desire to please everybody. On a paper napkin in an Indian restaurant one day he sketched out for me a plausible account of all the sexual selection theories working in concert.

Each male trait begins as a chance mutation. If it happens to hit a sensory bias of the female, it starts to spread. As it spreads, the Fisher effect takes over, and both the trait and the preference are exaggerated. Eventually the point is reached where the trait has spread to all males, and there is no point in females following the fashion anymore. It starts to fade again, under pressure from the fact that there is now a cost to female choice; if nothing else, it is a waste of females' time and effort to compare different males. The Fisher effect fades more slowly when that cost is small—for example, in lekking species where the males can all be viewed at once. But some traits do not fade because it so happens that they reflect the underlying health of their possessors—they change color if the male is infected with parasites, for example. And therefore females do not stop choosing the best males at all. They keep picking (or being seduced by) the fanciest male because if they do, they will have disease-resistant offspring. In other words, condition-reflecting traits will not be the only ones brought to an exaggerated state, but they will be the ones that persist the longest. And all the Fisher-exaggerated traits remain in lekking species as well because the cost of choosing is so small. The most promiscuous species end up a collage of different handicaps, ornaments, and gaudy blotches. Pomiankowski has since begun to confirm his intuition (based on the symmetry idea discussed earlier) that multiple traits on polygamous birds, such as the many adornments of a peacock, are Fisher ornaments, while single features on monogamous birds, such as the swallow's forked tail, are Good-gene ornaments, or condition-revealing handicaps.[80]

The next time you visit a zoo in the spring, try to watch a male Lady Amherst pheasant from China posturing before a hen. He is a riot of color: His face is pale green, his crest scarlet, his throat iridescent green, his back emerald, his rump orange, and his belly pristine white. Around his neck is a white ruff trimmed with black, and at the base of his tail are five pairs of vermillion feathers. His tail, white barred with black, is longer than his body. A dull or damaged feather would stand out anywhere on his body. He is one great advertisement for good genes, handicapped by the need to keep clean, healthy, and out of danger, a walking illustration of his mate's evolved sensory biases.

THE HUMAN PEACOCK

The antics of peacocks and guppies are interesting enough in themselves to naturalists; to students of evolution they are intriguing as test cases; but to the rest of us what makes them worth studying is pure self-centeredness. We want to know what lessons they teach us about human affairs. Are some men successful with women because their appearance sends an honest signal of their handicapping good genes and their ability to resist disease?

The idea is ridiculous. Men succeed with women for much more varied and subtle reasons: They are kind or clever or witty or rich or good-looking or just available. Humans are simply not a lekking species. Men do not gather in groups to display for passing women. Most men do not abandon women immediately after copulation. Men are not equipped with gorgeous ornaments or stereotyped courtship rituals, however it may look in the average discotheque. When a woman chooses a man to mate with, she is less concerned with whether he can father sexy sons or disease-resistant daughters than whether he would make a good husband. A man choosing a wife uses equally mundane considerations, though he is perhaps more of a sucker for beauty. Both genders use criteria that bear on parental abilities. They are more like terns, who choose mates that can fish well, than sage grouse hens, who copy one another's choice of a fast-displaying male. So the Red Queen race between the genders over seduction and sales resistance that follows from pure Good-gene choice does not happen.

And yet we cannot be so categorical. There are species of

mammal in which the effects of sexual selection are few and small. It is hard to argue that the average rat has been endowed with conspicuous display ornaments by the preferences of ancestral females. Even our closest relatives, the chimpanzees, are little touched by the effects of female choice: Males look much like females, and courtship is somewhat simple. But we should pause before dismissing the effects of sexual selection on human beings. People, after all, are universally interested in beauty. Lipstick, jewelry, eye shadow, perfume, hair dyes, high heels—people are just as willing to exaggerate or lie about their sexually alluring traits as any peacock or bowerbird. And as the list above makes clear, it seems as if men seek female beauty rather more than women seek male beauty. The human being, in other words, may be the victim of generations of male choice even more than female choice. If we are to apply sexual selection theory to man, it is male choice for female genes that we should examine. But it makes little difference. When one gender is being choosy, all the consequences of sexual selection theory inevitably flow. It is quite possible, even likely, as the next few chapters will reveal, that some parts of the human body and psyche have been sexually selected.

Chapter 6

POLYGAMY AND THE NATURE OF MEN

If women didn't exist, all the money in the world would have no meaning.

—Aristotle Onassis

Power is a great aphrodisiac.

—Henry Kissinger

In the ancient empire of the Incas, sex was a heavily regulated industry. The sun-king Atahualpa kept fifteen hundred women in each of many "houses of virgins" throughout his kingdom. They were selected for their beauty and were rarely chosen after the age of eight—to ensure their virginity. But they did not all remain virgins for long: They were the emperor's concubines. Beneath him, each rank of society afforded a harem of a particular legal size. Great lords had harems of more than seven hundred women. "Principal persons" were allowed fifty women; leaders of vassal nations, thirty; heads of provinces of 100,000 people, twenty; leaders of 1,000 people, fifteen; administrators of 500 people, twelve; governors of 100 people, eight; petty chiefs over 50 men, seven; chiefs of 10 men, five; chiefs of 5 men, three. That left precious few for the average male Indian whose enforced near-celibacy must have driven him to desperate acts, a fact attested to by the severity of the penalties that followed any cuckolding of his seniors. If a man violated one of Atahualpa's women, he, his wife, his children, his relatives, his servants, his fellow villagers, and all his lamas would be put to death, the village would be destroyed, and the site strewn with stones.

As a result, Atahualpa and his nobles had, shall we say, a majority holding in the paternity of the next generation. They systematically dispossessed less privileged men of their genetic share of posterity. Many of the Inca people were the children of powerful men.

In the kingdom of Dahomey in West Africa, all women were

at the pleasure of the king. Thousands of them were kept in the royal harem for his use, and the remainder he suffered to "marry" the more favored of his subjects. The result was that Dahomean kings were very fecund, while ordinary Dahomean men were often celibate and barren. In the city of Abomey, according to one nineteenth-century visitor, "it would be difficult to find Dahomeans who were not descended from royalty."

The connection between sex and power is a long one.[1]

MANKIND, AN ANIMAL

So far this book has taken only a few, sideways glances at human beings. This is deliberate. The principles I have been trying to establish are better illustrated by aphids, dandelions, slime molds, fruit flies, peacocks, and elephant seals than they are by one peculiar ape. But the peculiar ape is not immune to those principles. Human beings are a product of evolution as much as any slime mold, and the revolution of the last two decades in the way scientists now think about evolution has immense implications for mankind as well. To summarize the argument so far, evolution is more about reproduction of the fittest than survival of the fittest; every creature on earth is the product of a series of historical battles between parasites and hosts, between genes and other genes, between members of the same species, between members of one gender in competition for members of the other gender. Those battles include psychological ones, to manipulate and exploit other members of the species; they are never won, for success in one generation only ensures that the foes of the next generation are fitter to fight harder. Life is a Sisyphean race, run ever faster toward a finish line that is merely the start of the next race.

This chapter begins to follow the logic of these arguments into the heart of human behavior. Those who think this unjustified on the grounds that human beings are unique usually advance one of two arguments: that in humans everything about behavior is learned, and nothing is inherited; or inherited behavior is inflexible behavior, and human beings are clearly flexible. The first argument is an exaggeration, the second false. A man does not experience lust because he learned it at his father's knee; a person does not feel hunger or anger because she was taught it. They are human nature. We are born with the potential to develop lust, hunger, and anger. We learn to direct hunger at hamburgers, anger at delayed trains, and lust at the object of our affection—when appropriate. So we have "changed" our "nature." Inherited tendencies permeate everything we do, and they are flexible. There is no nature that exists devoid of nurture; there is no nurture that develops without nature. To say otherwise is like saying that the area of a field is determined by its length but not its width. Every behavior is the product of an instinct trained by experience.

The study of human beings remained resolutely unreformed by these ideas until a few years ago. Even now, most anthropologists and social scientists are firmly committed to the view that evolution has nothing to tell them. Human bodies are products of natural selection; but human minds and human behavior are products of "culture," and human culture does not reflect human nature, but the reverse. This restricts social scientists to investigating only differences between cultures and between individuals—and to exaggerating them. Yet what is most interesting to me about human beings is the things that are the same, not what is different—things like grammatical language, hierarchy, romantic love, sexual jealousy, long-term bonds between the genders ("marriage," in a sense). These are trainable instincts peculiar to our species and are just as surely the products of evolution as eyes and thumbs.[2]

THE POINT OF MARRIAGE

For a man, women are vehicles that can carry his genes into the next generation. For a woman, men are sources of a vital substance (sperm) that can turn their eggs into embryos. For each gender the other is a sought-after resource to be exploited. The question is, *how*? One way to exploit the other gender is to round up as many as possible of them and persuade them to mate with you, then desert them, as bull elephant seals do. The opposite extreme is to find one individual and share all the duties of parenthood equally, as albatrosses do. Every species falls somewhere on that spectrum, with its own characteristic "mating system." Where does humanity fall?

There are five ways to find out. One is to study modern people directly and describe what they do as the human mating system. The answer is usually monogamous marriage. A second way is to look at human history and divine from our past what sexual arrangements are typical of our species. But history teaches a dismal lesson: A common arrangement from our past was that rich and powerful men enslaved concubines in large harems. A third way is to look at people living in simple societies with Stone Age technologies and conjecture that they live much as our ancestors lived ten millennia ago. They tend to fall between the extremes: less polygamous than early civilizations, less monogamous than modern society. The fourth technique is to look at our closest relatives, the apes, and compare our behavior and anatomy with theirs. The answer that emerges is that men's testicles are not large enough for a system of promiscuity like the chimpanzee's, men's bodies are not big enough for a system of harem polygamy like the gorilla's (there is an iron link between harem polygamy in a species and a large size differential between male and female), and men are not as antisocial and adjusted to fidelity as the monogamous gibbon. We are somewhere in between. The fifth method is to compare humans with other animals that share our highly social habits: with colonial birds, monkeys, and dolphins. As we shall see, the lesson they teach is that we are designed for a system of monogamy plagued by adultery.

It is at least possible to rule out some options. There are characteristically human things that we do, such as form lasting bonds between sexual partners, even when polygamous. We are not like sage grouse whose marriages last for minutes. Nor are we polyandrous, like the jacana or lily-trotter, a tropical water bird that has big fierce females that control harems of small domesticated males. There is only one truly polyandrous society on Earth; it is in

Tibet and consists of women who marry two or more brothers simultaneously in an attempt to put together a family unit that is economically viable in a harsh land where men herd yaks to support women. The junior brother's ambition is to leave and obtain his own wife, so polyandry is plainly a second-best outcome for him.[3] Nor are we like the robin or the gibbon, which are strictly territorial, each pair monopolizing and defending a home range sufficient to live their whole lives within. We build garden fences, but even our homes are often shared with lodgers or fellow apartment dwellers, and most of our lives are spent on some form of common ground, at work, shopping, traveling, entertaining ourselves. People live in groups.

None of this is much help, then. Most people live in monogamous societies, but this may only tell us what democracy usually prescribes, not what human nature seeks. Relax the antipolygamy laws and it flourishes. Utah has a tradition of theologically sanctioned polygamy and in recent years has been less forceful about prosecuting polygamists, so the habit has reemerged. Although the most populous societies are monogamous, about three-quarters of all tribal cultures are polygamous, and even the ostensibly monogamous ones are monogamous in name only. Throughout history powerful men have usually had more than one mate each, even if they have had only one legitimate wife. However, that is for the powerful. For the rest, even in openly polygamous societies, most men have only one wife and virtually all women have only one husband. That leaves us precisely nowhere. Mankind is a polygamist and a monogamist, depending on the circumstances. Indeed, perhaps it is foolish even to talk of humans having a mating system at all. They do what they want, adapting their behavior to the prevailing opportunity.[4]

WHEN MALES POUNCE AND FEMALES FLIRT

Until recently, evolutionists had a fairly simple view of mating systems based on the essential differences between males and females.

If powerful men had their way, women would probably live in harems like seals; that is certainly the lesson of history. If most women had their way, men would be as faithful as albatrosses. Although research has modified this supposition, it is nonetheless true that males are generally seducers and females the seduced. Humanity shares this profile of ardent, polygamist males and coy, faithful females with about 99 percent of all animal species, including our closest relatives, the apes.

Consider, for example, the question of marriage proposals. In no society on earth do they usually come from the woman or her family. Even among the most liberated of Westerners, men are expected to ask and women to answer. The tradition of women asking men on Leap Year's Day reinforces the very paucity of their opportunities: They get one day to pop the question for every 1,460 that men can do so. It is true that many modern men do not go down on one knee but "discuss" the matter with their girlfriends as equals. Yet even so, the subject is usually first raised by the man. And in the matter of seduction itself, once more it is the male who is expected to make the first move. Women may flirt, but men pounce.

Why should this be? Sociologists will blame it on conditioning, and they are partly right. But that is not a sufficient answer because in the great human experiment called the 1960s much conditioning was rejected yet the pattern survives. Besides, conditioning usually reinforces instinct rather than overrides it. Since an insight of Robert Trivers's in 1972,[5] biologists have had a satisfying explanation for why male animals are usually more ardent suitors than females and why there are exceptions to the rule. There seems to be no reason why it should not also apply to people. The gender that invests the most in creating and rearing the offspring, and so forgoes most opportunities for creating and rearing other offspring, is the gender that has the least to gain from each extra mating. A peacock grants a peahen one tiny favor: a batch of sperm and nothing else. He will not guard her from other peacocks, feed her, protect a food supply for her, help her incubate her eggs, or help her bring up the chicks. She will do all the work.

Therefore, when she mates with him, it is an unequal bargain. She brings him the promise of a gigantic single-handed effort to make his sperm into new peacocks; he brings just the tiniest—though seminal—contribution. She could choose any peacock she likes and has no need to choose more than one. At the margin, he loses nothing and gains much by mating with every female who comes along; she loses time and energy for a futile gain. Every time he seduces a fresh female, he wins the jackpot of her investment in his sons and daughters. Every time she seduces a fresh peacock, she wins a little extra sperm that she probably does not need. No wonder he is keen on quantity of mates, and she on quality.

In more human terms, men can father another child just about every time they copulate with a different woman, whereas women can bear the child of only one man at a time. It is a fair bet that Casanova left more descendants than the Whore of Babylon.

This basic asymmetry between the genders goes right back to the difference in size of a sperm and an egg. In 1948 a British scientist named A. J. Bateman allowed fruit flies to mate with one another at will. He found that the most successful females were not much more prolific than the least successful, but the most prolific males were far more successful than the least prolific males.[6] The asymmetry has been greatly enhanced by the evolution of female parental care, which reaches its zenith in mammals. A female mammal gives birth to a gigantic baby that has been nurtured inside her for a long time; a male can become a father in seconds. Women cannot increase their fecundity by taking more mates; men can. And the fruit fly rule holds. Even in modern monogamous societies, men are far more likely to have lots of children than women are. For instance, men who marry twice are more likely to sire children by two wives than women who marry twice are to have children by both husbands.[7]

Infidelity and prostitution are special cases of polygamy in which no marriage bond forms between the partners. This puts a man's wife and his mistresses in different categories with respect to the investment that he is likely to make in his children. The man who can sufficiently arrange his business affairs to make time, opportunity, and money available for supporting two families is as rich as he is rare.

FEMINISM AND PHALAROPES

The rule that parental investment dictates which gender will attempt polygamy can be tested by looking at its exceptions. In sea horses the female has a sort of penis that she uses to inject eggs into the male's body, neatly reversing the usual method of mating. The eggs develop there, and as the theory predicts, it is the female sea horse who courts the male. There are about thirty species of birds, of which the phalaropes and jacanas are the best-known examples, in which the small dowdy male is courted by the large, aggressive female, and it is the male that broods the eggs and rears the chicks.[8]

Phalaropes and other seducer-female species are the exceptions that prove the rule. I remember watching a whole flock of female phalaropes badgering a poor male so intensely he almost drowned. And why? Because their mates were quietly sitting on their eggs for them, so these females had nothing better to do than look for second mates. Where males invest more time or energy in the care of the young, females take the initiative in courtship, and vice versa.[9]

In humans, the asymmetry is clear enough: nine months of pregnancy set against five minutes of fun. (I exaggerate.) If the balance of such investment determines sex roles in seduction, then it comes as no surprise that men seduce women rather than vice versa. This fact suggests that a highly polygamous human society represents a victory for men, whereas a monogamous one suggests a victory for women. But this is misleading. A polygamous society primarily represents a victory for one or a few men over all other men. Most men in highly polygamous societies are condemned to celibacy.

In any case, no moral conclusions of any kind can be drawn from evolution. The asymmetry in prenatal sexual investment between the genders is a fact of life, not a moral outrage. It is "natural." It is terribly tempting, as human beings, to embrace such an evolutionary scenario because it "justifies" a prejudice in favor of male philandering, or to reject it because it "undermines" the pressure for sexual equality. But it does neither. It says absolutely nothing about what is right and wrong. I am trying to describe the nature of humans, not prescribe their morality. That something is natural does not make it right. Murder is "natural" in the sense that our ape relatives commit it regularly, as apparently did our human ancestors. Prejudice, hate, violence, cruelty—all are more or less part of our nature, and all can be effectively countered by the right kind of nurture. Nature is not inflexible but malleable. Moreover, the most natural thing of all about evolution is that some natures will be pitted against others. Evolution does not lead to Utopia. It leads to a land in which what is best for one man may be the worst for another man, or what is best for a woman may be the worst for a man. One or the other will be condemned to an "unnatural" fate. That is the essence of the Red Queen's message.

In the pages that follow I will again and again be trying to guess what is "natural" for humanity. Perhaps my own moral prejudices will occasionally intrude as wishful thinking, but they will do so unconsciously. And even where I am wrong about human nature, I am not wrong that there is such a nature to be sought.

THE MEANING OF HOMOSEXUAL PROMISCUITY

Most prostitutes are female for the simple reason that the demand for female prostitutes is greater than for male ones. If the existence of female prostitutes reveals the male sexual appetite in its nakedness, then so, too, does the phenomenon of male homosexuality. Before the advent of AIDS, practicing male homosexuals were far more promiscuous than heterosexual men. Many gay bars were, and are, recognized places for picking up partners for one-night stands. The bathhouses of San Francisco catered to orgies and feats of repeated sex, assisted by stimulants, that boggled the mind when publicly discussed during the early years of the AIDS epidemic. A

Kinsey Institute study of gay men in the San Francisco Bay area found that 75 percent had had more than one hundred partners; 25 percent had had more than one thousand.[10]

This is not to deny that there are many homosexuals who were and are less promiscuous than many heterosexuals. But even homosexual activists admit that, before AIDS arrived, homosexuals were generally more promiscuous than heterosexuals. There is no single convincing explanation of this. Activists would say that homosexual promiscuity is caused largely by society's disapproval. Illegitimate, "shameful" activities tend to be indulged to excess when indulged at all. The legal and social difficulty of forming gay "marriages" mitigates against stable relationships.

But this is not persuasive. Promiscuity is not confined to those who indulge in gay sex clandestinely. Infidelity is acknowledged to be a greater problem in male gay "marriages" than in heterosexual ones, and society's disapproval is far greater of casual than of stable homosexual relations. Many of the same arguments apply to lesbians, who show a striking contrast: Lesbians rarely tend to indulge in sex with strangers but instead form partnerships that persist for many years with little risk of infidelity. Most lesbians have fewer than ten partners in their lifetimes.[11]

Donald Symons of the University of California at Santa Barbara has argued that the reason male homosexuals on average have more sexual partners than male heterosexuals, and many more than female homosexuals, is that male homosexuals are acting out male tendencies or instincts unfettered by those of women.

> Although homosexual men, like most people, usually want to have intimate relationships, such relationships are difficult to maintain, largely owing to the male desire for sexual variety; the unprecedented opportunity to satisfy this desire in a world of men; and the male tendency toward sexual jealousy. . . . I am suggesting that heterosexual men would be as likely as homosexual men to have sex most often with strangers, to participate in anonymous orgies in public baths, and to stop

off in public restrooms for five minutes of fellatio on the way home from work if women were interested in these activities.[12]

That is not to say that homosexuals do not long for stable intimacy or even that many are morally repelled by anonymous sex. But Symons's point is that the desire for monogamous intimacy with a life companion and the desire for casual sex with strangers are not mutually incompatible instincts. Indeed, they are characteristic of heterosexual men, as proven by the existence of a thriving call girl or "escort" industry that, at a price, supplies happily married businessmen with sexual diversions while they are traveling. Symons is commenting not on homosexual men but on men—average men. As he says, homosexual men behave like men, only more so; homosexual women behave like women, only more so.[13]

HAREMS AND WEALTH

In the chess game of sex, each gender must respond to the other's moves. The resulting pattern, whether polygamous or monogamous, is a stalemate rather than a draw or a victory. In elephant seals and sage grouse, the game reaches the point where males care only about the quantity of mates and females only about the quality. Each pays a heavy price, the males battling and exhausting themselves and dying in the often vain attempt to be the senior bull or master cock, the females entirely forgoing any practical help from the fathers in rearing their children.

The chess game reaches a very different stalemate in the case of the albatross. Every female gets her model husband; courtship is a mutual affair, and they share equally the chores of raising the chick. Neither gender seeks quantity of mates, but both are after quality: the hatching and rearing of one solitary chick that is pampered and fed for many months. Given that male albatrosses have the same genetic incentives as male elephant seals, why do they behave so differently?

The answer, as John Maynard Smith was the first to see, can be supplied by game theory, a technique borrowed from economics. Game theory is different from other forms of theorizing because it recognizes that the outcome of a transaction often depends on what other people are doing. Maynard Smith tried pitting different genetic strategies against each other in the same way that economists do with different economic strategies. Among the problems that were suddenly rendered soluble by this technique was the question of why different animals have such different mating systems.[14]

Imagine a population of ancestral albatrosses in which the males were highly polygamous and spared no time to help rear the young. Imagine that you were a junior male with no prospect of becoming a harem master. Suppose that instead of striving to be a polygamist, you married one female and helped rear her offspring. You would not have hit the jackpot, but at least you would have done better than most of your more ambitious brothers. Suppose, too, that by helping your wife to feed the baby, you greatly increased the chance that the baby survived. Suddenly, females in the population have two options: to seek a faithful mate like yourself or to seek a polygamist. Those that seek a faithful mate leave behind more young, so in each generation the number willing to join harems declines, and the rewards of becoming a polygamist fall with it. The species is "taken over" by monogamy.[15]

It works in reverse as well. The male lark bunting of Canada sets up a territory in a field and tries to attract several females to breed with him. By joining a male that already has a mate, a female forfeits the chance to make use of his skills as a father. But if his territory is sufficiently richer in food than his neighbor's, it still pays her to choose him. When the advantage of choosing a bigamist for his territory or genes exceeds the advantage of choosing a monogamist for his parental care, polygamy ensues. This so-called polygyny threshold model seems to explain how so many marshland birds in North America became polygamous.[16]

Both of these models could apply easily to humans. We became monogamous because the advantage that a junior father could supply in feeding the family outweighed the disadvantage in not being mated to the chief. Or we became polygamous because of the discrepancies in wealth between males. "Which woman would not rather be John Kennedy's third wife than Bozo the Clown's first?" said one (female) evolutionist.[17]

There is some evidence that the polygyny threshold does apply to human beings. Among the Kipsigis of Kenya, rich men have more cattle and more wives. Each wife of a rich man is at least as well off as the single wife of a poor man, and she knows it. According to Monique Borgehoff Mulder of the University of California at Davis, who has studied the Kipsigis, polygamy is willingly chosen by the women. A Kipsigis woman is consulted by her father when her marriage is arranged, and she is only too aware that being the second wife of a man with plenty of cattle is a better fate than being the first wife of a poor man. There is companionship and a sharing of the burden between co-wives. The polygyny threshold model holds for Kipsigis fairly well.[18]

There are two difficulties with this theory, however. The first is that it says nothing about the first wife's views. There is little advantage to a first wife in sharing her husband and his wealth with others. Among the Mormons of Utah it is well known that first wives resent the arrival of second wives. The Mormon church officially abandoned polygamy more than a century ago, but in recent years a few fundamentalists have resumed the practice and have even begun to campaign openly for its acceptance. In Big Water, Utah, the mayor, Alex Joseph, had nine wives and twenty children in 1991. Most of the wives were career women who were happy with their lot, but they do not all see eye to eye. "The first wife does not like it when the second wife comes along," said the third Mrs. Joseph, "and the second wife doesn't care for the wife who came first. So you can get some fighting and bad feeling."[19]

Supposing that first wives usually object to sharing their husbands, what can the husband do about it? He can force her to accept the arrangement, as presumably many despots did in times past, or he can bribe her to accept it. The legitimacy a first wife's children usually has compared with those from a second wife is a bonus that must go some way toward mollifying the former. In parts of Africa it is written into the law that the first wife inherits 70 percent of the husband's wealth.

Incidentally, the polygyny threshold leads me to ask the question In whose interest is it that polygamy be outlawed in our society? We automatically assume it is in the interest of women. But consider; it would presumably be illegal, as it is now, for people to be forced to marry against their will, so second wives would be choosing their lot voluntarily. A woman who wants a career would surely find a ménage à trois more, not less, convenient; she would have two partners to help share the chores of child care. As a Mormon lawyer put it recently, there are "compelling social reasons" that make polygamy "attractive to the modern career woman."[20] But think of the effect on men. If many women chose to be second wives of rich men rather than first wives of poor men, there would be a shortage of unmarried women, and many men would be forced to remain unhappily celibate. Far from being laws to protect women, antipolygamy statutes may really do more to protect men.[21]

Let us erect the four commandments of mating system theory. First, if females do better by choosing monogamous and faithful males, monogamy will result—unless, second, men can coerce them. Third, if females do no worse by choosing already-mated males, polygamy will result—unless fourth already-mated females can prevent their males from mating again, in which case monogamy will result. The surprising conclusion of game theory is therefore that males, despite their active role in seduction, may be largely passive spectators at their marital fate.

WHY PLAY SEXUAL MONOPOLY?

But the polygamy threshold is a bird-centric view. Those who study mammals take a rather different view, for virtually all mammals lie so far above the polygamy threshold that the four commandments are irrelevant. Male mammals can be of so little use to their mates during pregnancy that it need not concern the females whether the males have already married. Humanity is a startling exception to this rule. Because children are fed by their parents for so long, they are more like baby birds than baby mammals. The female can do a great deal better by choosing an unmarried wimp of a husband who will stay around to help rear the young than by marrying a philandering chief if she has to do all the work herself. That is a point to which I shall return in the next chapter. For the moment, forget people and think about deer.

A female deer has little need of a monopolized male. He cannot produce milk or bring grass to the young. So the mating system of a deer is determined by the battle among males, which in turn is determined by how females decide to distribute themselves. Where females live in herds (for example, elk), males can be harem masters. Where females live alone (white-tailed deer), males are territorial and mostly monogamous. Each species has its own pattern, depending on the behavior of the females.

In the 1970s zoologists began to investigate these patterns to try to find out what determined a species' mating system. They coined a new term, "socioecology," in the process. Its most successful forays were into antelope and monkey society. Two studies concluded that the mating system of an antelope or a primate could be safely predicted from its ecology. Small forest antelopes are selective feeders and, as a consequence, are solitary and monogamous. Middle-sized, open-woodland ones live in small groups and form harems. Big plains antelopes, such as the eland and African buffalo, live in great herds and are promiscuous. At first a very similar system seemed to apply to monkeys and apes. Small nocturnal bush babies are solitary and monogamous; leaf-eating indris live in harems; forest-fringe-dwelling gorillas live in small harems; tree-savanna chimps live in large promiscuous groups; grassland baboons live in large harems or multimale troops.[22]

It began to look as if such ecological determinism was on to something. The logic behind it was that female mammals set out to distribute themselves without regard to sex, living alone or in small groups or in large groups according to the dictates of food and safety. Males then set out to monopolize as many females as possible either by guarding groups of females directly or by defending a territory in which females lived. Solitary, widely dispersed females gave a male only one option: to monopolize a single female's home range and be her faithful husband (for instance, the gibbon). Females that were solitary but less far apart gave him the chance to monopolize the home ranges of two or more separate females (for instance, the orangutan). Small groups of females gave him the chance to monopolize the whole group and call it his harem (for instance, the gorilla). He would have to share large groups with other males (for instance, the chimp).

That picture has been complicated by one factor: A species' recent history can influence what mating system it ends up with. Or, to put it more simply, the same ecology can produce two different mating systems depending on the route taken to get there. On Northumbrian moors the red grouse and the black grouse live in virtually identical habitats. The black grouse prefers bushy areas and places that are not too heavily grazed by sheep, but apart from that, they are ecological brothers. Yet the black grouse gather in spring at spectacular leks where all the females mate with just one or two males, those that have most impressed them with their displays. They then rear their young without any help from the males. The nearby red grouse are territorial and monogamous; the cock is almost as attentive to the chicks as the hens. The two species share the same food, habitat, and enemies, yet have entirely different mating systems. Why? My preferred explanation, and that of most biologists who have studied them, is that they have different histories. Black grouse are the descendants of forest dwellers, and it was in the forest that their maternal ancestors developed the habit of choosing males according to genetic quality rather than territory.[23]

HUNTERS OR GATHERERS

The lesson for humanity is obvious. To determine our mating system we need to know our natural habitat and our past. We have lived mostly in cities for less than one thousand years. We have been agricultural for less than ten thousand. These are mere eye blinks. For more than a million years before that we were recognizably human and living, mostly in Africa, probably as hunter-gatherers, or foragers, as anthropologists now prefer to say. So inside the skull of a modern city dweller there resides a brain designed for hunting and gathering in small groups on the African savanna. Whatever humanity's mating system was then is what is "natural" for him now.

Robert Foley is an anthropologist at Cambridge University who has tried to piece together the history of our social system. He starts with the fact that all apes share the habit of females leaving their natal group, whereas all baboons share the habit of males leaving their natal group. It seems to be fairly hard for a species to switch from female exogamy to male exogamy, or vice versa. On average, human beings are typical apes in this respect even today. In most societies women travel to live with their husbands, whereas men tend to remain close to their relatives. There are many exceptions, though: In some but not most traditional human societies, men move to women.

Female exogamy means that apes are largely devoid of mechanisms for females to build coalitions of relatives. A young female chimpanzee generally must leave her mother's group and join a strange group dominated by unfamiliar males. To do so, she must gain favor with the females that already live in her new tribe. A male, by contrast, stays with his group and allies himself with powerful relatives in the hope of inheriting their status later.

So much for the ape's legacy to mankind. What about the habitat in which he lived? Toward the end of the Miocene era, some 25 million years ago, Africa's forests began to contract. Drier, more seasonal habitats—grasslands, scrublands, savannas—began to spread. About 7 million years ago the ancestors of mankind began to diverge from the ancestors of modern chimpanzees. Even more than chimps and much more than gorillas, mankind's ancestors moved into these new dry habitats and gradually adapted to them. We know this because the earliest fossils of manlike apes (the australopithecines) were living in places that at the time were not covered by forest—at Hadar in Ethiopia and Olduvai in Tanzania. Presumably, these relatively open habitats favored larger groups as they did for chimps and baboons, the two other open-country primates. As socioecologists find again and again, the more open the habitat, the bigger the group, both because big groups can be more vigilant in spotting predators and because the food is usually found in a patchier pattern. For reasons that are not especially persuasive (principally the apparently great size difference of males and females), most anthropologists believe the early australopithecines lived in single-male harems, like gorillas and some species of baboon.[24]

But then, sometime around 3 million years ago, the hominid lineage split in two (or more). Robert Foley believes the increasingly seasonal pattern of rainfall made the life-style of the original ape-man untenable, for its diet of fruit, seeds, and perhaps insects became increasingly rare in dry seasons. One line of its descendants developed especially robust jaws and teeth to deal with a diet increasingly dominated by coarse plants. *Australopithecus robustus*, or nutcracker man, could then subsist on coarse seeds and leaves during lean seasons. Its anatomy supplies meager clues, but Foley guesses that nutcrackers lived in multimale groups, like chimps.[25]

The other line, however, embarked on an entirely different path. The animals known as *Homo* took to a diet of meat. By 1.6 million years ago, when *Homo erectus* was living in Africa, he was without question the most carnivorous monkey or ape the world had ever known. That much is clear from the bones he left at his campsites. He may have scavenged them from lion kills or perhaps begun to use tools to kill game himself. But increasingly, in lean seasons, he could rely on a supply of meat. As Foley and P. C. Lee put it, "While the causes of meat-eating are ecological, the consequences would be distributional and social." To hunt, or even more, to seek lion kills, required a man to range farther from home and to rely on his companions for coordinated help. Whether as a result of this or coincidentally, his body embarked on a series of coordinated gradual changes. The shape of the skull began to retain more juvenile shape into adulthood, with a bigger brain and a smaller jaw. Maturity was gradually delayed so that children grew slowly into adulthood and depended on their parents longer.[26]

Then for more than a million years people lived in a way that couldn't have changed much. They inhabited grasslands and woodland savannas, first in Africa, later in Eurasia, and eventually in Australasia and the Americas. They hunted animals for food, gathered fruits and seeds, and were highly social within each tribe but hostile toward members of other tribes. Don Symons refers to this combination of time and place as the "environment of evolutionary adaptedness," or EEA, and he believes it is central to human psychology. People cannot be adapted to the present or the future; they can only be adapted to the past. But he readily admits that it is hard to be precise about exactly what lives people lived in the EEA. They probably lived in small bands; they were perhaps nomadic; they ate both meat and vegetable matter; they presumably shared the features that are universal among modern humans of all cultures: a pair bond as an institution in which to rear children, romantic love, jealousy and sexually induced male-male violence, a female preference for men of high status, a male preference for young females, warfare between bands, and so on. There was almost certainly a sexual division of labor between hunting men and gathering women, something unique to people and a few birds of prey. To this day, among the Aché people of Paraguay, men specialize in acquiring those foods that a woman encumbered with a baby could not manage to—meat and honey, for example.[27]

Kim Hill, at the University of New Mexico, argues that there was no consistent EEA, but he nonetheless agrees that there were universal features of human life that are not present today but that have hangover effects. Everybody knew or had heard of nearly all the people they were likely to meet in their lives: There were no strangers, a fact that had enormous importance for the history of trade and crime prevention, among other things. The lack of anonymity meant that charlatans and tricksters could rarely get away with their deceptions for long.

Another group of biologists at Michigan rejects these EEA arguments altogether with two arguments. First, the most critical feature of the EEA is still with us. It is other people. Our brains grew so big not to make tools but to psychologize one another. The lesson of socioecology is that our mating system is determined not by ecology but by other people—by members of the same gender and by members of the other gender. It is the need to outwit and dupe and help and teach one another that drove us to be ever more intelligent.

Second, we were designed above all else to be adaptable. We were designed to have all sorts of alternative strategies to achieve our ends. Even today, existing hunter-gatherer societies show enormous ecological and social variation, and they are probably an unrepresentative sample because they mostly occupy deserts and forests, which were not mankind's primary habitat. Even in the time of *Homo erectus*, let alone more modern people, there may have been specialized fishing, shore-dwelling, hunting, or plant-gathering cultures. Some of these may well have afforded opportunities for wealth accumulation and polygamy. In recent memory there was a preagricultural culture among the salmon-fishing Indians of the Pacific Northwest of America that was highly polygamous. If the local hunter-gathering economy favored it, men were capable of being polygamous and women were capable of joining harems over the protests of the preceding co-wives. If not, then men were capable of being good fathers and women jealous monopolizers. In other words, mankind has many potential mating systems, one for each circumstance.[28]

This is supported by the fact that larger, more intelligent and more social animals are generally more flexible in their mating systems than smaller, dumber, or more solitary ones. Chimps go from small feeding bands to big groups depending on the nature of the food supply. Turkeys do the same. Coyotes hunt in packs when their food is deer but hunt alone when their food is mice. These food-induced social patterns themselves induce slightly different mating patterns.

MONEY AND SEX

But if humanity is a flexible species, then the EEA is in a sense still with us. Where people in twentieth-century societies act adaptively or where power raises reproductive success, it could be because adaptations shaped in the EEA (wherever and whenever that was) are still working. The technological problems of suburban life may be a million miles from those of the Pleistocene savanna, but the human ones are not. We are still consumed by gossip about people we know or have heard about. Men are still obsessed with power-seeking and building or dominating male-male coalitions. Human institutions cannot be understood without understanding their internal politics. Modern monogamy may be just one of the many tricks in our mating-system repertoire, like harem polygamy in ancient China or gerontocratic polygamy in modern Australian aborigines, where men wait years to marry and then in their dotage enjoy huge harems.

If so, then the "sex drive" that we all acknowledge within us may be much more specific than we realize. Given the fact that men can always increase their reproductive success by philandering, whereas women cannot, we should suspect that men are apt to be behaviorally designed to take advantage of opportunities for polygamy and that some of the things they do have that end in mind.

There is broad agreement among evolutionary biologists that most of our ancestors lived in a condition of only occasional polygamy during the Pleistocene period (the two million years of modern human existence before agriculture). Societies that hunt and gather today are not much different from modern Western society. Most men are monogamous, many are adulterous, and a few manage to be polygamous, sharing perhaps up to five wives in extreme cases. Among the Aka pygmies of the Central African Republic, who hunt for food in the forest using nets, 15 percent of men have more than one wife, a pattern typical of foraging societies.[29]

One of the reasons hunting and gathering cannot support much polygamy is that luck, more than skill, plays a large part in hunters' success. Even the best hunter would often return empty-handed and would be reliant on his fellow men to share what they had killed. This equitable sharing of hunted food is characteristic of these people (in most other social hunting species there is a free-for-all) and is the clearest example of a habit of "reciprocal altruism" on which the whole of society sometimes appears to be based. A lucky hunter kills more than he can eat, so he loses little by sharing it with his companions but instead gains a lot because next time, if he is unlucky, the favor will be repaid by those with whom he shared now. Trading favors in this way was the ancient ancestor of the monetary economy. But because meat could not be stored and because luck did not last, hunter-gatherer societies did not allow the accumulation of wealth.[30]

With the invention of agriculture, the opportunity for some males to be polygamous arrived with a vengeance. Farming opened the way for one man to grow much more powerful than his peers by accumulating a surplus of food, whether grain or domestic animals, with which to buy the labor of other men. The labor of other men allowed him to increase his surplus still more. For the first time having wealth was the best way to get wealth. Luck does not determine why one farmer reaps more than his neighbor to the same degree that it determines the success of a hunter. Agriculture suddenly allowed the best farmer in the band to have not only the largest hoard of food but the most reliable supply. He had no need to share it freely, for he needed no favor in return. Among the //Gana San people of Namibia, who have given up their !Kung San neighbors' hunting life for farming, there is less food sharing and more political dominance within each band. Now, by owning the best or biggest fields or by working harder or by having an extra ox or by being a craftsman with a rare skill, a man could grow ten times as rich as his neighbor. Accordingly, he could acquire more wives. Simple agricultural societies often see harems of up to one hundred women per top man.[31]

Pastoral societies are, almost without exception, tradition-ally polygamous. It is not hard to see why. A herd of cattle or sheep is almost as easy to tend if it contains fifty animals as twenty-five.

Such scale economies allow a man to accumulate wealth at an ever-increasing rate. Positive feedback leads to inequalities of wealth, which leads to inequalities of sexual opportunity. The reason some Mukogodo men in Kenya have higher reproductive success than others is that they are richer; being richer enables them to marry early and marry often.[32]

By the time "civilization" had arrived, in six different parts of the globe independently (from Babylon in 1700 B.C. to the Incas in A.D. 1500), emperors had thousands of women in their harems. Hunting and warrior skills had previously earned a man an extra wife or two, then wealth had earned him ten or more. But wealth had another advantage, too. Not only could it buy wives directly, it could also buy "power." It is noteworthy that it is hard to distinguish between wealth and power before the time of the Renaissance. Until then there was no such thing as an economic sector independent of the power structure. A man's livelihood and his allegiance were owed to the same social superior.[33] Power is, roughly speaking, the ability to call upon allies to do your bidding, and that depended strictly on wealth (with a little help from violence).

Power seeking is characteristic of all social mammals. Cape buffalo rise within the hierarchy of the herd to positions of dominance that bring sexual rewards. Chimpanzees, too, strive to become "alpha male" in the troop and in so doing increase the number of matings they perform. But like men, chimps do not rise entirely on brute strength. They use cunning, and above all they form alliances. The tribal warfare between groups of chimps is both a cause and a consequence of the male tendency to build alliances. In Jane Goodall's studies the males of one chimp group were well aware when they were outnumbered by the males of another group and deliberately sought opportunities to single out individual males from the enemy. The bigger and more cohesive the male alliance, the more effective it was.[34]

Coalitions of males are found in a number of species. In turkeys, brotherhoods of males display competitively on a lek. If they win, the females will mate with the senior brother. In lions, brotherhoods combine to drive out the males from a pride and take it over themselves; they then kill the babies to bring the lionesses back into season, and all the brothers share the reward of mating with all the females. In acorn woodpeckers, groups of brothers live with groups of sisters in a free-love commune that controls one "granary tree," into which holes have been drilled that hold up to thirty thousand acorns to see the birds through the winter. The young, who are nieces and nephews of all the birds of whom they are not daughters and sons, must leave the group, form sisterhoods and brotherhoods themselves, and take over some other granary tree, driving out the previous owners.[35]

The alliances of males and females need not be based on relatedness. Brothers tend to help one another because they are related; what's good for your brother's genes is good for yours since you share half your genes with him. But there is another way to ensure that altruism pays: reciprocity. If an animal wants help from another, he could promise to return the favor in the future. As long as his promise is credible—in other words, as long as individuals recognize each other and live together long enough to collect their debts—a male can get other males to help him in a sexual mission. This seems to be what happens in dolphins, whose sex life is only just becoming known. Thanks to the work of Richard Connor, Rachel Smolker, and their colleagues, we now know that groups of male dolphins kidnap single females, bully them and display to them with choreographed acrobatics, then enjoy sexual monopoly over them. Once the female has given birth, the alliances of males lose interest in her, and she is free to return to an all-female group. These male alliances are often temporary and stitched together on a you-help-me-and-I'll-help-you basis.[36]

The more intelligent the species and the more fluid the coalitions, the less an ambitious male need be limited by his strength. Buffalos and lions win power in trials of strength. Dolphins and chimpanzees must not be weak if they are to win power but can rely much more on their ability to form winning coalitions of males. In people there is virtually no connection between strength and power, at least not since the invention of action-at-a-distance weapons such as the slingshot, as Goliath learned the hard way. Wealth, cunning, political skill, and experience lead to power among men. From Hannibal to Bill Clinton, men gain power by putting together coalitions of allies. In mankind, wealth became a way of putting together such alliances of power. The rewards, for other animals, are largely sexual. For men?

HIGHLY SEXED EMPERORS

In the late 1970s an anthropologist in California, Mildred Dickemann, decided to try to apply some Darwinian ideas to human history and culture. She simply set out to see if the kinds of predictions that evolutionists were making for other animals also applied to human beings. What she found was that in the highly stratified Oriental societies of early history, people seemed to behave exactly as you would expect them to if they knew that their goal on Earth was to leave as many descendants as possible. In other words, men tended to seek polygamy, whereas women strove to marry upward with men of high status. Dickemann added that a lot of cultural customs—dowries, female infanticide, the claustration of women so that their virginity could not be damaged—were consistent with this pattern. For example, in India, high castes practiced more female infanticide than low castes because there were fewer opportunities to export daughters to still higher castes. In other words, mating was a trade: male power and resources for female reproductive potential.[37]

About the same time as Dickemann's studies, John Hartung of Harvard University began to look at patterns of inheritance. He hypothesized that a rich person in a polygamous society would tend to leave his or her money to a son rather than a daughter because a rich son could provide more grandchildren than a rich daughter. This is because the son can have children by several wives, whereas a daughter cannot increase the number of her children even if she takes many husbands. Therefore, the more polygamous a society, the more likely it will show male-biased inheritance. A survey of four hundred societies found overwhelming support for Hartung's hypothesis.[38]

Of course, that *proves* nothing. It could be a coincidence that evolutionary arguments predict what does happen. There is a cautionary tale that scientists tell one another about a man who cuts the legs off a flea to test his theory that fleas' ears are on their legs. He then tells the flea to jump and it does not, so he concludes that he was right; fleas' ears are in their legs.

Nonetheless, Darwinians began to think that perhaps human history might be illuminated by a beam of evolutionary light. In the mid 1980s, Laura Betzig set out to test the notion that people are sexually adapted to exploit whatever situation they encounter. She had no great hopes of success, but she believed that the best way to test the conjecture was simply to postulate the simplest prediction she could make: that men would treat power not as an end in itself but as a means to sexual and reproductive success. Looking around the modern world, she was not encouraged; powerful men are often childless. Hitler was so consumed by ambition that he had little time left for philandering.[39]

But when she examined the record of history, Betzig was stunned. Her simplistic prediction that power is used for sexual success was confirmed again and again. Only in the past few centuries in the West has it failed. Not only that, in most polygamous societies there were elaborate social mechanisms to ensure that a powerful polygamist left a polygamous heir.

The six independent "civilizations" of early history—Babylon, Egypt, India, China, Aztec Mexico, and Inca Peru—were remarkable less for their civility than for their concentration of power. They were all ruled by men, one man at a time, whose power was arbitrary and absolute. These men were despots, meaning they could kill their subjects without fear of retribution. Without exception, that vast accumulation of power was always translated into prodigious sexual productivity. The Babylonian king Hammurabi had thousands of slave "wives" at his command. The Egyptian pharaoh Akhenaten procured 317 concubines and "droves" of consorts. The Aztec ruler Montezuma enjoyed 4,000 concubines. The

Indian emperor Udayama preserved sixteen thousand consorts in apartments ringed by fire and guarded by eunuchs. The Chinese emperor Fei-ti had ten thousand women in his harem. The Inca Atahualpa, as we have seen, kept virgins on tap throughout the kingdom.

Not only did these six emperors, each typical of his predecessors and successors, have similarly large harems, but they employed similar techniques to fill and guard them. They recruited young (usually prepubertal) women, kept them in highly defensible and escape-proof forts, guarded them with eunuchs, pampered them, and expected them to breed the emperor's children. Measures to enhance the fertility of the harem were common. Wet nurses, who allow women to resume ovulation by cutting short their breast-feeding periods, date from at least the code of Hammurabi in the eighteenth century B.C.; they were sung about in Sumerian lullabies. The Tang Dynasty emperors of China kept careful records of dates of menstruation and conception in the harem so as to be sure to copulate only with the most fertile concubines. Chinese emperors were also taught to conserve their semen so as to keep up their quota of two women a day, and some even complained of their onerous sexual duties. These harems could hardly have been more carefully designed as breeding machines, dedicated to the spread of emperors' genes.[40]

Nor were emperors anything more than extreme examples. Laura Betzig has examined 104 politically autonomous societies and found that "in almost every case, power predicts the size of a man's harem."[41] Small kings had one hundred women in their harems; great kings, one thousand, and emperors, five thousand. Conventional history would have us believe that such harems were merely one among many of the rewards that awaited the successful seeker of power, along with all the other accoutrements of despotism: servants, palaces, gardens, music, silk, rich food, and spectator sports. But women are fairly high on the list. Betzig's point is that it is one thing to find that powerful emperors were polygamous but quite another to discover that they each adopted similar measures to enhance their reproductive success within the harem:

wet nursing, fertility monitoring, claustration of the concubines, and so on. These are not the measures of men interested in sexual excess. They are the measures of men interested in producing many children.

However, if reproductive success was one of the perks of despotic power, one peculiar feature stands out: All six of the early emperors were monogamously married. In other words, they always raised one mate above all the others as a "queen." This is character-istic of human polygamous societies. Wherever there are harems, there is a senior wife who is treated differently from the others. She is usually noble-born, and crucially, she alone is allowed to bear legitimate heirs. Solomon had a thousand concubines and one queen.

Betzig investigated imperial Rome and found the distinc-tion between monogamous marriage and polygamous infidelity extending from the top to the bottom of Roman society. Roman emperors were famous for their sexual prowess, even while marrying single empresses. Julius Caesar's affairs with women were "com-monly described as extravagant" (Suetonius). Of Augustus, Sueto-nius wrote, "The charge of being a womanizer stuck, and as an elderly man he is said to have still harbored a passion for deflower-ing girls—who were collected for him by his wife." Tiberius's "criminal lusts" were "worthy of an oriental tyrant" (Tacitus). Caligula "made advances to almost every woman of rank in Rome" (Dio), including his sisters. Even Claudius was pimped for by his wife, who gave him "sundry housemaids to lie with" (Dio). When Nero floated down the Tiber, he "had a row of temporary brothels erected on the shore" (Suetonius). As in the case of China, though not so methodically, breeding seems to have been a principal func-tion of concubines.

Nor were emperors special. When a rich patrician named Gordian died leading a rebellion in favor of his father against the emperor Maximin in A.D. 237, Gibbon commemorated him thus: "Twenty-two acknowledged concubines and a library of sixty-two thousand volumes attested to the variety of his inclinations, and from the productions which he left behind him, it appears that both the one and the other were designed for use rather than osten-tation."

"Ordinary" Roman nobles kept hundreds of slaves. Yet, while virtually none of the female slaves had jobs around the house, female slaves commanded high prices if sold in youth. Male slaves were usually forced to remain celibate, so why were the Roman nobles buying so many young female slaves? To breed other slaves, say most historians. Yet that should have made pregnant slaves command high prices; they did not. If a slave turned out not to be a virgin, the buyer had a legal case against the seller. And why insist on chastity among the male slaves if breeding is the function of female slaves? There is little doubt that those Roman writers who equate slaves with concubines were telling the truth. The unrestricted sexual availability of slaves "is treated as a commonplace in Greco-Roman literature from Homer on; only modern writers have managed largely to ignore it."[42]

Moreover, Roman nobles freed many of their slaves at suspiciously young ages and with suspiciously large endowments of wealth. This cannot have been an economically sensible decision. Freed slaves became rich and numerous. Narcissus was the richest man of his day. Most slaves who were freed had been born in their masters' homes, whereas slaves in the mines or on farms were rarely freed. There seems little doubt that Roman nobles were freeing their illegitimate sons, bred of female slaves.[43]

When Betzig turned her attention to medieval Christendom, she discovered that the phenomenon of monogamous marriage and polygamous mating was so entrenched that it required some disinterring. Polygamy became more secret, but it did not expire. In medieval times the census shows a sex ratio in the countryside that was heavily male-biased because so many women were "employed" in the castles and monasteries. Their jobs were those of serving maids of various kinds, but they formed a loose sort of "harem" whose size depended clearly on the wealth and power of the castle's owner. In some cases historians and authors were more or less explicit in admitting that castles contained "gynoeciums," where lived the owner's harem in secluded luxury.

Count Baudouin, patron of a literary cleric named Lambert, "was buried with twenty-three bastards in attendance as well as ten legitimate daughters and sons." His bedchamber had access to the servant girls' quarters and to the rooms of adolescent girls upstairs. It had access, too, to the warming room, "a veritable incubator for suckling infants." Meanwhile, many medieval peasant men were lucky to marry before middle age and had few opportunities for fornication.[44]

THE REWARDS OF VIOLENCE

If reproduction has been the reward and goal of power and wealth, then it is little wonder that it has also been a frequent cause and reward of violence. This is presumably the reason that the early Church became so obsessed with matters of sex. It recognized sexual competition to be one of the principal causes of murder and mayhem. The gradual synonymy of sex and sin in Christendom is surely based more on the fact that sex often leads to trouble rather than that there is anything inherently sinful about sex.[45]

Consider the case of the Pitcairn Islanders. In 1790 nine mutineers from HMS *Bounty* landed on Pitcairn along with six male and thirteen female Polynesians. Thousands of miles from the nearest habitation, unknown to the world, they set about building a life on the little island. Notice the imbalance: fifteen men and thirteen women. When the colony was discovered eighteen years later, ten of the women had survived and only one of the men. Of the other men, one had committed suicide, one had died, and twelve had been murdered. The survivor was simply the last man left standing in an orgy of violence motivated entirely by sexual competition. He promptly underwent a conversion to Christianity and prescribed monogamy for Pitcairn society. Until the 1930s the colony prospered and good genealogical records were kept. Studies of these show that the prescription worked. Apart from rare and occasional adultery, the Pitcairners were and remain monogamous.[46]

Monogamy, enforced by law, religion, or sanction, does seem to reduce murderous competition between men. According to Tacitus, the Germanic tribes that so frustrated several Roman emperors attributed their success partly to the fact that they were a monogamous society and therefore able to direct their aggression outward (though no such explanation applied to the polygamous and successful Romans). No man was allowed more than one wife, so no man had an incentive to kill a fellow tribesman to take his wife. Not that socially imposed monogamy need extend to captive slaves. In the nineteenth century in Borneo, one tribe, the Iban, dominated the tribal wars of the island. Unlike their neighbors, the Iban were monogamous, which both prevented the accumulation of sullen bachelors in their ranks and motivated them to feats of great daring with the prize of foreign female slaves as reward.[47]

One of the legacies of being an ape is intergroup violence. Until the 1970s primatologists were busy confirming our prejudices about peaceable apes living in nonviolent societies. Then they began to observe the rare but more sinister side of chimp life. The males of a chimpanzee "tribe" sometimes conduct violent campaigns against the males of another tribe, seeking out and killing their enemies. This habit is very different from the territoriality of many animals, who are content to expel intruders. The prize may be to seize the enemy territory, but that is a small reward for so dangerous a business. A far richer reward awaits the successful male alliance: young females of the defeated group join the victors.[48]

If war is something we inherited directly from the hostility between groups of male apes over female apes, with territory as merely a means to the end—sex—then it follows that tribal people must be going to war over women rather than territory. For a long time anthropologists insisted that war was fought over scarce material resources, in particular protein, which was often in short supply. So when Napoleon Chagnon, trained in this tradition, went to Venezuela to study the tribal Yanomamö in the 1960s, he was in for a shock: "These people were not fighting over what I was trained to believe they were fighting over—scarce resources. They were fighting over women."[49] Or at least so they said. There is a tradition in anthropology that you should not believe what people tell you, so Chagnon was ridiculed for believing them. Or as he puts it, "You are allowed to admit the stomach as a source of war but not the gonads." Chagnon went back again and again and eventually accumulated a terrifying set of data that proves beyond doubt that men who kill other men (*unokais*) have more wives, independent of their social standing, than men who do not become murderers.[50]

Among the Yanomamö, war and violence are both primarily about sex. War between two neighboring villages breaks out over the abduction of a woman or in retaliation for an attack that had such a motive, and it always results in women changing hands. The most common cause of violence within a village is also sexual jealousy; a village that is too small is likely to be raided for women, but a village that is too large usually breaks up over adultery. Women are the currency and reward of male violence in the Yanomamö, and death is common. By the age of forty, two-thirds of the people have lost a close relative to murder—not that this dulls the pain and fear of murder. To Yanomamö who leave their forests, the existence in the outside world of laws that prevent chronic murder is miraculous and tremendously desirable. Likewise, the Greeks fondly remembered the replacement of revenge by justice as a milestone, through the legend of the trial of Orestes. According to Aeschylus, Orestes killed Clytemnestra for killing Agamemnon, but the Furies were persuaded by Athena to accept the court's verdict and end the system of blood feuds.[51] Thomas Hobbes did not exaggerate when he listed among the features of life of primitive mankind "continual fear and danger of violent death"; though he was much less correct in the second and more familiar part of the sentence: "and the life of man, solitary, poor, nasty, brutish, and short."

Chagnon now believes that the conventional wisdom—people only fight over scarce resources—misses the point. If resources are scarce, then people fight over them. If not, they do not. "Why bother," he says, "to fight for mangango nuts when the only point of having mangango nuts is so that you can have women. Why not fight over women?" Most human societies, he believes, are not touching some ceiling of resource limitation. The Yanomamö could easily clear larger gardens from the forest to grow more plantain trees, but then they would have too much to eat.[52]

There is nothing especially odd about the Yanomamö. All

studies of preliterate societies done before national governments were able to impose their laws upon them revealed routinely high levels of violence. One study estimated that one-quarter of all men were killed in such societies by other men. As for the motives, sex is dominant.

The founding myth of Western culture, Homer's *Iliad*, is a story that begins with a war over the abduction of a woman, Helen. Historians have long considered the abduction of Helen to Troy to be no more than a pretext for territorial confrontation between the Greeks and the Trojans. But can we be so confidently condescending? Perhaps the Yanomamö really do go to war over women, as they say they do. Perhaps Agamemnon's Greeks did, too, as Homer said they did. The Iliad opens with and is dominated by a quarrel between Achilles and Agamemnon, the cause of which is Agamemnon's insistence on confiscating a concubine, Briseis, from Achilles in compensation for having to give back his own concubine, Chryseis, to her priest-father who has enlisted Apollo's aid against the Greeks. This dissension in the ranks, caused by a dispute over a woman, nearly loses the Greeks the whole war, which in turn has been caused by a dispute over a woman.

In preagricultural societies, violence may well have been a route to sexual success, especially in times of turmoil. In many different cultures the captives taken in war have tended to be women rather than men. But echoes reach into modern times. Armies have often been motivated as much by the opportunities that victory would present for rape as they have been by patriotism or fear. Generals, recognizing this, turned blind eyes to the excesses of their troops and were sure to provide camp followers. Even in this century, access to prostitutes has been a more or less recognized purpose of shore leave in navies. And rape accompanies war still. In Bangladesh, during a nine-month occupation by west Pakistani troops in 1971, up to 400,000 women may have been raped by soldiers.[53] In Bosnia in 1992, the reports of organized rape camps for Serbian soldiers became too frequent to ignore. Don Brown, an anthropologist in Santa Barbara, recalls his days in the army: "Men talked about sex night and day; they never talked about power."[54]

MONOGAMOUS DEMOCRATS

The nature of the human male, then, is to take opportunities, if they are granted him, for polygamous mating and to use wealth, power, and violence as means to sexual ends in the competition with other men—though usually not at the expense of sacrificing a secure monogamous relationship. It is not an especially flattering picture, and it depicts a nature that is very much at odds with modern ethical preferences—for monogamy, fidelity, equality, justice, and freedom from violence. But my task is description, not prescription. And there is nothing inevitable about human nature. In *The African Queen*, Katharine Hepburn said to Humphrey Bogart, "Nature, Mr Allnutt, is what we are put in this world to rise above."

Besides, the long interlude of human polygamy, which began in Babylon nearly four thousand years ago, has largely come to an end in the West. Official concubines became unofficial mistresses, and mistresses became secrets kept from wives. In 1988, political power, far from being a ticket to polygamy, was jeopardized by any suggestion of infidelity. Whereas the Chinese emperor Fei-ti once kept ten thousand women in his harem, Gary Hart, running for the presidency of the most powerful nation on earth, could not even get away with two.

What happened? Christianity? Hardly. It coexisted with polygamy for centuries, and its strictures were as cynically self-interested as any layman's. Women's rights? They came too late. A Victorian woman had as much and as little say in her husband's affairs as a medieval one. No historian can yet explain what changed, but guesses include the idea that kings came to need internal allies enough that they had to surrender despotic power. Democracy, of a sort, was born. Once monogamous men had a chance to vote against polygamists (and who does not want to tear down a competitor, however much he might also like to emulate him?), their fate was sealed.

Despotic power, which came with civilization, has faded again. It looks increasingly like an aberration in the history of humanity. Before "civilization" and since democracy, men have been unable to accumulate the sort of power that enabled the most successful of them to be promiscuous despots. The best they could hope for in the Pleistocene period was one or two faithful wives and a few affairs if their hunting or political skills were especially great. The best they can hope for now is a good-looking younger mistress and a devoted wife who is traded in every decade or so. We're back to square one.

This chapter has kept its focus resolutely on the male. In doing so it may seem to have trampled on the rights of women by ignoring them and their wishes. But then so did men for many generations after the invention of agriculture. Before agriculture and since democracy, such chauvinism was impossible; the mating system of humans, like that of other animals, was a compromise between the strategies of males and females. And it is a curious truth that the monogamous marriage bond survived right through despotic Babylon, lascivious Greece, promiscuous Rome, and adulterous Christendom to emerge as the core of the family in the industrial age. Even in the most despotic and polygamous moment of human history, mankind was faithful to the institution of monogamous marriage, quite unlike any other polygamous animal. Even despots usually had one queen and many concubines. Explaining the human fascination with monogamous marriage requires us to understand the female strategy as closely as we have understood the male one. When we do, an extraordinary insight into human nature will emerge. That is what the next chapter is about.

Chapter 7

MONOGAMY AND THE NATURE OF WOMEN

SHEPHERD: *Echo, I ween, will in the wood reply,*

And quaintly answer questions: shall I try?

ECHO: *Try.*

What must we do our passion to express?

Press.

How shall I please her who never loved before?

Be Fore.

What most moves women when we them address?

A dress.

Say, what can keep her chaste whom I adore?

A door.

If music softens rocks, love tunes my lyre.

Liar.

Then teach me, Echo, how shall I come by her?

Buy her.

—Jonathan Swift, "A Gentle Echo on Woman"

In an astonishing study recently undertaken in Western Europe, the following facts emerged: Married females choose to have affairs with males who are dominant, older, more physically attractive, more symmetrical in appearance, and married; females are much more likely to have an affair if their mates are subordinate, younger, physically unattractive, or have asymmetrical features; cosmetic surgery to improve a male's looks doubles his chances of having an adulterous affair; the more attractive a male, the less attentive he is as a father; roughly one in three of the babies born in Western Europe is the product of an adulterous affair.

If you find these facts disturbing or hard to believe, do not worry. The study was not done on human beings but on swallows, the innocent, twittering, fork-tailed birds that pirouette prettily around barns and fields in the summer months. Human beings are entirely different from swallows. Or are they?[1]

THE MARRIAGE OBSESSION

The harems of ancient despots revealed that men are capable of making the most of opportunities to turn rank into reproductive success, but they cannot have been typical of the human condition for most of its history. About the only way to be a harem-guarding potentate nowadays is to start a cult and brainwash potential concubines about your holiness. In many ways modern people probably live in social systems that are much closer to those of their hunter-gatherer ancestors than they are to the conditions of early history.

No hunter-gatherer society supports more than occasional polygamy, and the institution of marriage is virtually universal. People live in larger bands than they used to, but within those bands the kernel of human life is the nuclear family: husband, wife, and children. Marriage is a child-rearing institution; wherever it occurs, the father takes at least some part in rearing the child even if only by providing food. In most societies men strive to be polygamists but few succeed. Even in the polygamous societies of pastoralists, the great majority of marriages are monogamous ones.[2]

It is our usual monogamy, not our occasional polygamy, that sets us apart from other mammals, including apes. Of the four other apes (gibbons, orangutans, gorillas, and chimpanzees), only the gibbon practices anything like marriage. Gibbons live in faithful pairs in the forests of Southeast Asia, each pair living a solitary life within a territory.

If men are opportunists-polygamists at heart, as I argued in the last chapter, then where does marriage come from? Although men are fickle ("You're afraid of commitment, aren't you?" says the stereotypical victim of a seducer), they are also interested in finding wives with whom to rear families and might well be very set on sticking by them despite their own infidelity ("You're never going to leave your wife for me, are you?" says the stereotypical mistress).

The two goals are contradictory only because women are not prepared to divide themselves neatly into wives and whores. Woman is not the passive chattel that the tussles of despots, described in the last chapter, have implied. She is an active adversary in the sexual chess game, and she has her own goals. Women are and always have been far less interested in polygamy than men, but that does not mean they are not sexual opportunists. The eager male/coy female theory has a great deal of difficulty answering a simple question: Why are women ever unfaithful?

THE HEROD EFFECT

In the 1980s a number of women scientists, led by Sarah Hrdy, now of the University of California at Davis, began to notice that the promiscuous behavior of female chimpanzees and monkeys sat awkwardly alongside the Trivers theory that heavily female biased parental investment leads directly to female choosiness. Hrdy's own studies of langurs and the studies of macaques by her student Meredith Small seemed to reveal a very different kind of female from the stereotype of evolutionary theory: a female who sneaked away from the troop for assignations with males; a female who actively sought a variety of sexual partners; a female who was just as likely as a male to initiate sex. Far from being choosy, female primates seemed to be initiators of much promiscuity. Hrdy began to suggest that there was something wrong with the theory rather than the females. A decade later it is suddenly clear what: A whole new light has been shed on the evolution of female behavior by a group of ideas known as "sperm competition theory."[3]

The solution to Hrdy's concern lay in her own work. In her study of the langurs of Abu in India, Hrdy discovered a grisly fact: The murder of baby monkeys by adult male monkeys was routine. Every time a male takes over a troop of females, he kills all the infants in the group. Exactly the same phenomenon had been discovered in lions a short time later: When a group of brothers wins a pride of females, the first thing they do is slaughter the innocents. In fact, as subsequent research revealed, infanticide by males is common in rodents, carnivores, and primates. Even our closest relatives, the chimpanzees, are guilty. Most naturalists, reared on a diet of sentimental natural history television programs, were inclined to believe they were witnessing a pathological aberration, but Hrdy and her colleagues suggested otherwise. The infanticide, they said, was an "adaptation"—an evolved strategy. By killing their stepchildren the males would halt the females' milk production and so bring forward the date on which the mother could conceive once more. An alpha male langur or a pair of brother lions has only a short time at the top, and infanticide helps these animals to father the maximum number of offspring during that time.[4]

The importance of infanticide in primates gradually helped scientists to understand the mating systems of the five species of apes because it suddenly provided a reason for females to be loyal to one or a group of males—and vice versa: to protect their genetic investment in each other from murderous rival males. Broadly speaking, the social pattern of female monkeys and apes is determined by the distribution of their food, while the social pattern of males is determined by the distribution of females. Thus, female orangutans choose to live alone in strict territories, the better to exploit their scarce food resources. Males also live alone and try to monopolize the territories of several females. The females that live within his territory expect their "husband" to come rushing to their aid if another male appears.

Female gibbons also live alone. Male gibbons are capable of defending the home ranges of up to five females, and they could easily practice the same kind of polygamy as orangutans: one male can patrol the territories of five females and mate with them all. What is more, male gibbons are of little use as fathers. They do not feed the young, they do not protect them from eagles, they do not even teach them much. So why do they stick with one female faithfully? The one enormous danger to a young gibbon that its father can guard against is murder by another male gibbon. Robin Dunbar of Liverpool University believes that male gibbons are monogamous to prevent infanticide.[5]

A female gorilla is as faithful to her husband as any gibbon; she goes where he goes and does what he does. And he is faithful to her in a manner of speaking. He stays with her for many years and watches her raise his children. But there is one big difference: He has several females in his harem and is, as it were, equally faithful to each. Richard Wrangham of Harvard University believes the gorilla social system is largely designed around the prevention of infanticide but that for females there is safety in numbers. (For fruit-eating gibbons there is not enough food in a territory to feed more than one female.) So a male keeps his harem safe from the attentions of rival males and pays his children the immense favor of preventing their murder.[6]

The chimpanzee has further refined the anti-infanticide strategy by inventing a rather different social system. Because they eat scattered but abundant food such as fruit and spend more time on the ground and in the open, chimps live in larger groups (a big group has more pairs of eyes than a small group) that regularly fragment into smaller groups before coming back together. These "fission-fusion" groups are too large and too flexible for a single male to dominate. The way to the top of the political tree for a male chimp is by building alliances with other males, and chimp troops contain many males. So a female is now accompanied by many dangerous stepfathers. Her solution is to share her sexual favors more widely with the effect that all the stepfathers might be the father. As a result, there is only one circumstance in which a male chimp can be certain an infant he meets is *not* his: when he has never seen the female before. And as Jane Goodall found, male chimps attack strange females that are carrying infants and kill the infants. They do not attack childless females.[7]

Hrdy's problem is solved. Female promiscuity in monkeys and apes can be explained by the need to share paternity among many males to prevent infanticide. But does it apply to mankind?

The short answer is no. It is a fact that stepchildren are sixty-five times more likely to die than children living with their true parents,[8] and it is inescapable that young children often have a terror of new stepfathers that is hard to overcome. But neither of these facts is of much relevance, for both apply to older children, not to suckling infants. Their deaths do not free the mother to breed again.

Moreover, the fact that we are apes can be misleading. Our sex lives are very different from those of our cousins. If we were like orangutans, women would live alone and apart from one another. Men, too, would live alone but each would visit several women (or none) for occasional sex. If two men ever met, there would be an almighty, violent battle. If we were gibbons, our lives would be unrecognizable. Every couple would live miles apart and fight to the death any intrusion into their home range—which they would never leave. Despite the occasional antisocial neighbor, that is not how we live. Even people who retreat to their sacred suburban homes do not pretend to remain there forever, let alone keep out all strangers. We spend much of our lives on common territory, at work, shopping, or at play. We are gregarious and social.

We are not gorillas, either. If we were, we would live in seraglios, each dominated by one giant middle-aged man, twice the weight of a woman, who would monopolize sexual access to all the women in the group and intimidate the other men. Sex would be rarer than saints' days, even for the great man, who would have sex once a year, and would be all but nonexistent for the other males.[9]

If we were hairless chimpanzees, our society would still look fairly familiar in some ways. We would live in families, be very social, hierarchical, group-territorial, and aggressive toward other groups than those we belong to. In other words, we would be family-based, urban, class-conscious, nationalist, and belligerent, which we are. Adult males would spend more time trying to climb the political hierarchy than with their families. But when we turn to sex, things would begin to look very different. For a start, men would take no part at all in rearing the young, not even paying child support; there would be no marriage bonds at all. Most women would mate with most men, though the top male (the president, let us call him) would make sure he had *droit du seigneur* over the most fertile women. Sex would be an intermittent affair, indulged in to spectacular excess during the woman's estrus but totally forgotten by her for years at a time when pregnant or rearing a young child. This estrus would be announced to everybody in sight by her pink and swollen rear end, which would prove irresistibly fascinating to every male who saw it. They would try to monopolize such females for weeks at a time, forcing them to go away on a "consortship" with them; they would not always succeed and would quickly lose interest when the swelling went down. Jared Diamond of the University of California at Los Angeles has speculated on how disruptive this would be to society by imagining the effect on the average office of a woman turning up for work one day irresistibly pink.[10]

If we were pygmy chimps or bonobos, we would live in groups much like those of chimps, but there would be roving bands of dominant men who visited several groups of women. As a consequence, women would have to share the possibility of paternity still more widely, and female bonobos are positively nymphomaniac in their habits. They have sex at the slightest suggestion and in a great variety of ways (including oral and homosexual) and are sexually attractive to males for long periods. A young female bonobo who arrives at a tree where others of the species are feeding will first mate with each of the males in turn—including the adolescents—and only then get on with eating. Mating is not wholly indiscriminate, but it is very catholic.

Whereas a female gorilla will mate about ten times for every baby that is born, a female chimp will mate five hundred to a thousand times and a bonobo up to three thousand times. A female bonobo is rarely harassed by a nearby male for mating with a more junior male, and mating is so frequent that it rarely leads to conception. Indeed, the whole anatomy of male aggression is reduced in bonobos: Males are no larger than females and spend less energy trying to rise in the male hierarchy than ordinary chimps. The best strategy for a male bonobo intent on genetic eternity is to eat his greens, get a good night's sleep, and prepare for a long day of fornication.[11]

THE BASTARD BIRDS

Compared to our ape cousins, we, the most common of the great apes, have pulled off a surprising trick. We have somehow reinvented monogamy and paternal care without losing the habit of living in large multimale groups. Like gibbons, men marry women singly and help them to rear their young, confident of paternity, but like chimpanzees, those women live in societies where they have continual contact with other men. There is no parallel for this among apes. It is my contention, however, that there is a close parallel among birds. Many birds live in colonies but mate monogamously within the colony. And the bird parallel brings an altogether different explanation for females to be interested in sexual variety. A female human being does not have to share her sexual favors with many males to prevent infanticide, but she may have a good reason to share them with one well-chosen male apart from her husband.

This is because her husband is, almost by definition, usually not the best male there is—else how would he have ended up married to her? His value is that he is monogamous and will therefore not divide his child-rearing effort among several families. But why accept his genes? Why not have his parental care and some other male's genes?

In describing the human mating system, it is hard to be precise. People are immensely flexible in their habits, depending on their racial origin, religion, wealth, and ecology. Nonetheless, some universal features stand out. First, women most commonly seek monogamous marriage—even in societies that allow polygamy. Rare exceptions notwithstanding, they want to choose carefully and then, as long as he remains worthy, monopolize a man for life, gain his assistance in rearing the children, and perhaps even die with him. Second, women do not seek sexual variety per se. There are exceptions, of course, but fictional and real women regularly deny that nymphomania holds any attraction for them, and there is no reason to disbelieve them. The temptress interested in a one-night stand with a man whose name she does not know is a fantasy fed by male pornography. Lesbians, free of constraints imposed by male nature, do not suddenly indulge in sexual promiscuity; on the contrary, they are remarkably monogamous. None of this is surprising: Female animals gain little from sexual opportunism, for their reproductive ability is limited not by how many males they mate with but how long it takes to bear offspring. In this respect men and women are very different.

But third, women are sometimes unfaithful. Not all adultery is caused by men. Though she may rarely or never be interested in casual sex with a male prostitute or a stranger, a woman, in life as in soap operas, is perfectly capable of accepting or provoking an offer of an affair with one man whom she knows, even if she is "happily" married at the time. This is a paradox. It can be resolved in one of three ways. We can blame adultery on men, asserting that the persuasive powers of seducers will always win some hearts, even the most reluctant. Call this the "Dangerous Liaisons" explanation. Or we can blame it on modern society and say that the frustrations and complexities of modern life, of unhappy marriages and so on, have upset the natural pattern and introduced an alien habit into human females. Call this the "Dallas" explanation. Or we can suggest that there is some valid biological reason for seeking sex outside marriage without abandoning the marriage—some instinct in women not to deny themselves the option of a sexual "plan B" when plan A does not work out so well. Call this the "Emma Bovary" strategy.

I am going to argue in this chapter that adultery may have played a big part in shaping human society because there have often been advantages to both sexes from within a monogamous marriage in seeking alternative sexual partners. This conclusion is based on studies of human society, both modern and tribal, and on comparisons with apes and birds. By describing adultery as a force that shaped our mating system, I am not "justifying" it. Nothing is more "natural" than people evolving the tendency to object to being cuckolded or cheated on, so if my analysis were to be interpreted as justifying adultery, it would be even more obviously interpreted as justifying the social and legal mechanisms for discouraging adultery. What I am claiming is that adultery and its disapproval are both "natural."

In the 1970s, Roger Short, a British biologist who later moved to Australia, noticed something peculiar about ape anatomy. Chimpanzees have gigantic testicles; gorillas have minuscule ones. Although gorillas are four times the weight of chimps, chimps' testicles weigh four times as much as gorillas'. Short wondered why that was and suggested that it might have something to do with the mating system. According to Short, the bigger the testicles, the more polygamous the females.[12]

The reason is easy to see. If a female animal mates with several males, then the sperm from each male competes to reach her eggs first; the best way for a male to bias the race in his favor is to produce more sperm and swamp the competition. (There are other ways. Some male damsel flies use their penis to scoop out sperm that was there first; male dogs and Australian hopping mice both "lock" their penis into the female after copulation and cannot free it for some time, thus preventing others from having a go; male human beings seem to produce large numbers of defective "kamikaze" sperm that form a sort of plug that closes the vaginal door to later entrants.)[13] As we have seen, chimpanzees live in groups where several males may share a female, and therefore there is a premium on the ability to ejaculate often and voluminously— he who does so has the best chance of being the father. This conjecture holds up across all the monkeys and across all rodents. The more they can be sure of sexual monopoly, as the gorilla can, the smaller their testes; the more they live in multimale promiscuous groups, the larger their testes.[14]

It began to look as if Short had stumbled on an anatomical clue to a species' mating system: Big testicles equals polygamous females. Could it be used to predict the mating system of species that had not been studied? For example, very little is known about the societies of dolphins and whales, but a good deal is known of their anatomy, thanks to whaling. They all have enormous testicles, even allowing for their size. The testicles of a right whale weigh more than a ton and account for 2 percent of its body weight. So, given the monkey pattern, it is reasonable to predict that female whales and dolphins are mostly not monogamous but will mate with several males. As far as is known, this is the case. The mating system of the bottle-nosed dolphin seems to consist of forcible "herding" of fertile females by shifting coalitions of males and sometimes even the simultaneous impregnation of such a female by two males at the same time—a case of sperm competition more severe than anything in the chimpanzee world.[15] Sperm whales, which live in harems like gorillas, have comparatively smaller testicles; one male has a monopoly over his harem and has no sperm competitors.

Let us now apply this prediction to man. For an ape, man's testicles are medium-sized—considerably bigger than a gorilla's. Like a chimpanzee's, human testicles are housed in a scrotum that hangs outside the body where it keeps the sperm that have already been produced cool, therefore increasing their shelf life, as it were.[16] This is all evidence of sperm competition in man.

But human testicles are not nearly as large as those of chimps, and there is some tentative evidence that they are not operating on full power (that is, they might once have been bigger in our ancestors): Sperm production per gram of tissue is unusually low in man. All in all, it seems fair to conclude that women are not highly promiscuous, which is what we expected to find.[17]

It is not just monkeys, apes, and dolphins that have large testicles when faced with sperm competition. Birds do, too. And it is from birds that the clinching clue comes about the human mating system. Zoologists have long known that most mammals are polygamous and most birds are monogamous. They put this down to the fact that the laying of eggs gives male birds a much earlier opportunity to help rear his children than a male mammal ever has. A male bird can busy himself with building the nest, with sharing the duties of incubation, with bringing food for the young; the only thing he cannot do is lay the eggs. This opportunity allows junior male birds to offer females a more paternal alternative than merely inseminating them, an offer that is accepted in species that have to feed their young, such as sparrows, and rejected in those that do not feed their young, such as pheasants.

Indeed, in some birds, as we have seen, the male does all these things alone, leaving his mate with the single duty of egg laying for her many husbands. In a mammal, by contrast, there is not much he can do to help even if wants to. He can feed his wife while she is pregnant and thereby contribute to the growth of the fetus, and he can carry the baby about when it is born or bring it food when it is weaned, but he cannot carry a fetus in his belly or feed it milk when it is born. The female mammal is left literally holding the baby, and with few opportunities to help her, the male is often better off expending his energy on an attempt to be a polygamist. Only when opportunities for further mating are few and his presence increases the baby's safety—as in gibbons—will he stay.

This kind of game-theory argument was commonplace by the mid-1970s, but in the 1980s when it became possible for the first time to do genetic blood testing of birds, an enormous surprise was in store for zoologists. They discovered that many of the baby birds in the average nest were not their ostensible father's offspring. Male birds were cuckolding one another at a tremendous rate. In the indigo bunting, a pretty little blue bird from North America that seemed to be faithfully monogamous, about 40 percent of the babies the average male feeds in his nest are bastards.[18]

The zoologists had entirely underestimated an important part of the life of birds. They knew it happened, but not on such a scale. It goes under the abbreviation EPC, for extra-pair copulation, but I will call it adultery, for that is what it is. Most birds are indeed monogamous, but they are not by any means faithful.

Anders Møller is a Danish zoologist of legendary energy whom we have already met in the context of sexual selection. He and Tim Birkhead from Sheffield University have written a book that summarizes what is now known about avian adultery, and it reveals a pattern of great relevance to human beings. The first thing they proved is that the size of a bird's testicles varies according to the bird's mating system. They are largest in polyandrous birds, where several males fertilize one female, and it is not hard to see why. The male who ejaculates the most sperm will presumably fertilize the most eggs.

That came as no surprise. But the testicles of lekking birds, such as sage grouse, where each male may have to inseminate fifty females in a few weeks, are unusually small. This puzzle is resolved by the fact that a female sage grouse will mate only once or twice and usually with only one male. That, remember, is the whole point of female choosiness at leks. So although the master cock may need to mate with many hens, he need not waste much sperm on each because those sperm will have no competitors. It is not how often a male bird copulates that determines the size of his testicles but how many other males he competes with.

The monogamous species lie in between. Some have fairly small testicles, implying little sperm competition; others have huge testicles, as big as those of polyandrous birds. Birkhead and Møller noticed that the ones with large testicles were mostly birds that lived in colonies: seabirds, swallows, bee eaters, herons, sparrows. Such colonies give females ample opportunity for adultery with the male from the nest next door, an opportunity that is not passed up.[19]

Bill Hamilton believes that adultery may explain why in so many "monogamous" birds the male is gaudier than the female. The traditional explanation, suggested by Darwin, is that the gaudiest males or the best songsters get the first females to arrive, and an early nest is a successful nest. That is certainly true, but it does not explain why song continues in many species long after a male has found a wife. Hamilton's suggestion is that the gaudy male is not trying to get more wives but more lovers. As Hamilton put it, "Why did Beau Brummel in Regency England dress up as he did? Was it to find a wife or to find an 'affair'?"[20]

EMMA BOVARY AND FEMALE SWALLOWS

What's in it for the birds? For the males it is obvious enough: Adulterers father more young. But it is not at all clear why the female is so often unfaithful. Birkhead and Møller rejected several suggestions: that she is adulterous because of a genetic side effect of the male adulterous urge, that she is ensuring some of the sperm she gets is fertile, that she is bribed by the philandering males (as seems to be the case in some human and ape societies). None of these fit the exact facts. Nor did it quite work to blame her infidelity on a desire for genetic variety. There seems to be little point in having more varied children than she would have anyway.

Birkhead and Møller were left with the belief that female birds benefit from being promiscuous because it enables them to have their genetic cake and eat it—to follow the Emma Bovary strategy. A female swallow needs a husband who will help look after her young, but by the time she arrives at the breeding site, she might find all the best husbands taken. Her best tactic is therefore to mate with a mediocre husband or a husband with a good territory and have an affair with a genetically superior neighbor. This theory is supported by the facts: Females always choose more dominant, older, or more "attractive" (that is, ornamented) lovers than their husbands; they do not have affairs with bachelors (presumably rejects) but with other females' husbands; and they sometimes incite competition between potential lovers and choose the winners. Male swallows with artificially lengthened tails acquired a mate ten days sooner, were eight times as likely to have a second brood, and had twice as high a chance of seducing a neighbor's wife as ordinary swallows.[21] (Intriguingly, when female mice choose to mate with males other than those they "live with," they usually choose ones whose disease-resistance genes are *different* from their own.)[22]

In short, the reason adultery is so common in colonial birds is that it enables a male bird to have more young and enables a female bird to have better young.

One of the most curious results to come out of bird studies in recent years has been the discovery that "attractive" males make inattentive fathers. Nancy Burley, whose zebra finches consider one another more or less attractive according to the color of their leg bands, first noticed this,[23] and Anders Moller has since found it to be true of swallows as well. When a female mates with an attractive male, he works less hard and she works harder at bringing up the young. It is as if he feels that he has done her a favor by providing superior genes and therefore expects her to repay him with harder work around the nest. This, of course, increases her incentive to find a mediocre but hardworking husband and cuckold him by having an affair with a superstud next door.[24]

In any case, the principle—marry a nice guy but have an affair with your boss or marry a rich but ugly man and take a handsome lover—is not unknown among female human beings. It is called having your cake and eating it, too. Flaubert's Emma Bovary wanted to keep both her handsome lover and her wealthy husband.

The work on birds has been conducted by people who knew little of human anthropology. In just the same way, a pair of British zoologists had been studying human beings in the late 1980s, largely in isolation from the bird work. Robin Baker and Mark Bellis of Liverpool University were curious to know if sperm competition happened inside women, and if it did, whether women had any control over it. Their results have led to a bizarre and astonishing explanation of the female orgasm.

What follows is the only part of this book in which the details of sexual intercourse itself are relevant to an evolutionary argument. Baker and Bellis discovered that the amount of sperm that is retained in a woman's vagina after sex varies according to whether she had an orgasm and when. It also depends on how long it was since she last had sex: The longer the period, the more sperm stays in, unless she has what the scientists call "a noncopulatory orgasm" in between.

So far none of this contained great surprises; these facts were unknown before Baker and Bellis did their work (which consisted of samples collected by selected couples and of a survey of four thousand people who replied to a questionnaire in a magazine), but they did not necessarily mean very much. But Baker and Bellis also did something rather brave. They asked their subjects about their extramarital affairs. They found that in faithful women about 55 percent of the orgasms were of the high-retention (that is, the most fertile) type. In unfaithful women, only 40 percent of the copulations with the partner were of this kind, but 70 percent of the copulations with the lover were of this fertile type. Moreover, whether deliberately or not, the unfaithful women were having sex with their lovers at times of the month when they were most fertile. These two effects combined meant that an unfaithful woman in their sample could have sex twice as often with her husband as with her lover but was still slightly more likely to conceive a child by the lover than the husband.

Baker and Bellis interpret their results as evidence of an evolutionary arms race between males and females, a Red Queen game, but one in which the female sex is one evolutionary step ahead. The male is trying to increase his chances of being the father in every way. Many of his sperm do not even try to fertilize her eggs but instead either attack other sperm or block their passage.

But the female has evolved a sophisticated set of techniques for preventing conception except on her own terms. Of course, women did not know this before now and therefore did not set out to achieve it, but the astonishing thing is that if the study by Baker and Bellis proves to be right, they are doing it anyway, perhaps quite unconsciously. This, of course, is typical of evolutionary explanations. Why do women have sex at all? Because they consciously want to. But why do they consciously want to? Because sex leads to reproduction, and being the descendants of those who reproduced, they are selected from among those who want things that lead to reproduction. This is merely a form of the same argument: The typical woman's pattern of infidelity and orgasm is exactly what you would expect to find if she were unconsciously trying to get pregnant from a lover while not leaving a husband.

Baker and Bellis do not claim to have found more than a tantalizing hint that this is so, but they have tried to measure the extent of cuckoldry in human beings. In a block of flats in Liverpool, they found by genetic tests that fewer than four in every five people were the sons of their ostensible fathers. In case this had something to do with Liverpool, they did the same tests in southern England and got the same result. We know from their earlier work that a small degree of adultery can lead to a larger degree of cuckoldry through the orgasm effect. Like birds, women may be—quite unconsciously—having it both ways by conducting affairs with genetically more valuable men while not leaving their husbands.

What about the men? Baker and Bellis did an experiment on rats and discovered that a male rat ejaculates twice as much sperm when he knows that the female he is mating with has been near another male recently. The intrepid scientists promptly set out to test whether human beings do the same. Sure enough, they do. Men whose wives have been with them all day ejaculate much smaller amounts than men whose wives have been absent all day. It is as if the males are subconsciously compensating for any opportunities for female infidelity that might be present. But in this particular battle of the sexes, the women have the upper hand because even if a man—again unconsciously—begins to associate his wife's lack of late orgasms with a desire not to conceive his child, she can always respond by faking them.[25]

CUCKOLDRY PARANOIA

The cuckold, however, does not stand by and accept his evolutionary lot even unto the extinction of his genes. Birkhead and Møller think that much of the behavior of male birds can be explained by the assumption that they are in constant terror of their wives' infidelity. Their first strategy is to guard the wife during the period when she is fertile (a day or so before each egg is laid). Many male birds do this. They follow their mates everywhere, so that a female bird who is building a nest is often accompanied on every trip by a male who never lends a hand; he just watches. The moment she is finished laying the clutch of eggs, he relaxes his vigil and begins to seek adulterous opportunities himself.

If a male swallow cannot find his mate, he often gives a loud alarm call, which causes all the swallows to fly into the air, effectively interrupting any adulterous act in progress. If the pair has just been reunited after a separation or if a strange male intrudes into the territory and is chased out, the husband will often copulate with the wife immediately afterward, as if to ensure that his sperm are there to compete with the intruder's.

Generally it works. Species that practice effective mate guarding keep the adultery rate low. But some species cannot guard their mates. In herons and birds of prey, for example, husband and wife spend much of the day apart, one guarding the nest while the other collects food. These species are characterized by extremely frequent copulation. Goshawks may have sex several hundred times for every clutch of eggs. This does not prevent adultery, but at least it dilutes it.[26]

Just like herons and swallows, people live in monogamous pairs within large colonies. Fathers help to rear the young if only by bringing food or money. And crucially, because of the sexual division of labor that characterized early human hunter-gathering societies (broadly speaking, men hunt, women gather), the sexes spend much time apart. So women have ample opportunities for adultery, and men have ample incentives to guard their mates or, failing that, to copulate frequently with them.

To demonstrate that adultery is a chronic problem throughout human society, rather than an aberration of modern apartment blocks in Britain, is paradoxically difficult: first, because the answer is so stunningly obvious that nobody has studied it, and second, because it is a universally kept secret and therefore almost impossible to study. It is easier to watch birds.

Nonetheless, attempts have been made. The 570 or so Aché people of Paraguay were hunter-gatherers until 1971, living in twelve bands. They then gradually came into contact with the outside world and were lured onto government reservations run by missionaries. Today, they no longer depend on hunted meat and gathered fruit but grow most of their own food in gardens. But when they still depended on men's hunting skills for much of their food, Kim Hill of the University of New Mexico found an intriguing pattern. Aché men would donate any spare meat they had caught to women with whom they wanted to have sex. They were not doing so to feed children they might have fathered but as direct payment for an affair. It was not easy to discover. Hill found that he was gradually forced to drop questions about adultery from his studies because the Aché, under missionary influence, became increasingly squeamish about discussing the subject. The chiefs and the head men were especially reluctant to talk about it, which is hardly surprising in view of the fact that they were the ones having the most affairs. Nonetheless, by relying on gossip Hill was able to piece together the pattern of adultery in the Aché. As expected, he found that high-ranking men were involved most, which is consistent with the idea of having your paternal-genetic cake and eating it. However, unlike in birds, it was not just the wives of low-ranking men who indulged. It is true that Aché adulterers frequently ply their mistresses with gifts of meat, but Hill thinks the most important motive is that Aché women are constantly preparing for the possibility that they will be deserted by their husbands. They are building up alternative relationships and are more likely to be unfaithful if the marriage is going badly. That is, of course, a double-edged sword: The marriage could break up because the affair is discovered.[27]

Whatever the motive for women, Hill and others believe that adultery has been much underemphasized as an influence in the evolution of the human mating system. In hunter-gatherer societies the male opportunist streak would have been far more easily satisfied by adultery than by polygamy. In only two known hunter-gatherer societies is polygamy either common or extreme. In the rest it is rare to find a man with more than one wife and very rare to find a man with more than two. The two exceptions prove the rule. One is among the Indians of the Pacific Northwest of America, who depended on abundant and reliable supplies of salmon and were more like farmers than hunter-gatherers in their ability to stockpile surpluses. The other is certain tribes of Australian aborigines, which practice gerontocratic polygamy: Men do not marry until they are forty, and by the age of sixty-five they have usually accumulated up to thirty wives. But this peculiar system is far from what it seems. Each old man has younger assistant men whose help, protection, and economic support he purchases by, among other things, turning a blind eye to their affairs with his wives. The old man looks the other way when the helpful nephew carries on with one of his junior wives.[28]

Polygamy is rare in hunter-gatherer societies, but adultery is common wherever it has been looked for. By analogy with monogamous colonial birds, therefore, one would expect to find human beings practicing either mate guarding or frequent copulation. Richard Wrangham has speculated that human beings practice mate guarding in absentia. Men keep an eye on their wives by proxy. If the husband is away hunting all day in the forest, he can ask his mother or his neighbor whether his wife was up to anything during the day. In the African pygmies Wrangham studied, gossip was rife and a husband's best chance of deterring his wife's affairs was to let her know that he kept abreast of the gossip. Wrangham went on to observe that this was impossible without language, so he speculated that the sexual division of labor, the institution of child-rearing marriages, and the invention of language—three of the most fundamental human characteristics shared with no other ape—all depend on another.[29]

WHY THE RHYTHM METHOD DOES NOT WORK

What happened before language allowed proxy mate guarding? Here, anatomy provides an intriguing clue. Perhaps the most startling difference between the physiology of a woman and that of a female chimpanzee is that it is impossible for anybody, including the woman herself, to determine precisely when in the menstrual cycle she is fertile. Whatever doctors, old wives' tales, and the Roman Catholic Church may say, human ovulation is invisible and unpredictable. Chimpanzees become pink; cows smell irresistible to bulls; tigresses seek out tigers; female mice solicit male mice— throughout the mammal order, the day of ovulation is announced with fanfare. But not in man. A tiny change in the woman's temperature, undetectable before thermometers, and that is all. Women's genes seem to have gone to inordinate lengths to conceal the moment of ovulation.

With concealed ovulation came continual sexual interest. Although women are more likely to initiate sex, masturbate, have an affair with a lover, or be accompanied by their husband on the day of ovulation than on other days,[30] it is nonetheless true that human beings of both sexes are interested in sex at all times of the menstrual cycle; both men and women have intercourse whenever they feel like it, without reference to hormonal events. Compared with many animals, we are astonishingly hooked on copulation. Desmond Morris called mankind "the sexiest primate alive"[31] (but that was before anybody studied bonobos). Other animals that copulate frequently—lions, bonobos, acorn woodpeckers, goshawks, white ibises—do so for reasons of sperm competition. Males of the first three species live in groups that share access to females, so every male must copulate as frequently as he can or risk another male's sperm reaching the egg first. Goshawks and white ibises do so to swamp any sperm that might have been received by the female while the male was away at work. Since it is clear that humanity is not a promiscuous species—even the most carefully organized free-love commune soon falls apart under the pressure of jealousy and possessiveness—the case of the ibis is the most pertinent for man: a monogamous colonial animal driven by the threat of adultery into the habit of frequent copulation. At least the male ibis need only keep his sex-six-times-a-day routine up for a few days each season before egg laying. Men must keep up sex twice a week for years.[32]

But concealed ovulation in women cannot have evolved for the convenience of the man. In the late 1970s there was a flurry of speculative theorizing about the evolutionary cause of concealed ovulation. Many of the ideas apply only to human beings. An example is Nancy Burley's suggestion that ancestral women with unconcealed ovulation learned to be celibate when fertile because of the uniquely painful and dangerous business of human childbirth; but such women left behind no descendants, so the rare exceptions who could not detect their own ovulation mothered the human race. Yet concealed ovulation is a habit we share with some monkeys and at least one ape (the orangutan). It is also a habit we share with nearly all birds. Only our absurdly parochial anthropocentrism has allowed us to think that silent ovulation is special.

Nonetheless, it is worth going through the attempted explanations of what Robert Smith once called human "reproductive inscrutability" because they shed an interesting light on the theory of sperm competition. They come in two kinds: those suggesting concealed ovulation as a way of ensuring that fathers did not desert their young, and those suggesting the exact opposite. The first kind of argument went as follows: Because he does not know when his wife is fertile, a husband must stay around and have sex with her often to be sure of fathering her children. This keeps him from mischief and ensures he is still around to help rear the babies.[33]

The second kind of argument went this way: If females wish to be discriminating in their choice of partner, it makes little sense to advertise their ovulation. Conspicuous ovulation will have the effect of attracting several males, who will either fight over the right to fertilize her, or share her. If a female wishes (is designed) to be promiscuous in order to share paternity, as chimps do, or if she wishes to set up a competition so that the best male wins her, as buffalo and elephant seals do, then it pays to advertise the moment of ovulation. But if she wishes to choose one mate herself for whatever reason, then she should keep it secret.[34]

This idea has several variants. Sarah Hrdy proposed that silent ovulation helps prevent infanticide because neither the husband nor the lover knows if he has been cuckolded. Donald Symons thinks women use perpetual sexual availability to seduce philanderers in exchange for gifts. L. Benshoof and Randy Thornhill suggested that concealed ovulation allows a woman to mate with a superior man by stealth without deserting or alerting her husband. If, as seems possible, ovulation is less concealed from her (or her unconscious) than it is from him, then it would help her make each extramarital liaison more rewarding since she is more likely to "know" when to have sex with her lover, whereas her husband does not know when she is fertile. In other words, silent ovulation is a weapon in the adultery game.[35]

This intriguingly sets up the possibility of an arms race between wives and mistresses. Genes for concealed ovulation make both adultery and fidelity easier. It is a peculiar thought, and there is at present no way of knowing if it is right, but it throws into stark contrast the fact that there can be no genetic feminine solidarity. Women will often be competing with women.

SPARROW FIGHTS

It is this competition between females that provides the final clue to the reason adultery, rather than polygamy, has probably been the most common way for men to have many mates. Red-winged blackbirds, which nest in marshes in Canada, are polygamous; the males with the best territories each attract several females to nest in them. But the males with the biggest harems are also the most successful adulterers, fathering the most babies in their neighbors' territories, too. Which raises the question of why the males' lovers do not simply become extra wives.

There is a small owl called Tengmalm's owl that lives in Finnish forests. In years when mice are abundant, some of the male owls have two mates, one in each of two territories, while other males go without a mate at all. The females that are married to polygamous males rear noticeably fewer young than the females married to monogamous males, so why do they put up with it?

Why not leave for one of the nearby bachelors? A Finnish biologist believes that the polygamists are deceiving their victims. The females judge potential suitors by how many mice they can catch to feed them during courtship. In a good year for mice a male can catch so many mice that he can simultaneously give two females the impression that he is a fine male; he can provide each with more mice than he could catch for one in a normal year.[36]

Nordic forests seem to be full of deceitful adulterers, for a similar habit by a deceptively innocent-looking little bird led to a long-running dispute in the scientific literature of the 1980s. Some male pied flycatchers in the forests of Scandinavia manage to be polygamous by holding two territories, each with a female in it, like the owls or like Sherman McCoy in Tom Wolfe's *Bonfire of the Vanities* who keeps an expensive wife on Park Avenue and a beautiful mistress in a rent-controlled apartment across town. Two teams of researchers have studied the birds and come to different conclusions about what is going on. The Finns and Swedes say that the mistress is deceived into believing the male is unmarried. The Norwegians say that since the wife sometimes visits the mistress's nest and may try to drive her away, the mistress can be under no illusions. She accepts the fact that her mate may desert her for his wife but hopes that if things go wrong at the wife's nest—they often do—he will come back to help her raise her young. He gets away with it only when the two territories are so far apart that the wife cannot visit the mistress's territory often enough to persecute her. In other words, according to the Norwegians, men deceive their wives about their affairs, not their mistresses.[37]

It is not clear, therefore, whether the wife or the mistress is the victim of treachery, but one thing is certain: The bigamous male pied flycatcher has pulled off a minor triumph, fathering two broods in one season. The male has fulfilled his ambition of bigamy at the expense of a female. The wife and the mistress would both have done better had each monopolized a male rather than shared him.

To test the suggestion that it is better to cuckold a faithful husband than leave him to become the second wife of a bigamist, José Veiga studied house sparrows breeding in a colony in Madrid. Only about 10 percent of the males in the colony were polygamous. By selectively removing certain males and females he tested various theories about why more males did not have multiple wives. First, he rejected the notion that males were indispensable to the rearing of young. Females in bigamous marriages reared as many young as those in monogamous ones, though they had to work harder. Second, by removing some males and observing which males the widows chose to remarry, he rejected the idea that females preferred to mate with unmated males; they were happy to choose already mated males and to reject bachelors. Third, he rejected the idea that males could not find spare females; 28 percent of males remated with a female who had not bred in the previous year. Then he tried putting nest boxes closer together to make it easier for the male to guard two at once; he found that it entirely failed to increase the amount of polygamy. That left him with one explanation for the rarity of polygamy in sparrows: The senior wives do not stand for it. Just as male birds guard their mates, so female birds chase away and harass their husbands' chosen second fiancées. Caged females are attacked by mated female sparrows. They do so presumably because even though they could rear the chicks on their own, it is a great deal easier with the husband's undivided help.[38]

It is my contention that man is just like an ibis or a swallow or a sparrow in several key respects. He lives in large colonies. Males compete with one another for places in a pecking order. Most males are monogamous. Polygamy is prevented by wives who resent sharing their husbands lest they also share his contributions to child rearing. Even though they could bring up the children unaided, the husband's paycheck is invaluable. But the ban on polygamous marriage does not prevent the males from seeking polygamous matings. Adultery is common. It is most common between high-ranking males and females of all ranks. To prevent it males try to guard their wives, are extremely violent toward their wives' lovers, and copulate with their wives frequently, not just when they are fertile.

That is the life of the sparrow anthropomorphized. The life of man sparrowmorphized might read like this: The birds live and breed in colonies called tribes or towns. Cocks compete with one another to amass resources and gain status within the colony; it is known as "business" and "politics." Cocks eagerly court hens, who resent sharing their males with other hens, but many cocks, especially senior ones, trade in their hens for younger ones or cuckold other cocks by having sex with their (willing) wives in private.

The point does not lie in the details of the sparrow's life. There are significant differences, including the fact that human beings tend to have a much more uneven distribution of dominance, power, and resources within the colony. But they still share the principal feature of all colonial birds: monogamy, or at least pair bonds, plus rife adultery rather than polygamy. The noble savage, far from living in contented sexual equanimity, was paranoid about becoming, and intent on making his neighbor into, a cuckold. Little wonder that human sex is first and foremost in all societies a private thing to be indulged in only in secret. The same is not true of bonobos, but it is true of many monogamous birds. One reason the high bastard rates of birds came as such a shock was that few naturalists had ever witnessed an adulterous affair between two birds—they do it in private.[39]

THE GREEN-EYED MONSTER

Cuckoldry paranoia is deep-seated in men. The use of veils, chaperones, purdah, female circumcision, and chastity belts all bear witness to a widespread male fear of being cuckolded and a widespread suspicion that wives, as well as their potential lovers, are the ones to distrust. (Why else circumcise them?) Margo Wilson and Martin Daly of McMaster University in Canada have studied the phenomenon of human jealousy and come to the conclusion that the facts fit an evolutionary interpretation. Jealousy is a "human universal," and no culture lacks it. Despite the best efforts of anthropologists to find a society with no jealousy and so prove that it is an emotion

introduced by pernicious social pressure or pathology, sexual jealousy seems to be an unavoidable part of being a human being.

> *The Demon, Jealousy, with Gorgon frown*
> *Blasts the sweet flowers of pleasure not his own,*
> *Rolls his wild eyes, and through the shuddering grove*
> *Pursues the steps of unsuspecting Love.*[40]

Wilson and Daly believe that a study of human society
reveals a mindset whose manifestations are diverse in detail but
"monotonously alike in the abstract." They are "socially recognized
marriage, the concept of adultery as a property violation, the valua-
tion of female chastity, the equation of 'protection' of women with
protection from sexual contact, and the special potency of infidelity as a provocation to violence." In short, in every age and in every
place, men behave as if they owned their wives' vaginas.[41]

Wilson and Daly reflect on the fact that love is an admired
emotion, whereas jealousy is a despised one, but they are plainly
two sides of the same coin—as anybody who has been in love can
testify. They are both part of a sexual proprietary claim. As many a
modern couple knows, the absence of jealousy, far from calming a
relationship, is itself a cause of insecurity. If he or she is not jealous when I pay attention to another man or woman, then he or she
no longer cares whether our relationship survives. Psychologists
have found that couples who lack moments of jealousy are less likely to stay together than jealous ones.

As Othello learned, even the suspicion of infidelity is enough
to drive a man to such rage that he may kill his wife. Othello was fictional, but many a modern Desdemona has paid with her life for her
husband's jealousy. As Wilson and Daly said: "The major source of
conflict in the great majority of spouse killings is the husband's
knowledge or suspicion that his wife is either unfaithful or intending
to leave him." A man who kills his wife in a fit of jealousy can rarely
plead insanity in court because of the legal tradition in Anglo-Amer-
ican common law that such an act is "the act of a reasonable man."[42]

This interpretation of jealousy probably seems astonishingly banal. After all, it is only putting an evolutionary slant on what everybody knows about everyday life. But among sociologists and psychologists it is heretical nonsense. Psychologists have tended to see jealousy as a pathology to be discouraged and generally thought shameful—as something that has been imposed by that eternal villain "society" to corrupt the nature of man. Jealousy shows low self-esteem, they say, and emotional dependency. Indeed it does, and that is exactly what the evolutionary theory would predict. A man held in low esteem by his wife is exactly the kind of person in danger of being cuckolded, for she has the motive to seek a better father for her children. This may even explain the extraordinary and hitherto baffling fact that husbands of rape victims are more likely to be traumatized and, despite themselves, to resent their raped wives if the wife was *not* physically hurt during the rape. Physical hurt is evidence of her resistance. Husbands may have been programmed by evolution to be paranoidly suspicious that their wives were not raped at all, or "asked for it."[43]

Cuckoldry is an asymmetrical fate. A woman loses no genetic investment if her husband is unfaithful, but a man risks unwittingly raising a bastard. As if to reassure fathers, research shows that people are strangely more apt to say of a baby, "He (or she) looks just like his father," than to say, "He (or she) looks just like his mother"—and that it is the mother's relatives who are most likely to say this.[44] It is not that a woman need not mind about her husband's infidelity; it might lead to his leaving her or wasting his time and money on his mistress or picking up a nasty disease. But it does imply that men are likely to mind even more about their wives' infidelity than vice versa. History and law have long reflected just that. In most societies adultery by a wife was illegal and punished severely, while adultery by a husband was condoned or treated lightly. Until the nineteenth century in Britain, a civil action could be brought against an adulterer by an aggrieved husband for "criminal conversation."[45] Even among the Trobriand islanders, who were celebrated by Bronislaw Malinowski in 1927 as a sexually uninhibited people, females who committed adultery were condemned to die.[46]

This double standard is a prime example of the sexism of society and is usually dismissed as no more than that. Yet the law has not been sexist about other crimes: Women have never been punished more severely than men for theft or murder, or at least the legal code has never prescribed that they be so. Why is adultery such a special case? Because man's honor is at stake? Then punish the adulterous man as harshly, for that is just as effective a deterrent as punishing the woman. Because men stick together in the war of the sexes? They do not do so in anything else. The law is quite explicit on this: All legal codes so far studied define adultery "in terms of the marital status of the woman. Whether the adulterous man was himself married is irrelevant."[47] And they do so because "it is not adultery per se that the law punishes but only the possible introduction of alien children into the family and even the uncertainty that adultery creates in this regard. Adultery by the husband has no such consequences."[48] When, on their wedding night, Angel Clare confessed to his new wife, Tess, in Thomas Hardy's *Tess of the D'Urbervilles*, that he had sown his wild oats before marriage, she replied with relief by telling the story of her own seduction by Alec D'Urberville and the short-lived child she bore him. She thought the transgressions balanced.

> "Forgive me as you are forgiven! *I* forgive *you*, Angel."
> "You—yes you do."
> "But do you not forgive me?"
> "O Tess, forgiveness does not apply to the case! You were one person; now you are another. My God—how can forgiveness meet such a grotesque—prestidigitation as that!"
> Clare left her that night.

COURTLY LOVE

Human mating systems are greatly complicated by the fact of inherited wealth. The ability to inherit wealth or status from a parent is not unique to man. There are birds that succeed to the ownership of their parents' territories by staying to help them rear subsequent broods. Hyenas inherit their dominance rank from their mothers (in hyenas, females are dominant and often larger); so do many monkeys and apes. But human beings have raised this habit to an art. And they usually show a much greater interest in passing on wealth to sons than to daughters. This is superficially odd. A man who leaves his wealth to his daughters is likely to see that wealth left to his certain granddaughters. A man who leaves his wealth to his sons is likely to see the wealth left to what may or may not be his grandsons. In the few matrilineal societies there is indeed such promiscuity that men are not sure of paternity, and in such societies it is uncles that play the role of father to their nephews.[49]

Indeed, in more stratified societies the poor often favor their daughters over their sons. But this is not because of certainty of paternity but because poor daughters are more likely to breed than poor sons. A feudal vassal's son had a good chance of remaining childless, while his sister was carted off to the local castle to be the fecund concubine of the resident lord. Sure enough, there is some evidence that in the fifteenth and sixteenth centuries in Bedfordshire, peasants left more to their daughters than to their sons.[50] In eighteenth-century Ostfriesland in Germany, farmers in stagnant populations had oddly female-biased families, whereas those in growing populations had male-biased families. It is hard to avoid the conclusion that third and fourth sons were a drain on the family unless there were new business opportunities, and they were dealt with accordingly at birth, resulting in female-biased sex ratios in the stagnant populations.[51]

But at the top of society, the opposite prejudice prevailed. Medieval lords banished many of their daughters to nunneries.[52] Throughout the world rich men have always favored their sons and often just one of them. A wealthy or powerful father, by leaving his status or the means to achieve it to his sons, is leaving them the wherewithal to become successful adulterers with many bastard sons. No such advantage could accrue to wealthy daughters.

This has a curious consequence. It means that the most

successful thing a man or a woman can do is beget a legitimate heir to a wealthy man. Logic such as this suggests that philanderers should not be indiscriminate. They should seduce the women with the best genes and also the women with the best husbands, who therefore have the potential to produce the most prolific sons. In medieval times this was raised to an art. The cuckolding of heiresses and the wives of great lords was considered the highest form of courtly love. Jousting was little more than a way for potential philanderers to impress great ladies. As Erasmus Darwin put it:

> Contending boars with tusks enamel'd strike,
> And guard with shoulder shield the blow oblique;
> While female bands attend in mute surprise,
> And view the victor with admiring eyes.
> So Knight on Knight, recorded in romance,
> Urged the proud steed, and couch'd the extended lance;
> He, whose dread prowess with resistless force,
> Bless'd, as the golden guerdon of his toils,
> Bow'd to the Beauty, and receiv'd her smiles.[53]

At a time when the legitimate eldest son of a great lord would inherit not only his father's wealth but also his polygamy, the cuckolding of such lords was sport indeed. Tristan expected to inherit the kingdom of his uncle, King Mark, in Cornwall. While in Ireland he ignored the attentions of the beautiful Isolde until she was summoned by King Mark to be his wife. Panic-struck at the thought of losing his inheritance but determined to save it at least for his son, he suddenly took an enormous interest in Isolde. Or at least so Laura Betzig retells the old story.[54]

Betzig's analysis of medieval history includes the idea that the begetting of wealthy heirs was the principal cause of Church-state controversies. A series of connected events occurred in the tenth century or thereabouts. The power of kings declined and the power of local feudal lords increased. As a consequence, noblemen gradually became more concerned with producing legitimate heirs to succeed to their titles, as the seigneurial system of primogeniture was established. They divorced barren wives and left all to the firstborn son. Meanwhile, resurgent Christianity conquered its rivals to become the dominant religion of northern Europe. The early Church was obsessively interested in matters of marriage, divorce, polygamy, adultery, and incest. Moreover, in the tenth century the Church began to recruit its monks and priests from among the aristocracy.[55]

The Church's obsessions with sexual matters were very different from St. Paul's. It had little to say about polygamy or the begetting of many bastards, although both were commonplace and against doctrine. Instead, it concentrated on three things: first, divorce, remarriage, and adoption; second, wet nursing, and sex during periods when the liturgy demanded abstinence; and third, "incest" between people married to within seven canonical degrees. In all three cases the Church seems to have been trying to prevent lords from siring legitimate heirs. If a man obeyed the doctrines of the Church in the year 1100, he could not divorce a barren wife, he certainly could not remarry while she lived, and he could not adopt an heir. His wife could not give her baby daughter to a wet nurse and be ready to bear another in the hope of its being a son, and he could not make love to his wife "for three weeks at Easter, four weeks at Christmas, and one to seven weeks at Pentecost; plus Sundays, Wednesdays, Fridays, and Saturdays—days for penance or sermons; plus miscellaneous feast days." He also could not bear a legitimate heir by any woman closer than a seventh cousin—which excluded most noble women within three hundred miles. It all adds up to a sustained attack by the Church on the siring of heirs, and "it was not until the Church started to fill up with the younger brothers of men of state that the struggle over inheritance—over marriage—between them began." Individuals in the Church (disinherited younger sons) were manipulating sexual mores to increase the Church's own wealth or even regain property and titles for themselves. Henry VIII 's dissolution of the monasteries, following his break with Rome, which followed Rome's disapproval of his divorcing the sonless Catherine of Aragon, is a sort of parable for the whole history of Church-state relations.[56]

Indeed, the Church-state controversy was just one of many historical instances of wealth-concentration disputes. The practice of primogeniture was a good way to keep wealth—and its polygamy potential—intact through the generations. But there were other ways, too. First among them was marriage itself. Marrying an heiress was always the quickest way to wealth. Of course, strategic marriage and primogeniture work against each other: If women inherit no wealth, then there is nothing to be gained from marrying a rich man's daughter. Among the royal dynasties of Europe, though, in most of which women could inherit thrones (in default of male heirs), eligible marriages were often possible. Eleanor of Aquitaine brought Britain's kings a large chunk of France. The War of the Spanish Succession was fought solely to prevent a French king from inheriting the throne of Spain as the result of a strategic marriage. Right down to the Edwardian practice of English aristocrats marrying the daughters of American robber barons, the alliances of great families have been a force to concentrate wealth.

Another way, practiced commonly among slave-owning dynasties in the American South, was to keep marriage within the family. Nancy Wilmsen Thornhill of the University of New Mexico has shown how in such families more often than not men married their first cousins. By tracing the genealogies of four southern families, she found that fully half of all marriages involved kin or sister exchange (two brothers marrying two sisters). By contrast, in northern families at the same time, only 6 percent of marriages involved kin. What makes this result especially intriguing is that Thornhill had predicted it before she found it. Wealth concentration works better for land, whose value depends on its scarcity, than for business fortunes, which are made and lost in many families in parallel.[57]

Thornhill went on to argue that just as some people have an incentive to use marriage to concentrate wealth, so other people have an incentive to prevent them from doing exactly that. And kings, in particular, have both the incentive and the power to achieve their wishes. This explains an otherwise puzzling fact: that prohibitions on "incestuous" marriages between cousins are fierce and numerous in some societies and absent in others. In every case it is the more highly stratified society that most regulates marriage. Among the Trumai of Brazil, an egalitarian people, marriage between cousins is merely frowned upon. Among the Maasai of East Africa, who have considerable disparities of wealth, such marriage is punished with "a severe flogging." Among the Inca people, anybody having the temerity to marry a female relative (widely defined) had his eyes gouged out and was cut into quarters. The emperor was, of course, an exception: His queen was his full sister, and Pachacuti began a tradition of marrying all his half sisters as well. Thornhill concludes that these rules had nothing to do with incest but were all about rulers trying to prevent wealth concentration by families other than their own; they usually excepted themselves from such laws.[58]

DARWINIAN HISTORY

This kind of science goes by the name of Darwinian history, and it has been greeted with predictable ridicule by real historians. For them, wealth concentration requires no further explanation. For Darwinians, it must once have been (or must still be) the means to a reproductive end. No other currency counts in natural selection.

When we study sage grouse or elephant seals in their natural habitat, we can be fairly sure that they are striving to maximize their long-term reproductive success. But it is much more difficult to make the same claim for human beings. People strive for something, certainly, but it is usually money or power or security or happiness. The fact that they do not translate these into babies is raised as evidence against the whole evolutionary approach to human affairs.[59] But the claim of evolutionists is not that these measures of success are today the tickets to reproductive success but that they once were. Indeed, to a surprising extent they still are. Successful men remarry more frequently and more widely than unsuccessful ones, and even with contraception preventing this from being turned into reproductive success, rich people still have as many or more babies as poor people.[60]

Yet Western people conspicuously avoid having as many

children as they could. William Irons of Northwestern University in Chicago has tackled this problem. He believes that human beings have always taken into account the need to give a child a "good start in life." They have never been prepared to sacrifice quality of children for quantity. Thus, when an expensive education became a prerequisite for success and prosperity, around the time of the demographic transition to low birthrates, people were able to readjust and lower the number of children they had in order to be able to afford to send them to school. Exactly this reason is given today by Thai people for why they are having fewer children than their parents.[61]

There has been no genetic change since we were hunter-gatherers, but deep in the mind of the modern man is a simple male hunter-gatherer rule: Strive to acquire power and use it to lure women who will bear heirs; strive to acquire wealth and use it to buy other men's wives who will bear bastards. It began with a man who shared a piece of prized fish or honey with an attractive neighbor's wife in exchange for a brief affair and continues with a pop star ushering a model into his Mercedes. From fish to Mercedes, the history is unbroken: via skins and beads, plows and cattle, swords and castles. Wealth and power are means to women; women are means to genetic eternity.

Likewise, deep in the mind of a modern woman is the same basic hunter-gatherer calculator, too recently evolved to have changed much: Strive to acquire a provider husband who will invest food and care in your children; strive to find a lover who can give those children first-class genes. Only if she is very lucky will they be the same man. It began with a woman who married the best unmarried hunter in the tribe and had an affair with the best married hunter, thus ensuring her children a rich supply of meat. It continues with a rich tycoon's wife bearing a baby that grows up to resemble her beefy bodyguard. Men are to be exploited as providers of parental care, wealth, and genes.

Cynical? Not half as cynical as most accounts of human history.

Chapter 8

SEXING THE MIND

No woman, no cry.

—Bob Marley

O, the trouble, the trouble with women,

I repeat it again and again

From Kalamazoo to Kamchatka

The trouble with women is—men.

—Ogden Nash and Kurt Weill

The pine vole, *Microtus pinetorum*, is a monogamous species of mouse. The males help the females look after the babies. Male and female pine voles have similar brains. In particular, the hippocampus of both the male and female is much the same size. When required to run a maze, male and female pine voles prove equally good at the task. The meadow vole, *Microtus pennsylvanicus*, is a different story altogether. It is polygamous. Males, which must visit the scattered holes of their several wives, travel farther than females every day. Male meadow voles have bigger hippocampi than females and are better at finding and remembering their way through mazes. Their brains are simply better at such spatial tasks.[1]

Just like meadow voles, men are better at spatial tasks than women. When asked to compare the shapes of two objects seen from different angles and judge whether they are the same shape or to judge whether two glasses of different shape are equally full or any such task that involves spatial judgments, men generally do better than women. Polygamy and spatial skills seem to go together in several species.

EQUALITY OR IDENTITY?

Men and women have different bodies. The differences are the direct result of evolution. Women's bodies evolved to suit the demands of bearing and rearing children and of gathering plant food. Men's bodies evolved to suit the demands of rising in a male hierarchy, fighting over women, and providing meat to a family.

Men and women have different minds. The differences are the direct result of evolution. Women's minds evolved to suit the demands of bearing and rearing children and of gathering plant food. Men's minds evolved to suit the demands of rising in a male hierarchy, fighting over women, and providing meat to a family.

The first paragraph is banal; the second inflammatory. The proposition that men and women have evolved different minds is anathema to every social scientist and politically correct individual. Yet I believe it to be true for two reasons. First, the logic is impeccable. As the last two chapters have demonstrated, over long stretches of evolutionary time men and women have faced different evolutionary pressures, so the ones who succeeded will have been those whose brains produced behaviors well suited to those pressures. Second, the evidence is overwhelming. Gingerly, reluctantly, but with increasing conviction, physiologists and psychologists have begun to probe the differences between male and female brains. Often they have done so determined to find none. Yet again and again they come back with good evidence that there are such differences. Not everything is different; most things, in fact, are identical between the sexes. Much of the folklore about differences is merely convenient sexism. And there are enormous overlaps. Although it is a fair generalization to say that men are taller than women, nonetheless the tallest woman in a large group of people is usually taller than the shortest man. In the same way, even if the average woman is better at some mental task than the average man, there are many women who are worse at the task than the best man, or vice versa. But the evidence for the average male brain differing in certain ways from the average female brain is now all but undeniable.

Evolved differences are by definition "genetic," and any suggestion that men and women have genetically different minds horrifies the modern conscience for it seems to justify prejudice. How can we strive to build an equal society when men are given "scientific" support for their sexism? Give men an inch of inequality and they will claim a mile of bias. The Victorians believed men and women were so different that women should not even have the vote; in the eighteenth century some men thought women incapable of reason.

These concerns are fair. But just because people have exaggerated sexual differences in the past does not mean they cannot exist. There is no a priori reason for assuming that men and women have identical minds and no amount of wishing it were so will make it so if it is not so. Difference is not inequality. Boys are interested in guns, girls in dolls. That may be conditioning, or it may be genes, but neither is "better" than the other. As anthropologist Melvin Konner put it: "Men are more violent than women and women are more nurturant, at least toward infants and children, than men. I am sorry if this is a cliché; that cannot make it less factual."[2] Moreover, suppose there are differences in the mentalities of men and women. Is it then fair to assume and act as if there were none? Suppose that boys are more competitive than girls. Would that not suggest that girls would be better educated apart from boys? The evidence suggests that girls are indeed more successful after education in a single-sex school. Sex-blind education may be unfair education.

In other words, to assume the sexes are mentally identical in the face of evidence that they are not is just as unfair as to assume sexual differences in the face of evidence that they are the same. We have always assumed that the burden of proof must rest with those who believe there are innate mental differences between the sexes. We may have been wrong.

MEN AND MAP READING

With that out of the way, let us examine the evidence. There are three reasons to expect evolution to have produced different mentalities in men and women. The first is that men and women are mammals, and all mammals show sexual differences in behavior. As Charles Darwin put it, "No one disputes that the bull differs in disposition from the cow, the wild boar from the sow, the stallion from the mare."[3] The second is that men and women are apes, and in all apes there are great rewards for males that show aggression toward other males, for males that seek mating opportunities, and for females that pay close attention to their babies. The third is that men and women are human beings, and human beings are mammals with one highly unusual characteristic: a sexual division of labor. Whereas a male and a female chimp seek the same sources of food, a male and a female human being, in virtually every preagricultural society, set about gathering food in different ways. Men look for sources that are mobile, distant, and unpredictable (usually meat), while women, burdened with children, look for sources that are static, close, and predictable (usually plants).[4]

In other words, far from being an ape with fewer than usual sex differences, the human being may prove to be an ape with more than usual sex differences. Indeed, mankind may be the mammal with the greatest division of sexual labor and the greatest of mental differences between the sexes. Yet, though mankind may have added division of labor to the list of causes of sexual dimorphism, he has subtracted the effect of male parental care.

Of the many mental features that are claimed to be different between the sexes, four stand out as repeatable, real, and persistent in all psychological tests. First, girls are better at verbal tasks. Second, boys are better at mathematical tasks. Third, boys are more aggressive. Fourth, boys are better at *some* visuo-spatial tasks and girls at others. Put crudely, men are better at reading a map and women are better judges of character and mood—on average.[5] (And, interestingly, gay men are more like women than heterosexual men in some of these respects.)[6]

The case of the visuo-spatial tasks is intriguing because it has been used to argue that men are naturally polygamous[7] by analogy with the case of the mice quoted at the beginning of this chapter. Crudely put, polygamist mice need to know their way from one wife's house to another—and it is certainly true that in many polygamous animals, including our relatives, orangutans, males patrol an area that includes the territories of several wives. When people are asked to rotate a diagram of an object mentally to see if it is the same as another object, only about one in four women scores as highly as the average man. This difference grows during childhood. Mental rotation is the essence of map reading, but it seems a huge jump to argue that men are polygamous because they are better at map reading just because the same is true of mice.

Besides, there are spatial skills that women perform better than men. Irwin Silverman and Marion Eals at York University in Toronto reasoned that the male skill at mental-rotation tasks probably reflected not some parallel with polygamous male mice patrolling broad territories to visit many females but a much more particular fact about human history: that during the Pleistocene period, when early man was an African hunter-gatherer for a million years or more, men were the hunters. So men needed superior spatial skills to throw weapons at moving targets, to make tools, to find their way home to camp after a long trek, and so on.

Much of this is conventional wisdom, but Silverman and Eals then asked themselves: What special spatial skills would women gatherers need that men would not? One thing they predicted was that women would need to notice things more—to spot roots, mushrooms, berries, plants—and would need to remember landmarks so as to know where to look. So Silverman and Eals did a series of experiments that required students to memorize a picture full of objects and then recall them later, or to sit in a room for three minutes and then recall what objects were where in the room. (The students were told they were merely being asked to wait in the room until a different experiment was ready.) On every measure of object memory and location memory, the women students did 60–70 percent better than the men. The old jokes about women noticing things and men losing things about the house and having to ask their wives are true. The difference appears around puberty, just as the social and verbal skills of women begin to exceed those of men.[8]

When a family in a car gets lost, the woman wants to stop and ask the way, while the man persists in trying to find his way by map or landmarks. So pervasive is this cliché that there must be some truth to it. And it fits with what we know of the sexes. To a man, stopping to ask the way is an admission of defeat, something status-conscious males avoid at all costs. To a woman, it is common sense and plays on her strengths in social skills.

NURTURE WITH NATURE

These social skills may also have had Pleistocene origins. A woman is dependent on her social intuition and skills for success at making allies within the tribe, manipulating men into helping her, judging potential mates, and advancing the cause of her children. Now this is not to claim that the difference is purely genetic. It could well be true—it is in my marriage—that men read maps more and women read novels more. So perhaps it is all a matter of training: Women think about character more, and so their brains get more practice at it. Yet where does the preference come from? Perhaps it comes from conditioning. Women learn to imitate their mothers who are more interested in character than maps. So where did the mothers get the interest? From their mothers? Even if you suggest that the original Eve took an arbitrary step in deciding to be more interested in character than Adam, you cannot escape genetic change, for Eve's female descendants, concentrating on one another's character, would have thrived in proportion to their skill at judging character and mood, and so genes for better ability to judge character and mood would have spread. If such skill was genetically influenced, people could not avoid being influenced by genes for preferring what they were genetically good at, and so genetic differences would be reinforced with cultural conditioning.

This phenomenon—that people specialize in what they are good at and so create conditions that suit their genes—is known as the Baldwin effect since it was first described by one James Mark Baldwin in 1896. It leads to the conclusion that conscious choice and technology can both influence evolution, an idea explored at length by Jonathan Kingdon in his recent book *Self-Made Man and His Undoing*.[9] It is impossible in the end to deny that even a highly conditioned trait can be without some basis in biology—or vice versa. Nurture always reinforces nature; it rarely fights it. (An exception may be aggressiveness, which develops more in boys despite frequent parental discouragement.) I find it very hard to believe that the fact that 83 percent of murderers and 93 percent of drunken drivers in America are male is due to social conditioning alone.[10]

It is hard for a nonscientist to realize how revolutionary the implications of these ideas were when men like Don Symons first began to sketch them out in the late 1970s.[11] What Symons was saying—that men and women have different minds because they have had different evolutionary ambitions and rewards—accords easily with common sense, but the overwhelming majority of the research that social scientists had done on human sexuality was infused with the assumption that there are no mental differences. To this day many social scientists assume—not conclude, assume—that all differences are learned from parents and peers by identical brains. Listen, for instance, to Liam Hudson and Bernadine Jacot, authors of a book called *The Way Men Think*: "At the heart of men's psychology is a 'wound,' a developmental crisis experienced by infant boys as they distance themselves from mother's love and establish themselves as male. This makes men adept at abstract reasoning but vulnerable to insensitivity, misogyny, and perversion."[12] Through their assumption that the cause must lie in a childhood experience, the authors are condemning 49 percent of the human race as "wounded" perverts. How much more generous it would be if, instead of writing parables about childhood wounds, psychologists were to accept that some differences between the sexes just are, that they are in the nature of the beasts, because each sex has an evolved tendency to develop that way in response to experience. Deborah Tannen, author of a fascinating book about men's and women's styles of conversation called *You Just Don't Understand*, while not considering the possibility that men's and women's natures are on average innately different, at least has the courage to argue that the differences are better recognized and lived with than condemned and blamed on personality: "When sincere attempts to communicate end in stalemate, and a beloved partner seems irrational and obstinate, the different languages men and women speak can shake the foundations of our lives. Understanding the other's ways of talking is a giant leap across the communication gap between women and men, and a giant step toward opening lines of communication."[13]

HORMONES AND BRAINS

There is, nonetheless, a sense in which sexual differences cannot be strictly left to the genes. If a gene appeared in a Pleistocene man for, say, better sense of direction at the expense of poorer social intuition, it might have been of benefit to him. But, as well as his sons, his daughters would have inherited it from him. In them the gene might have been positively disadvantageous because it left them less socially intuitive. So its net effect, over time, would be neutral, and it would not spread.[14]

The genes that would spread would therefore be ones that responded to signals of gender: if in a male, improve the sense of direction; if in a female, improve the social intuition. And this is precisely what we find. There is no evidence of genes for different brains, but there is ample evidence of genes for altering brains in response to male hormones. (For reasons of historical accident, the "normal brain" is female unless masculinized.) So the mental differences between men and women are caused by genes that respond to testosterone.

We last met the steroid hormone testosterone in fish and birds where it was rendering them more vulnerable to parasites by exaggerating their sexual ornaments. In recent years more and more evidence has been found that testosterone affects not just ornaments and bodies but also brains. Testosterone is an ancient chemical, found in much the same form throughout the vertebrates. Its concentration determines aggressiveness so exactly that in birds with reversed sex roles, such as phalaropes and in female-dominated hyena clans, it is the females that have higher levels of testosterone in the blood. Testosterone masculinizes the body; without it, the body remains female, whatever its genes. It also masculinizes the brain.

Among birds it is usually only the male that sings. A zebra finch will not sing unless there is sufficient testosterone in its blood. With testosterone, the special song-producing part of its brain grows larger and the bird begins to sing. Even a female zebra finch will sing as long as she has been exposed to testosterone early in life and as an adult. In other words, testosterone primes the young zebra finch's brain to be responsive later in life to testosterone again and so develop the tendency to sing. Insofar as a zebra finch can be said to have a mind, the hormone is a mind-altering drug.

Much the same applies to human beings. Here the evidence comes from a series of natural and unnatural experiments. Nature has left some men and women with abnormal hormonal doses, and in the 1950s doctors changed the hormonal conditions of some wombs by injecting some pregnant women with certain hormones. Women with a condition known as Turner's syndrome (they are born without ovaries) have even less testosterone in their blood than do women who have ovaries. (Ovaries produce some testosterone, though not as much as testicles do.) They are exaggeratedly feminine in their behavior, with typically a special interest in babies, clothes, housekeeping, and romantic stories. Men with less than usual testosterone in their blood as adults—eunuchs, for example—are noted for their femininity of appearance and attitude. Men exposed to less than usual testosterone as embryos—for example, the sons of diabetic women who took female hormones during pregnancy—are shy, unassertive, and effeminate. Men with too much testosterone are pugnacious. Women whose mothers were injected with progesterone in the 1950s (to avert miscarriage) later described themselves as having been tomboys when young; progesterone is not unlike testosterone in its effects. Girls who were born with an unusual condition called either adrenogenital syndrome or congenital adrenal hyperplasia are equally tomboyish. This disorder causes the adrenal gland, near the kidney, to produce a hormone that acts like testosterone instead of cortisol, its usual product.[15]

Somewhat like in the zebra finches, there are two periods when testosterone levels rise in male children: in the womb, from about six weeks after conception, and at puberty. As Anne Moir and David Jessel put it in a recent book, *Brain Sex*, the first pulse of hormone exposes the photographic negative; the second develops it.[16] This is a crucial difference from the way the hormone affects the body. The body is masculinized by testosterone from the testicles at puberty, whatever its womb experience. But not the mind. The mind is immune to testosterone unless it was exposed to a sufficient concentration (relative to female hormones) in the womb. It would be easy to engineer a society with no sex difference in attitude between men and women. Inject all pregnant women with the right dose of hormones, and the result would be men and women with normal bodies but identical feminine brains. War, rape, boxing, car racing, pornography, and hamburgers and beer would soon be distant memories. A feminist paradise would have arrived.

SUGAR AND SPICE

The effect of this double-barreled burst of testosterone on the male brain is dramatic. The first dose produces a baby that is mentally different from a girl baby from its first day on the planet. Baby girls are more interested in smiling, communicating, and in people, boys in action and things. Shown cluttered pictures, boys select objects, girls people. Boys are instantly obsessed with dismantling, assembling, destroying, possessing, and coveting things. Girls are fascinated by people and treat their toys as surrogate people. Hence, to suit their mentalities, we have invented toys that suit each sex. We give boys tractors and girls dolls. We are reinforcing the stereotypical obsessions that they already have, but we are not creating them.

This is something every parent knows. Despairingly they watch their son turn every stick into a sword or gun, while their daughter cuddles even the most inanimate object as if it were a doll. A woman wrote to the *Independent*, a British national newspaper, on November 2, 1992, as follows: "I would be interested if any of your more learned readers could tell me why, from the time my twins could reach for toys and were put on a rug together with a mixed selection of 'boys' and 'girls' baby toys, he would inevitably select the car/train items and she the doll/teddy ones."

The genes cannot be denied. And yet, of course, there are no genes for liking guns or dolls, there are only genes for channeling male instincts into imitating males and female instincts into imitating female behavior. There are natures that respond to some nurtures and not others.

At school, boys are fidgety, difficult, inattentive, and slow to learn, compared to girls. Nineteen out of every twenty hyperactive children are boys. Four times as many boys as girls are dyslexic and learning disabled. "Education is almost a conspiracy against the aptitudes and inclinations of a schoolboy," wrote psychologist Dianne McGuiness, a sentiment to which almost every man with a memory of school will raise a hearty cry of assent.[17]

But another fact begins to emerge at school. Girls are simply better at linguistic forms of learning, boys at mathematical and some spatial skills. Boys are more abstract, girls more literal. Boys with an extra X chromosome (XXY instead of the normal XY) are much more verbal than other boys. Girls with Turner's syndrome (no ovaries) are even worse at spatial tasks than other girls but just as good at verbal ones. Girls who were exposed to male hormones in the womb are better at spatial tasks. Boys who were exposed to female hormones are worse at spatial tasks. These facts have been first disputed and then actively suppressed by the educational establishment, which continues to insist that there are no differences in learning ability between boys and girls. According to one researcher, such suppression has done both boys and girls far more harm than good.[18]

And the brain itself begins to show strange differences. Brain functions become more diffuse in girls, whereas they take up specific locations in the heads of boys. The two hemispheres of the brain become more different and more specialized in boys. The corpus callosum, which connects the two, grows larger in girls. It is as if testosterone begins to isolate the boy's right hemisphere from colonization by verbal skills from the left.

These facts are far too few and unsystematic to be regarded as anything more than hints of what actually happens, but the role of language acquisition must be critical. Language is the most human and therefore most recent of our mental skills—the one we share with no other ape. Language seems to come into the brain like an invading Goth, taking the place of other skills, and testosterone appears to resist this. Whatever actually happens, it is indisputable fact that at the age of five, when they first arrive at school, the average boy has a very different brain from the average girl.

Yet at five the testosterone levels in the average boy are identical to those of the average girl, and a fraction of what they were at birth. The pulse of testosterone in the womb is a distant memory, and there will be little difference between the sexes in testosterone levels until the age of eleven or twelve. A boy of eleven is far more similar to a girl of the same age than he has been before or will be again. He is academically her equal for the first time, and his interests are not so far apart. Indeed, there is one piece of medical evidence that at this age a person can still grow up to be, mentally, either a typical man or a typical woman, despite the hormonally induced differences of childhood. This evidence comes from thirty-eight cases of a rare congenital disorder in the Dominican Republic. Called 5-alpha-reductase deficiency, this disorder causes its male possessor to be unusually insensitive to the effect of testosterone before birth. As a result, such people are born with female genitalia and reared as girls. Suddenly, at puberty, their testosterone level rises, and they turn into almost normal men. (The main difference is that they ejaculate through a hole at the base of the penis.) Yet, despite their childhood as girls, these men have for the most part adapted fairly easily to male roles in their society, which suggests either that their brains were masculinized even as their genitalia were not or that their brains were still adaptable at puberty.[19]

Puberty strikes a young man like a hormonal thunderbolt. His testicles descend, his voice breaks, his body becomes hairier and leaner, and he begins to grow like a weed. The cause of all this is a veritable flood of testosterone from his testicles. He now has twenty times as much of it in his blood as a girl of the same age. The effect is to develop the mental photograph laid down in his head by the womb's dose and to make his mind into that of an adult man.[20]

SEXISM AND THE KIBBUTZ LIFE

Asked about their ambitions, men from six different cultures replied with much the same answer. They wanted to be practical, shrewd, assertive, dominating, competitive, critical, and self-controlled. They sought power and independence above all. Women from the same cultures wanted to be loving, affectionate, impulsive, sympathetic, and generous. They sought to serve society above all.[21] Studies of male conversation find it to be public (that is, men clam up at home), domineering, competitive, status-obsessed, attention-seeking, factual, and designed to reveal knowledge and skill. Female conversation tends to be private (that is, women clam up in big groups), cooperative, rapport-establishing, reassuring, empathetic, egalitarian, and meandering (that is, to include talk for talk's sake).[22]

There are, of course, exceptions and overlaps. Just as there are women who are taller than men, so there are women who want to be assertive and men who want to be sympathetic. But just as it is still valid to make the generalization that men are taller than women, so it is valid to conclude that the adjectives listed above are fairly typical of the natures of men and women. Some must be related to the differences between hunting and gathering, the most uniquely human of the sex differences. For example, it cannot be a coincidence that men enjoy hunting, fishing, and eating meat much more than women do. Some may be more recent, reflecting social norms that the sexes have imposed on themselves through peer pressure and education (which was not always as sex blind as it strives to be today). For example, the male desire to be self-controlled may be a modern attribute, a recognition that he has a nature that needs controlling. Others may be more ancient, reflecting basic patterns that all apes share and that baboons do not, such as the fact that a woman generally leaves her group on marriage and lives with her offspring among what were hitherto strangers, whereas a man lives among kin. Others may be more ancient still and shared with all mammals and many birds, such as the fact that women nurture babies while men compete with other men for access to women. It surely cannot be a coincidence that men are obsessed with status in hierarchies and that male chimpanzees compete for status in strict hierarchies of dominance.

The Israeli kibbutz system has proved to be a large natural

experiment in the persistence of sex roles. Men and women were initially encouraged to drop all sex roles in kibbutzim: Haircuts and clothes were unisex; boys were encouraged to be peaceful and sensitive, while girls were treated like tomboys; men did household chores and women went out to work. Yet three generations later, the attempt has largely been abandoned, and kibbutz life is actually *more* sexist than life in the rest of Israel. People have returned to stereotypes. Men politick, while women tend the home; boys study physics and become engineers, while girls study sociology and become teachers and nurses. Women manage the morale, health, and education of the kibbutz, while men manage the finances, security, and business. To some this is easily explicable: People have simply rebelled against the eccentric pattern set by their parents. Yet that explanation is more condescending than one that treats them as agents of their own choice, choosing according to their natures. Women clean house in a kibbutz because, like women everywhere, they complain that men would not do it properly. Men do not clean house in a kibbutz because, like men everywhere, they complain that if they did, their wives would say it had not been done properly.[23]

Nor are the kibbutzim unique. Even in liberated Scandinavia, it is women who feed the family, wash the clothes, and care for the children. Even where women go to work, some professions remain male bastions (for instance, garage mechanics, air-traffic controllers, driving test examiners, architects), while others have become female bastions (for instance, bank tellers, elementary school teachers, secretaries, interpreters). It is getting gradually more implausible to maintain that in the most egalitarian Western societies women are prevented by social prejudice from becoming garage mechanics. Women rarely want to become garage mechanics. They do not want to become garage mechanics because the world of the garage mechanic is an uninviting "man's world" in which they would feel unwelcome. But why is it a man's world? Because it is a job that men have molded to suit their personalities, and male personalities are different from female ones.

FEMINISM AND DETERMINISM

The bizarre thing about this assertion of different natures is that it is a thoroughly feminist assertion. There is a contradiction at the heart of feminism, one that few feminists have acknowledged. You cannot say, first, that men and women are equally capable of all jobs and, second, that if jobs were done by women, they would be done differently. So feminism itself is anything but egalitarian. Feminists argue explicitly that if more women were in charge, more caring values would prevail. They begin from the presumption that women are by nature different beings. If women ran the world, there would be no war. When women run companies, cooperation, not competition, is the watchword. These are all explicit and firm assertions of sexism: that the personalities and natures of women are different from men. If women's personalities are different, is it not likely that they will prove better or worse at certain jobs than men? Differences cannot be appealed to when they suit and denied when they do not.

Nor does it help to appeal to social pressure as the source of personality differences. If social pressure is as powerful as social scientists would have us believe, then a person's nature is irrelevant; only his or her background counts. A man from a broken home who has led a life of crime is the product of that experience, and there is no spark of decent "nature" in his soul to redeem. Of course we scoff at such nonsense. We recognize him to be a product of both his background and his nature. It is the same with sex differences. To say that Western women do not enter politics in the same numbers as men because they have been conditioned to think of it as a man's career is to patronize women. Politics is all about status-seeking ambition, which many women have a healthy cynicism about. Women have their own minds. They are capable of deciding to enter politics if they want to, whatever society says (and Western society, if anything, now affirms that they should). One of the things that make a political career uninviting may well be the sexism of those around them, but it is absurd to assume it is the only thing.

I have argued that men and women are different and that some of these differences stem from an evolutionary past in which men hunted and women gathered. So I am dangerously close to arguing that a woman's place is in the home while her husband works as the breadwinner. Yet that conclusion does not at all follow from the logic presented here. The practice of going out to work in an office or a factory is foreign and novel to the psychology of a savanna-dwelling ape. It is just as foreign to a man as to a woman. If in the Pleistocene period men went off from the home base on long hunts while women went a shorter distance to gather plants, then maybe men are mentally better suited to long commutes. But neither is evolutionarily suited to sit at a desk all day and talk into a telephone or sit at a factory bench all day tightening screws. The fact that "work" became a male thing and "home" a female one is an accident of history: The domestication of cattle and the invention of the plow made food gathering a task that benefited from male muscle power. In societies where the land is tilled by hand, women do most of the work. The industrial revolution reinforced the trend, but the post industrial revolution—the recent growth of service industries—is reversing it again. Women are going "out to work" again as they did when they sought tubers and berries in the Pleistocene period.[24]

Therefore, there is absolutely no justification from evolutionary biology for the view that men should earn and women should darn their socks. There may be professions, such as car mechanic or big-game hunter, that men are psychologically more suited to than women, just as there are professions, such as doctor and nanny, that women are probably naturally better at. But there is no general support in biology for sexism about careers.

Indeed, in a curious way, an evolutionary perspective justifies affirmative action more than a more egalitarian philosophy would, for it implies that women have different ambitions and even more than different abilities. Men's reproductive success depended for generations on climbing political hierarchies. Women have rarely had an incentive to seek success of that kind, for their reproductive success depended on other things. Therefore, evolutionary thinking predicts that women often will not seek to climb political ladders, but it says nothing about how good they would be if they did. I suggest it is no accident that women have reached the top rung (as the prime minister in many countries) in numbers disproportionate to their strength on the lower rungs. I suggest that it is no accident that queens of Britain have a far more distinguished and consistent history than the kings. The evidence suggests that women are on average slightly better than men at running countries. The evidence supports the feminist assertion that men can only envy the female touches they bring to such jobs—intuition, character judgment, lack of self-worship. Since the bane of all organizations, whether they are companies, charities, or governments, is that they reward cunning ambition rather than ability (the people who are good at getting to the top are not necessarily the people who are best at doing the job) and since men are more endowed with such ambition than women, it is absolutely right that promotion should be biased in favor of women—not to redress prejudice but to redress human nature.

And also, of course, to represent the woman's point of view. Feminists believe that women need to be proportionally represented in Parliament and Congress because women have a different agenda. They are right if women are by nature different. If they were the same as men, there would be no reason for men not to represent women's interests as competently as they represent men's. To believe in sexual equality is just. To believe in sexual identity is a most peculiar and unfeminist thing to do.

Feminists who recognize this contradiction are pilloried for their pains. Camille Paglia, literary critic and gadfly, is one of the few who sees that feminism is trying an impossible trick: to change the nature of men while insisting that the nature of women is unchangeable. She argues that men are not closet women and women are not closet men. "Wake up," she cries. "Men and women are different."[25]

THE CAUSES OF MALE HOMOSEXUALITY

A man develops a sexual preference for women because his brain develops in a certain way. It develops in a certain way because testosterone produced by his genetically determined testicles alter the brain inside his mother's womb in such a way that later, at puberty, it will react to testosterone again. Miss out on the genes for testicles, the testosterone burst in the womb, or the testosterone burst at puberty—any one of the three—and you will not be a typical man. Presumably, a man who develops a preference for other men is a man who has a different gene that affects how his testicles develop or a different gene that affects how his brain responds to hormones or a different learning experience during the pubertal burst of testosterone—or some combination of these.

The search for the cause of homosexuality has begun to shed a great deal of light on the way the brain develops in response to testosterone. It was fashionable until the 1960s to believe that homosexuality was entirely a matter of upbringing. But cruel Freudian aversion therapy proved incapable of changing it, and the fashion then changed to hormonal explanations. Yet adding male hormones to the blood of gay men does not make them more heterosexual; it merely makes them more highly sexed. Sexual orientation has already been fixed before adulthood. Then, in the 1960s, an East German doctor named Gunter Dörner began a series of experiments on rats which seemed to show that in the womb the male homosexual brain releases a hormone, called luteinizing hormone, that is more typical of female brains. Dörner, whose motives have often been questioned on the grounds that he seemed to be searching for a way to "cure" homosexuality, castrated male rats at different stages of development and injected them with female hormones. The earlier the castration, the more likely the rat was to solicit sex from other male rats. Research in Britain, America, and Germany has all confirmed that a prenatal exposure to deficiency of testosterone increases the likelihood of a man becoming homosexual. Men with an extra X chromosome and men exposed in the womb to female hormones are more likely to be gay or effeminate, and effeminate boys do indeed grow up to be gay more often than other boys. Intriguingly, men who were conceived and born in periods of great stress, such as toward the end of World War II, are more often gay than men born at other times. (The stress hormone cortisol is made from the same progenitor as testosterone; perhaps it uses up the raw material, leaving less to be made into testosterone.) The same is true of rats: Homosexual behavior is more common in rats whose mothers were stressed during pregnancy. The things that male brains are usually good at gay brains are often bad at, and vice versa. Gays are also more often left-handed than heterosexuals, which makes a sort of sense because handedness is affected by sex hormones during development, but it is also odd because left-handed people are supposed to be better at spatial tasks than right-handers. This only demonstrates how sketchy our knowledge still is of the relationship between genes, hormones, brains, and skills.[26]

It is clear, however, that the cause of homosexuality lies in some unusual balance of hormonal influence in the womb but not later on, a fact that further supports the idea that the mentality of sexual preference is affected by prenatal sex hormones. This is not incompatible with the growing evidence that homosexuality is genetically determined. The "gay gene" that I will discuss in the next chapter is widely expected to turn out to be a series of genes that affect the sensitivity of certain tissues to testosterone.[27] It is both nature and nurture.

It is no different from genes for height. Fed on identical diets, two genetically different men will not grow to the same height. Fed on different diets, two identical twins will grow to different heights. Nature is the length of the rectangle, nurture the width. There can be no rectangle without both. The genes for height are really only genes for responding to diet by growing.[28]

WHY DO RICH MEN MARRY BEAUTIFUL WOMEN?

If homosexuality is determined by hormonal influences in the womb, then so, presumably, are heterosexual preferences. Throughout our evolutionary history, men and women have faced different sexual opportunities and constraints. For a man casual sex with a stranger carried only a small risk—infection, discovery by the wife—and a potentially enormous reward: a cheap addition of an extra child to his genetic legacy. Men who seized such opportunities certainly left behind more descendants than men who did not. Therefore, since we are by definition descended from prolific ancestors rather than barren ones, it is a fair bet that modern men possess a streak of sexual opportunism. Virtually all male mammals and birds do, even those that are mainly monogamous. This is not to say that men are irredeemably promiscuous or that every man is a potential rapist, it is just that men are more likely to be tempted by an opportunity for casual sex than women.

Women are likely to be different. Having sex with a stranger not only encumbered a Pleistocene woman with a possible pregnancy before she had won the man's commitment to help rear the child, but it also exposed her to probable revenge from her husband if she had one and to possible spinsterhood if she did not. These enormous risks were offset by no great reward. Her chances of conceiving were just as great if she remained faithful to one partner, and her chances of losing the child without a husband's help were greater. Therefore, women who accepted casual sex left fewer rather than more descendants, and modern women are likely to be equipped with suspicion of casual sex.

Without this evolutionary history in mind, it is impossible to explain the different sexual mentalities of men and women. It is fashionable to deny such differences and to maintain that only social repression prevents women from buying explicit pornography about men or that only socially paranoid machismo drives men to promiscuity. Yet this is to ignore the enormous social pressures now placed on men and women to disregard or minimize differences between them. A modern woman is exposed to pressure from men to be sexually uninihibited, but she is also exposed to the same pressure from other women. Likewise, men are under constant pressure to be more "responsible," sensitive, and faithful—from other men as well as from women. Perhaps more out of envy than morality, men are just as censorious of philanderers as they are of women; often more so. If men are sexual predators, it is despite centuries of social pressure not to be. In the words of one psychologist, "Our repressed impulses are every bit as human as the forces that repress them."[29]

But what exactly are the differences between men and women in their sexual mentalities? I argued in the last two chapters that men, for whom the reproductive stakes are higher, are likely to be more competitive with one another and therefore are more likely to end up wielding power, controlling wealth, and seeking fame. Consequently, women are more likely to have been rewarded for seeking power, wealth, or fame in a husband than men are in a wife. Women who did so probably left more descendants among modern women, so it follows from evolutionary thinking that women are more likely to value potential mates who are rich and powerful. Another way to look at it is to think of what a woman can most profitably seek in a husband that will increase the number and health of her children. The answer is not more sperm but more money or more cattle or more tribal allies or whatever resource counts.

A man, by contrast, is seeking a mate who will use his sperm and his money to produce babies. Consequently, he has always had an enormous incentive to seek youth and health in his mates. Those men who preferred to marry forty-year-old women rather than twenty-year-olds stood a small chance of begetting any children at all, let alone more than one or two. They also stood a large chance of inheriting a bunch of stepchildren from a previous marriage. They left fewer descendants than the men who always sought out the youngest, postpubertal females on offer. We would expect, therefore, that while women pay attention to cues of wealth and power, men pay attention to cues of health and youth.

This may seem a startlingly obvious thing to say. As Nancy Thornhill put it, "Surely no one has ever seriously doubted that men desire young, beautiful women and that women desire wealthy, high-status men."[30] The answer to her question is that sociologists do doubt it. Judging by their reaction to a recent study, only the most rigorous evidence will convince them. The study was done by David Buss of the University of Michigan, who asked a large sample of American students to rank the qualities they most preferred in a mate. He found that men preferred kindness, intelligence, beauty, and youth, while women preferred kindness, intelligence, wealth, and status. He was told that this may be the case in America, but it is not a universal facet of human nature.

So he repeated the study in thirty-seven different samples from thirty-three countries, asking over one thousand people, and found exactly the same result. Men pay more attention to youth and beauty, women to wealth and status. To which came this answer: Of course women pay more attention to wealth because men control it. If women controlled wealth, they would not seek it in their spouses. Buss looked again and found that American women who make more money than the average American woman pay *more* attention than average to the wealth of potential spouses, not less.[31] High-earning women value the earning capacity of their husbands more, not less, than low-earning women. Even a survey of fifteen powerful leaders of the feminist movement revealed that they wanted still more powerful men. As Buss's colleague Bruce Ellis put it, "Women's sexual tastes become more, rather than less, discriminatory as their wealth, power, and social status increase."[32]

Many of Buss's critics argued that he ignored context altogether. Different criteria of mate preference develop in different cultures at different times. To this Buss replied with a simple analogy. The amount of muscle on the average man is highly context-dependent: In the United States young men tend to be beefier about the shoulders than in Britain, perhaps partly because they eat better food and perhaps partly because their sports emphasize throwing strength rather than agility. Yet this does not negate the generalization that "men have more muscle on their shoulders than women." So, too, the fact that women pay more attention to men's wealth in one place than in another does not negate the generalization that women pay more attention to the wealth of potential mates than men do.[33]

The main difficulty with Buss's study is that it fails to distinguish between a partner chosen as a spouse and a partner chosen for a fling. Douglas Kenrick of Arizona State University asked a group of students to rank various attributes of potential mates according to four levels of intimacy. When seeking a marriage partner, intelligence is important to both sexes. When seeking a sexual partner for a one-night stand, intelligence matters much less, especially to men. There is little doubt that people of both sexes are sensible enough to value kindness, compatibility, and wit in those with whom they may spend the rest of their lives.[34]

The difficulty with measuring sexual preferences is that they are compromises. An aging ugly man does not mate with several young and beautiful women (unless he is very rich indeed). He settles for a faithful wife of the same age. A young woman does not mate faithfully with a wealthy tycoon. She chooses whatever is available, probably a slightly older man with no more money but a steady job. People lower their expectations according to their age, looks, and wealth. To discover just how different the sexual mentalities of men and women are, it is necessary to do a controlled experiment. Take an average man and an average women and give each the option of faithful marriage to a familiar partner or continual orgies with beautiful strangers. The experiment has not been done, and it is hard to imagine its getting a grant. But it need not be, for it is in effect possible to do exactly that experiment by looking inside people's heads and examining their fantasies.

Bruce Ellis and Don Symons gave 307 California students a questionnaire about their sexual fantasies. Had their subjects been Arabs or English people, the study would have been easily dismissed by social scientists because any sex differences that emerged could be attributed to social pressures from a sexist background. But there can be no people on Earth or in history so steeped in the politically correct ideology that there are no psychological sex differences as students at a university in California. Any differences that emerged could therefore be regarded as conservative estimates for the species as a whole.

Ellis and Symons found that two things showed no sex differences at all. The first was the students' attitudes toward their fantasies. Guilt, pride, and indifference were each as common among men as among women. And both sexes had a clear image of their fantasized partner's face during the fantasy. On every other measure there were substantial differences between the men and the women. Men had more sexual fantasies and fantasized about more partners. One in three men said they fantasized about more than one thousand partners in their lives; only 8 percent of women imagined so many partners. Nearly half the women said they never switch partners during a sexual fantasy; only 12 percent of men never switch. Visual images of the partner(s) were more important for men than touching, the partner's response, or any feelings and emotions. The reverse was true of women, who were more likely to focus on their own responses and less likely to focus on the partner. Women overwhelmingly fantasized about sex with a familiar partner.[35]

These results are not alone. Every other study of sexual fantasy has concluded that "male sexual fantasies tend to be more ubiquitous, frequent, visual, specifically sexual, promiscuous, and active. Female sexual fantasies tend to be more contextual, emotive, intimate, and passive."[36] Nor need we rely on such surveys alone. Two industries relentlessly exploit the sexual fantasizing of men and women: pornography and the publishing of romance novels. Pornography is aimed almost entirely at men. It varies little from a standard formula all over the world. "Soft porn" consists of pictures of naked or seminaked women in provocative positions. Such pictures are arousing to men, whereas pictures of naked (anonymous) men are not especially arousing to women. "A propensity to be aroused merely by the sight of males would promote random matings from which a female would have nothing to gain reproductively and a great deal to lose."[37]

"Hard porn," which depicts actual acts of sex, is almost invariably about the gratification of male lust by willing, easily aroused, varied, multiple, and physically attractive women (or men, in the case of gay porn). It is virtually devoid of context, plot, flirtation, courtship, and even much foreplay. There are no encumbering relationships, and the coupling duo are usually depicted as strangers. When two scientists showed heterosexual students pornographic films and measured their arousal by them, they found a consistent pattern of the kind common sense would suggest. First, men were more aroused than women. Second, men were aroused more by depictions of group sex than by films of a heterosexual couple, whereas for women it was the other way around. Third, women and men were both aroused by lesbian scenes, but neither was aroused by male homosexual sex. (Remember, all these students were heterosexual.) When watching pornography, men and women are both interested in the women actors. But porn is designed for, marketed to, and sought out by men, not women.[38]

The romance novel, by contrast, is aimed entirely at a female market. It, too, depicts a fictional world that has changed remarkably little except in adapting to female career ambitions and to a less inhibited attitude toward the description of sex. Authors adhere strictly to a formula provided by the publishers. Sexual acts play a small part in these novels; the bulk of each book is about love, commitment, domesticity, nurturing, and the formation of relationships. There is little promiscuity or sexual variety, and what sex there is, is described mainly through the heroine's emotional reaction to what is done to her—particularly the tactile things—and not to any detailed description of the man's body. His character is often discussed in detail but not his body.

Ellis and Symons claim that the romance novel and pornography represent the respective utopian fantasies of the two sexes. Their data on the sexual fantasies of California students would seem to support this contention. So does the repeated failure of magazines that try to repackage the male-porn formula for women (much of *Playgirl*'s readership is gay men), plus the burgeoning business of selling explicit novels about promiscuous sex at airports—for men. In any bookstore there are magazines for men with pictures of women on the covers, promising more inside, and magazines for women with pictures of women on the covers, promising hints about improving relationships inside. There are romance novels aimed at women with pictures of women on the covers and sexy novels aimed at men with pictures of women on the covers. The publishing industry, living by the market, not the prevailing ideology, has no doubts about the differences between men and women's attitudes toward sex.

As Ellis and Symons put it,

The data on sexual fantasy reported here, the scientific
literature on sexual fantasy, ... the consumer-driven
selective forces of a free market (which have shaped the
historically stable contrasts between male-oriented
pornography and female-oriented romance novels), the
ethnographic record on human sexuality, and the
ineluctable implications of an evolutionary perspective
on our species, taken together, imply the existence of a
profound sex difference in sexual psychology.[19]

This is a far more enlightened view than the peculiarly
uncharitable assumption among the politically correct that the reason women are not more turned on by nudity and pornography is
that they are repressed.

CHOOSY MEN

A paradox looms. Men are promiscuous opportunists at heart and
in their fantasies. Truly promiscuous opportunists would not be
too choosy, one would think. And yet men care about women's
looks more than women care about men's looks. A sports car and an
expense account can turn a frog into a prince for women, but even
a rich woman cannot afford to be ugly (although in these times of
cosmetic surgery, she can sometimes afford the means not to be
ugly). Advertisements for "escorts" emphasize looks. A man contemplating an affair should not restrict himself to what he consid-
ers a good-looking woman, yet he usually does. This is rather
unusual. A male gorilla or sage grouse does not refuse to mate with
a female because of her appearance. He takes every opportunity on
offer regardless of looks. Polygamous despots of ancient times may
have been promiscuous, but they were still choosy; their harems
were always recruited from among the young, the virginal, and the
beautiful.

The paradox is soluble. The degree to which an animal of
either sex is choosy correlates exactly with the degree to which it

invests in parental care. A black grouse, investing no more than sperm, is prepared to copulate with anything that even resembles a female: A stuffed bird or a model will do.[40] A male albatross, who will put all his best efforts into raising one female's young, is elaborately suspicious and selective, striving for the best female on offer. So man's choosiness reflects once more the fact that man does indeed form a pair bond and invest in his young, unlike some of his undiscriminating ape cousins. It is a legacy of his past monogamy: Choose well, for it may be the only chance you will get. Indeed, the overwhelming fascination of men with female youth argues that pair bonds have lasted lifetimes. In this we are quite unlike any other mammal. Chimpanzees find old females just as attractive as young ones as long as both are in estrus. The fact that men do prefer twenty-year-olds adds one piece of evidence to the theory that a Pleistocene man, like a modern man, married for life.

Anthropologist Helen Fisher has argued that there is a natural term to marriage, which is why divorce rates peak after four years of marriage. Four years is long enough to rear a single child beyond utter dependence, and Fisher believes that when each child reached four, Pleistocene women sought a fresh husband for the next child. She argues that therefore divorce is natural. But there are several problems with her case. The four-year peak is merely what statisticians call a mode and not a very prominent one at that: Divorce rates are bound to peak in one of the years after marriage. Moreover, her theory sits oddly alongside the fact that men consistently prefer younger women and that husbands contribute to their children's rearing long after the children reach four. A woman who divorced her husband four years after the birth of every child would be less attractive to new men every time, not only because she would be older but because she would bring a growing retinue of stepchildren. The male preference for young mates implies lifelong mateships.[41]

Even the most cursory inspection of the personal advertisements in a newspaper confirms what we all know: that men seek younger wives and women seek older husbands—despite the fact that they will almost certainly outlive them by a decade or more. In his survey Buss found that men seek women of about twenty-five, slightly past their maximum reproductive potential (they have already missed several breeding years) but close to their period of maximum fertility. However, this result may be misleading, as two of those commenting on Buss's data have suggested. First, as Don Symons points out, a twenty-five-year-old modern Westerner shows probably as much wear and tear as a twenty-year-old tribal woman. When asked what women they prefer, Yanomamö men do not hesitate to say *moko dude* women, meaning those between puberty and first child. Other things being equal, that is also the Western man's ideal.[42]

RACISM AND SEXISM

This chapter, obsessed with differences between the sexes, has ignored the differences between races, yet they are often thrown together in the demonology of modern prejudice. In an extraordinary equation, to insist on sexual differences is to insist on racial differences, too. Sexism is the sister of racism. I confess to being baffled by this. I think it is easy and, given the evidence, rational to believe that the differences between the natures of men of different races are trivial, while the differences between the natures of men and women of the same race are considerable.

Not that racial and cultural differences cannot exist. Just as a white man has different skin color from a black man, so it is quite possible that he also has a somewhat different mind. But given what we know of evolution, it is not very likely. The evolutionary pressures that have shaped the human mind—principally competitive relations with kin members, tribal allies, and sexual partners—are and have been the same for white and black men and were at work before the ancestors of whites left Africa 100,000 years ago. While skin color is affected by things such as climate, which differs markedly between Africa and northern Europe, the shape of the mind is affected only very marginally by nonhuman problems such as what kind of game to hunt or how to keep warm or cool. Infinitely more important is how to deal with fellow human beings, and that is the same problem everywhere—that is, the same for men everywhere and the same for women everywhere. But not the same for men *and* for women.

This is the essential difference between anthropology and Darwinism. Anthropologists insist that a Western urban man is far different in his habits and thoughts from a bushman tribesman than either is from his wife. Indeed, it is the foundation of their discipline that this is so, for anthropology consists of studying the differences between peoples. But this has led anthropologists to exaggerate the motes of racial difference and to ignore the beams of similarity. Men fight, compete, love, show off, and hunt all over the world. True, bushmen fight with spears and sticks, whereas Chicagoans fight with guns and lawsuits; bushmen strive to be headmen, whereas Chicagoans strive to become senior partners. The stuff of anthropology—the traditions, the myths, the crafts, the language, the rituals—is to me but the froth on the surface. Beneath lie giant themes of humanity that are the same everywhere and that are characteristically male and female. To a Martian an anthropologist studying the differences between races would seem like a farmer studying the differences between each of the wheat plants in his field. The Martian is much more interested in the typical wheat plant. It is the human universals, not the differences, that are truly intriguing.[43]

One of the most persistent of those universals is sexual role playing. As Edward Wilson put it: "In diverse cultures men pursue and acquire, while women are protected and bartered. Sons sow wild oats and daughters risk being ruined. When sex is sold, men are usually the buyers."[44] John Tooby and Leda Cosmides have put the challenge to cultural interpretations of this universal pattern even more baldly:

> The assertion that "culture" explains human variation will be taken seriously when there are reports of women war parties raiding villages to capture men as husbands, or of parents cloistering their sons but not their daugh-

ters to protect their sons' virtue, or when cultural distributions for preferences concerning physical attractiveness, earning power, relative age, and so on show as many cultures with bias in one direction as in the other.[45]

Just as it is foolish to deny the differences between the sexes in the face of the evidence presented here, so it is foolish to exaggerate them. In the matter of intelligence, for example, there is no reason to believe that men are dumber than women or vice versa—nothing in evolutionary thinking suggests as much, and no data test the proposition. As noted earlier, the data do suggest that men are probably better at abstract and spatial tasks, women at verbal and social ones, which vastly complicates the job of trying to design a test that is gender-neutral. Indeed, it helps to demolish the farcical notion of general, unitary intelligence altogether.

Nor does an appeal to sexual difference excuse anything. In the words of Anne Moir and David Jessel, "We do not consecrate the natural just because it is biologically true; men, for instance, have a natural disposition to homicide and promiscuity, which is not a recipe for the happy survival of society."[46]

People seem to forget easily that the word *is* is different from the word *should*. If we choose to redress the sexual differences between the minds of men and women through policy, we are going against nature, but no more than when we outlaw murder. But we should be clear that we are redressing a difference, not discovering an identity. Wishful thinking that they are the same will be mere propaganda and no favor to either sex.

Chapter 9

THE USES OF BEAUTY

Sigh no more, ladies, sigh no more,

Men were deceivers ever

One foot in sea, and one on shore,

To one thing constant never.

—Shakespeare, *Much Ado About Nothing*

In the early 1990s, there was a flurry of interest that a "gay gene" had been found on the X chromosome. The excitement faded as it proved hard to replicate the original study. But twin studies show that homosexuality is heritable, and one day the genes that can cause a man to be gay—perhaps in response to maternal genes expressed in his mother's womb—will be found.

The first implication will be political. Although it raises the possibility of selective abortion by mothers who do not wish to have gay sons, the theory of the gay gene has been largely welcomed in recent years by homosexual activists. The reason is that they find it will convince their more stubborn detractors that homosexuality is a condition into which they were born rather than a choice they made. In the eyes of disapproving heterosexuals, it exculpates them, their parents, and their education for their sexual proclivity. It also relieves parents of any anxiety they might have that their son might be led into homosexuality simply because his favorite rock group consists of homosexuals or because he has been seduced by a homosexual during adolescence.

The second implication is moral. The gay gene would at last demolish the myth that there is something "better" and less evil about theories that ascribe conditions to nurture or environment than theories that ascribe them to innate nature. In the name of Freudian nurture theories, gays were once treated with aversion therapy—electric shocks and emetics accompanied by homoerotic images. The most compelling of the new evidence for the gay gene is that fraternal twins, carried in the same womb and reared in the same household, have only a one-in-four chance of being gay. Identical twins, on the other hand, with the same nurture and the same nature, have a one-in-two chance of being gay. If one identical twin is gay, the chances that his brother is also gay are 50 percent. There is also good evidence that the gene is inherited from the mother and not from the father.[1]

How could such a gene survive, given that gay men generally do not have children? There are two possible answers. One is that the gene is good for female fertility when in women, to the same extent that it is bad for male fertility when in men. The second possibility is more intriguing. Laurence Hurst and David Haig of Oxford University believe that the gene might not be on the X chromosome after all. X genes are not the only genes inherited through the female line. So are the genes of mitochondria, described in chapter 4, and the evidence linking the gene to a region of the X chromosome is still very shaky statistically. If the gay gene is in the mitochondria, then a conspiracy theory springs to the devious minds of Hurst and Haig. Perhaps the gay gene is like those "male killer" genes found in many insects. It effectively sterilizes males, causing the diversion of inherited wealth to female relatives. That would (until recently at least) have enhanced the breeding success of the descendants of those female relatives, which would have caused the gay gene to spread. If the sexual preferences of gay men are greatly influenced (not wholly determined) by a gene, then it is probable that so are the sexual preferences of heterosexuals. And if our sexual instincts are heavily determined by our genes, then they have evolved by natural and sexual selection, and that means they bear the imprint of design. They are adaptive. There is a reason that beautiful people are attractive. They are attractive because others have genes that cause them to find beautiful people attractive. People have such genes because those that employed criteria of beauty left more descendants than those that did not. Beauty is not arbitrary. The insights of evolutionary biologists are transforming our view of sexual attraction, for they have begun at last to suggest why we find some features beautiful and others ugly.

BEAUTY AS A UNIVERSAL

Botticelli's Venus is beautiful. Michelangelo's David is handsome. But were they always thus? Would a Neolithic hunter-gatherer have agreed? Do Japanese or Eskimo people agree? Will our great-grand-children agree? Is sexual attraction fashionable and evanescent or permanent and inflexible?

We all know how dated and frankly unattractive the fashions and the beauties of a decade ago look now, let alone those of a century ago. Men in doublet and hose may still seem sexy to some, but men in frock coats surely do not. It is hard to avoid the conclusion that a person's sense of what is beautiful and sexy is subtly educated to prefer the prevailing norms of fashion. Rubens would not have chosen Twiggy as a model. Moreover, beauty is plainly relative, as any prisoner who has spent months without seeing a member of the opposite sex can testify.

And yet this flexibility stays within limits. It is impossible to name a time when women of ten or forty were considered "sexier" than women of twenty. It is inconceivable that male paunches were ever actually attractive to women or that tall men were thought uglier than short ones. It is hard to imagine that weak chins were ever thought beautiful on either sex. If beauty is a matter of fashion, how is it that wrinkled skin, gray hair, hairy backs, and very long noses have never been "in fashion"? The more things change, the more they stay the same. The famous sculpture of Nefertiti's head and neck, 3,300 years old, is as stunning today as when Akhenaten first courted the real thing.

Incidentally, in this chapter on what makes people sexually attractive to one another, I am going to take almost all my examples from white Europeans, and from northern Europeans at that. By this I am not implying that white European standards of beauty are absolute and superior but merely that they are the only ones I know enough about to describe. There is no room for a separate investigation into the standards of beauty that black, Asian, or other people employ. But the problem that I am principally concerned with is universal to all people: Are standards of beauty cultural whims or innate drives? What is flexible and what endures? I will argue in this chapter that only by understanding how sexual attraction evolved is it possible to make sense of the mixture of culture and instinct, and understand why some features flow with the fashion while others resist. The first clue comes from the study of incest.

FREUD AND INCEST TABOOS

Very few men have sex with their sisters. Caligula and Cesare Borgia were notorious because they were (rumored to be) such exceptions. Even fewer men have sex with their mothers, in spite of what Freud tells us is an intense longing to do so. Sexual abuse by fathers of daughters is far more common. But it is still rare.

Compare two explanations of these facts. First, that people secretly desire incest but are able to overcome these desires with the help of social taboos and rules; second, that people do not find their very close relatives sexually arousing, that the taboo is in the mind. The first explanation is Sigmund Freud's. He argued that our first and most intense sexual attraction is toward our opposite-sex parent. That is why, he went on to say, all human societies impose on their subjects strict and specific taboos against incest. Since the taboo "is not to be found in the psychology of the individual," there is a "necessity for stern prohibitions." Without those taboos, he implied, we would all be dreadfully inbred and suffer from genetic abnormalities.[2]

Freud made three unjustified assumptions. First, he equated attraction with sexual attraction. A two-year-old girl may love her father, but that does not mean she lusts after him. Second, he assumed without proof that people have incestuous desires. Freudians say the reason very few people express these desires is that they have "repressed" them—which makes Freud's argument irrefutable. Third, he assumed that social rules about cousins marrying were "incest taboos." Until very recently scientists and laymen alike followed Freud in believing that laws forbidding marriages between cousins were enacted to prevent incest and inbreeding. They may not have been.

Freud's rival in this field was Edward Westermarck; in 1891 he suggested that men do not mate with their mothers and sisters not because of social rules but because they are simply not turned on by those they were reared with. Westermarck's idea was simple. Men and women cannot recognize their relatives as relatives, so they have no way of preventing inbreeding as such. (Curiously, quail are different; they can recognize their brothers and sisters even when reared apart.) But they can use a simple psychological rule that works ninety-nine times out of a hundred to avert an incestuous match. They can avoid mating with those whom they knew very well during childhood. Sexual aversion to one's closest relatives is thus achieved. True, this will not avert marriage between cousins, but then there is nothing much wrong with marriage between cousins: The chance of a recessive deleterious gene emerging from such a match is small, and the advantages of genetic alliance to preserve complexes of genes that are adapted to work with one another probably outweigh it. (Quail prefer to mate with first cousins rather than with strangers.) Westermarck did not know that, of course, but it strengthens his argument, for it suggests that the only incestuous relations a human being should avoid are the ones between brother and sister, and parent and child.[3]

Westermarck's theory leads to several simple predictions: Stepsiblings would generally not be found to marry unless they were brought up apart. Very close childhood friends would also generally not be found to marry. Here the best evidence comes from two sources: Israeli kibbutzim and an old Chinese marriage custom. In kibbutzim, children are reared in crèches with unrelated companions. Lifelong friendships are formed, but marriages between fellow kibbutz children are very rare. In Taiwan some families practice *"shimpua* marriage" in which an infant daughter is brought up by the family of the man she will marry. She is therefore effectively married to her stepbrother. Such marriages are often infertile, largely because the two partners find each other sexually unattractive. Conversely, two siblings reared apart are surprisingly likely to fall in love with each other if they meet at the right age.[4]

All of this adds up to a picture of sexual inhibition between people who saw a great deal of each other during childhood; fraternal incest, as Westermarck suggested, is therefore prevented by this instinctive aversion that siblings have for each other. But Westermarck's theory would also predict that if incest does occur, it will prove to be between parent and child, and specifically between father and daughter, because a father is past the age at which familiarity breeds aversion and because men usually initiate sex. That, of course, is the most common form of incest.[5]

This contradicts Freud's idea that incest taboos are there because people need to be told not to commit incest. Indeed, Freud's theory requires that evolutionary pressures have not just failed to generate some mechanism to avoid incest but have actually encouraged maladaptive incestuous instincts, which the taboos repress. Freudians have often criticized the Westermarck theory on the grounds that it would obviate the need for incest taboos at all. But in fact incest taboos that outlaw marriage within the nuclear family are rare. The taboos that Freud observed are nearly always concerned with outlawing marriage between cousins. In most societies there is no need to outlaw incest within the nuclear family because there is little risk of its happening.[6]

So why are the taboos there? Claude Lévi-Strauss invented a different theory called the "alliance theory," which stressed the importance of using women as bargaining chips between tribes and therefore not letting them marry within the tribe, but since no two anthropologists can agree on exactly what Levi-Strauss meant, it is hard to test his idea. Nancy Thornhill of the University of New Mexico has argued that the so-called incest taboos are actually rules about marriage customs invented by powerful men to prevent rivals from accumulating wealth by marrying their own cousins. They are not about incest at all but about power.[7]

TEACHING OLD CHAFFINCHES NEW TRICKS

The incest story neatly demonstrates the interdependence of nature and nurture. The incest avoidance mechanism is socially induced: You become sexually averse to your siblings during your childhood.

In that sense there is nothing genetic about it. And yet it is genetic, for it is not taught: It just develops within the brain. The instinct not to mate with childhood companions is nature, but the features by which you recognize them are nurture.

It is critical to Westermarck's argument that this aversion to mating with familiar people wear off for new acquaintances in later life. Otherwise, people would become averse to mating with their spouses within weeks of marrying them, which they plainly do not. Biologically, this is not hard to arrange. One of the most striking features of animal brains is the "critical period" of youth during which something can be learned and after which the learning is not erased or superseded. Konrad Lorenz discovered that chicks and goslings "imprint" on the first moving thing they meet, which is usually their mother and rarely an Austrian zoologist, and thereafter they prefer to follow that object. But chicks a few hours old will not imprint, nor will those two days old. They are at their most sensitive to imprinting at thirteen to sixteen hours old. During that sensitive period they will fix their preferred image of a parent in their heads.

The same is true of a chaffinch learning to sing. Unless it hears another chaffinch, it never learns the species's typical song. If it hears no chaffinch until it is fully grown, it never learns the right song but produces a feeble half-song. Nor will it learn the song if it hears another chaffinch only when it is a few days old. It must hear a chaffinch during a critical period in between—from two weeks to two months of age—and then it will learn to sing correctly; after that period it never modifies its song by imitation.[8]

It is not hard to find examples of critical-period learning in people. Few people change their accents after the age of about twenty-five, even if they move from, say, the United States to Britain. But if they move at ten or fifteen, they quickly adopt a British accent. They are just like white-crowned sparrows, which sing with the dialect of the place where they lived at two months old.[9] Likewise, children are remarkably good at picking up foreign languages just by exposure to them, whereas adults must laboriously learn them. We are not chicks or chaffinches, but we still have critical periods during which we acquire preferences and habits that are fairly hard to change.

This concept of the critical period is presumably what lies behind the Westermarck incest-avoiding instinct: We become sexually indifferent to those with whom we were reared during a critical period. Nobody is certain exactly what constitutes the critical period, but it is a plausible guess that it lasts from, say, eight to fourteen, the years before puberty. Common sense dictates that sexual orientation must be decided in such a fashion: A genetic predisposition meets examples during a critical period. Recall the fate of the baby chaffinch. For six weeks it is sensitive to learning chaffinch song. But during those six weeks of sensitivity, it hears all sorts of things: cars, telephones, lawn mowers, thunder, crows, dogs, sparrows, starlings. Yet it only imitates the song of chaffinches. It has a predilection to learn chaffinch song. (If it were a thrush or a starling, it could indeed imitate some of the other things. One bird in Britain learned the call of a telephone, causing havoc among backyard sunbathers.)[10] This is often the case with learning. Ever since the work of Nikolaas Tinbergen and Peter Marler in the 1960s, it has been well known that animals do not learn anything and everything; they learn what their brains "want" to learn. Men are instinctively attracted to women thanks to the interaction of their genes and hormones, but that tendency is much influenced in a critical period by role models, peer pressure, and free will. There is learning, but there are predispositions.

A heterosexual man emerges from puberty with more than a general sexual preference for all women. He emerges with a distinct notion of beauty and ugliness. He is "stunned" by some women, indifferent to others, and finds others sexually repulsive. Is this, too, something that he acquired by a mixture of genes, hormones, and social pressure? It must be, but the interesting question is how much of each. If social pressure is everything, then the images and lessons we give to the youth of both sexes, through films, books, advertisements, and by example, are crucially important. If not, then the fact that men prefer, say, thin women is fixed by the genes and hormones and not a passing fad.

Suppose you were a Martian interested in studying people as William Thorpe studied chaffinches. You want to know how men learn their standards of beauty, so you keep boys in cages. Some you expose to endless films of plump men admiring and being admired by plump women, while thin men and thin women are reviled; others you keep in total ignorance of womanhood so that their existence comes as a shock at the age of twenty.

It is revealing to speculate on what you think the outcome of the Martian's experiment would be because what follows is an attempt to piece together from much inferior experiments and facts the same result. What kind of woman would the men who had never seen women prefer once they got over the shock of seeing women for the first time? Old ones or young ones, fat ones or thin ones? And would the men reared to believe that fat was beautiful really prefer plump women to skinny models?

Bear in mind the reason we are concentrating on male preferences. As we saw in the last chapter, men care more about the physical appearance of women than vice versa, and for good reason: Youth and health are better clues to women's value as a mate and potential mother than to a man's. Women are not indifferent to youth and health, but they are more concerned than men with other features.

SKINNY WOMEN

But fashions change. If beauty is subject to fashion, however despotic, it can change. Consider a case where the definition of beauty does seem to have changed drastically in recent years: thinness. Wallis Simpson, later the Duchess of Windsor, is credited with the remark that a woman "can never be too rich or too thin," but even she might be surprised at the emaciated appearance of the average modern model. In the words of Roberta Seid, thinness became a "prejudice" in the 1950s, a "myth" in the 1960s, an "obsession" in the 1970s, and a "religion" in the 1980s.[11] Tom Wolfe coined the term "social X rays" for New York society women who starve themselves into fashionable shape. The weight of the Miss Americas falls steadily year after year. So does that of *Playboy* centerfolds. Both categories of women are 15 percent lighter than the average for their ages.[12] Slimming diets fill the newspapers and the wallets of charlatans. Anorexia and bulimia, diseases brought on by excessive dieting, maim and kill young women.

　　　One thing is painfully obvious: There is no preference for the average. Even allowing for the fact that abundant, cheap, refined food makes the average woman much plumper than was normal a millennium or two ago, women must go to extraordinary lengths to achieve the fashionable reedlike shape. Nor has it ever been sensible for men to pick the thinnest woman available. Today, as in the Pleistocene period, that is a sure way to choose the least fertile woman: A woman can be rendered infertile by a body fat content only 10–15 percent below normal. Indeed, one theory is that the widespread obsession of young women with their weight is an evolved strategy to avoid getting pregnant too early or before a man has committed himself to raising a family. But this does not help explain the male preference for skinniness, which seems positively maladaptive.[13]

　　　If the male preference for thinness is paradoxical, how much more puzzling is the fact that it seems to be new. There is ample evidence from sculpture and painting that Victorian beauties were not especially thin, and from sculpture and painting as far back as the Renaissance that beautiful women were plump women. There are exceptions. Nefertiti's neck was that of a thin, elegant woman. Botticelli's Venus was not exactly overweight. And for a time, Victorians worshiped at the shrine of wasp waists, so much so that some women allegedly removed a pair of ribs to make their waists slimmer. Lillie Langtry could enclose her eighteen-inch waist with two hands, but even the slimmest of today's models are twenty-two inches around the waist. And it is implausible that a Renaissance man would have found them ugly. Yet we need not rely only on our own culture for evidence that plump women can be more attractive than thin ones. There is a willingly expressed preference for plump female bodies among tribal people all over the world, and in many subsistence societies, thin women are shunned.

As Robert Smuts of the University of Michigan has argued, thinness was once all too common and was a sign of relative poverty. Nowadays, poverty-induced thinness is confined to the Third World. But in the industrialized nations, wealthy women are able to afford a diet low in fat and spend their money on dieting and exercise. Thinness has become what fatness was: a sign of status.

Smuts argues that male preferences, keying in on whatever signs of status prevailed, simply switched. They did this presumably by a switch of association. A young man growing up today is bombarded with correlations between thinness and wealth, from the fashion industry in particular. His unconscious mind begins to make the connection during his critical period, and when he is forming his idealized mental preference for a woman, he accordingly makes her slim.[14]

STATUS CONSCIOUSNESS

Unfortunately, this theory conflicts directly with the conclusions of the last chapter, so something has to give. It is women, not men, who are supposed to be especially sensitive to the social status of their potential spouses. Sociobiologists argue that the reason men notice women's looks is not as a proxy for their wealth but as a clue to their reproductive potential. Yet here we have men supposedly using women's waists as clues to their bank balances and positively panting after infertile emaciates.

Several studies have come to the unambiguous conclusion that beautiful women and rich men end up together far more than vice versa. In one study the physical attractiveness of a woman was a far better predictor of the occupational status of the man she married than her own socioeconomic status, intelligence, or education—a rather surprising fact when you consider how often people marry within their profession, class, and education brackets.[15] If men are using appearance as a proxy for status, why do they not use knowledge of status itself?

Unlike female thinness, male status symbols are generally "honest." If they were not, they would not remain status symbols.

Only the very best con man can fake conspicuous consumption or get away for long with boasts about his prowess or rank. Thinness is altogether trickier because poor, low-status women once found it easier to be thin than rich, high-status women. Even today when poor women can afford only junk food while rich women eat lettuce, it is hard to argue that every thin woman is rich and every fat one poor.[16]

So the argument that links status with skinniness is not persuasive. Skinniness is a very poor clue to wealth, and in any case men are not much interested in women's status or wealth. Indeed, the argument is circular: Social status and thinness are correlated because of a male preference for thinness. I find the explanation that men have cued in on a woman's thinness as a clue to her status unconvincing.

The trouble is, I am not sure what to suggest in place of it. Suppose it is true that in the days of Rubens men preferred plump women and that today they prefer thin women. Suppose between the plump matrons of Rubens's paintings and the "no woman can be too thin" days of Wallis Simpson, men stopped preferring the fattest or some half-plump ideal and started preferring the thinnest women available. Ronald Fisher's sexual selection theory suggests one way in which it may have been adaptive for men to like thin women. By preferring a thin female, a human male may have had thin daughters who would have attracted the attention of high-status males because other males also preferred thinness. In other words, even if a thin wife could bear fewer children than a fat one, her daughters would be more likely to marry well, and having married well, to be wealthy enough to rear more of the children they bore. So the man who marries a thin woman may have more grandchildren than the man who marries a fat one. Now imagine that a cultural sexual preference spreads by imitation and that young men learn the equation thin equals beautiful by watching others behave. That in itself would be adaptive because it would be one way for males to ensure that they did not flout the prevailing fashion (just as females copying each other in mate choice is adaptive in black grouse). Were they to ignore the cultural preference for plump or

for thin women, they would risk having spinster daughters as surely
as a peahen would risk having bachelor sons by choosing a short-
tailed mate. In other words, as long as the preference is cultural
and the preferred trait is genetic, Fisher's insight that fashion is
despotic still stands.[17]

I confess, however, that these ideas do not really convince
me. If fashions are despotic, they cannot easily be changed. The
puzzle is how men stopped liking plump women without depriving
themselves of eligible offspring by doing so. It is hard to escape
the conclusion that the fashion in men's preferences for women's
fatness cannot have changed adaptively. Either men's preferences
shifted spontaneously and for no good reason or men always preferred some ideal shape that was always quite thin.

WHY WAISTS MATTER

The solution to this puzzle may lie in the work of an ingenious
Indian psychologist named Dev Singh, who now works at the University of Texas in Austin. He observed that women's bodies, unlike
men's, go through two remarkable transitions between puberty and
middle age. At ten a girl has a figure not unlike what she will have
at forty. Then suddenly her vital statistics are transformed: The
ratio of her waist to her chest measurement and to her hips shrinks
rapidly. By thirty it is rising again as her breasts lose their firmness
and her waist its narrowness. That ratio, of waist to breasts and
hips, is not only known as the vital statistic but it is also the feature that, with a few brief exceptions, fashion has always empha-
sized above all else. Bodices, corsets, hoops, bustles, and crinolines
existed to make waists look smaller relative to bosom and bottom.
Bras, breast implants, shoulder pads (which make the waist look
smaller), and tight belts do the same today.

Singh noticed that however much the weight of *Playboy* centerfolds changed, one feature did not: the ratio of their waist width
to their hip width. Recall that Bobbi Low at the University of Michigan argued that fat on the buttocks and breasts mimics broad hip

bones and high mammary tissue content, while thin waists seem to indicate that these features could not be caused by fat. Singh's theory is slightly different but intriguingly parallel. He argues that, within reason, a man will find almost any weight of a woman attractive as long as her waist is much thinner than her hips.[18]

If that sounds foolish, consider the results of Singh's experiments. First, he showed men four versions of the same picture of the midriff of a young woman in shorts. Each picture was subtly touched up to alter slightly the waist-to-hip ratio: 0.6, 0.7, 0.8, and 0.9. Unerringly, men chose the thinnest-waist version as the most attractive. This was no great surprise, but he found a remarkable consistency among his subjects. Next he showed his subjects a range of drawings of female forms, which varied according to their weight and according to their waist-to-hip ratio. He found that a heavy woman with a low ratio of waist to hips was usually preferred to a thin woman with a high ratio. The ideal figure was the one with the lowest ratio, not the one with the thinnest torso.

Singh's interest is in anorexics, bulimics, and women obsessed with losing weight even when thin. He believes that because dieting in fairly thin women has no effect on the waist-to-hip ratio—if anything, it makes it larger by shrinking the hips—they are doomed never to feel more attractive.

Why does the waist-to-hip ratio matter? Singh observes that a "gynoid" fat distribution—more fat on the hips, less on the torso—is necessary for the hormonal changes associated with female fertility. An "android" fat distribution—fat on the belly, thin hips—is associated with the symptoms of male disabilities such as heart disease, even in women. But which is cause and which effect? It seems to me more likely that both the shape and the hormonal effects of it are sexually selected by generations of males rather than males preferring the shape because it is the only way the hormones can be made to work. The relatively brief period during which women have hourglass bodies—from fifteen to thirty-five, say—is a sexually selected phenomenon. It owes more to competition to attract men than to any other biological need. Men have been unconsciously acting as selective breeders of women.

Low provides one possible reason for the male preference for a low ratio—choosing broad-hipped women more able to give birth. Most apes give birth to babies whose brain is half-grown; human babies' brains are one-third grown at birth, and they spend far less time in the womb than is normal for a mammal, given the longevity of man. The reason is obvious: Were the hole in the pelvis through which we are born (the birth canal) commensurately larger, our mothers would be unable to walk at all. The width of human hips reached a certain point and could go no further; as brains continued to grow bigger, earlier birth was the only option left to the species. Imagine the evolutionary pressure of this process on female hip size. It was always wise for a man to choose the biggest-hipped woman he could find, generation after generation, for millions of years. At a certain point hips could get no bigger but men still had the preference, so women with slender waists who appeared to have larger hips by contrast were preferred instead.[19]

I do not know if I believe this tale or not. I cannot find the logical flaw in it (though on first reading there will seem to be many), nor can I quite match it to the male passion for thinness. I also have a nagging doubt about our assumptions that fashions have changed in the admiration of thinness. Suppose our assumptions are at fault, as in the story of the king and the goldfish. Suppose men always preferred slim women to fat ones because slimness meant youth and virginity. After all, as every cosmetic company and plastic surgeon knows, youth has always been the most reliable key to beauty. Perhaps men do not use slimness as a clue to status or childbearing ability but to youth.

YOUTH EQUALS BEAUTY?

A man cannot tell the age of a woman directly. He must infer it from her physical appearance, her behavior, and her reputation. It is intriguing to note that many of the most noticed features of female beauty decay rapidly with age: unblemished skin, full lips, clear eyes, upright breasts, narrow waists, slender legs, even blond hair, which, without chemical intervention, rarely lasts beyond the twenties except among the most Viking of people. These things are, in the sense developed in chapter 5, honest handicaps. They tell a tale of age that cannot be easily disguised without surgery, makeup, or veils.

That blond hair on a woman has been considered by Europeans more beautiful than brown or black has long been noted. In ancient Rome women dyed their hair blond. In medieval Italy fair hair and great beauty were inseparable. In Britain the words fair and beautiful were synonymous.[20] Blond adult hair may be a sexually selected honest handicap, just like a swallow's tail streamers. Blond hair in children is a fairly common gene among Europeans (and, curiously, Australian aborigines). So when a mutation arose in the not-so-distant past, somewhere near Stockholm, say, for that blondness to last into adulthood but not beyond the early twenties, any men with a genetic preference for blond women would have found themselves marrying only young women, which—in a heavily clothed civilization—others might not have done. They would therefore have left more descendants, and a preference for blond hair would have spread. This in turn would have increased the spread of the trait itself because it was indeed an honest indicator of female reproductive value. Hence, gentlemen prefer blondes.[21]

Of course, the part about the male genetic preference is optional or, if you like, a parable. It is more probable that the preference for blond hair among northern European men, if it exists, is a cultural trait instilled in them unconsciously by the association between blondness and youth—an association, incidentally, that the cosmetic industry is rapidly undermining. But the effect is the same: a genetic change brought about by a sexual preference. The alternative theory is to suggest some natural reason for blond hair's being advantageous—for example, that it goes with fair skin, and fair skin allows the absorption of ultraviolet light to help stave off vitamin D deficiency. But the skin is not much fairer in blond than in dark Swedes; truly fair skin goes with red hair, not blond.

Until recently, sexual selection was an argument of last resort, when appeals to natural selection by the "environment" had failed. But why should it be? Why is it more plausible to suggest that blond hair in Baltic people was selected by vitamin D deficiency than to suggest it was selected by sexual preferences? The evidence is beginning to accumulate that humanity is a highly sexually selected species and that this explains the great variations between races in hairiness, nose length, hair length, hair curliness, beards, eye color—variations that plainly have little to do with climate or any other physical factor. In the common pheasant, every one of forty-six isolated wild populations in central Asia has a different combination of male plumage ornaments: white collars, green heads, blue rumps, orange breasts. Likewise, in mankind, sexual selection is at work.[22]

The male obsession with youth is characteristically human. There is no other animal yet studied that shares this obsession quite as strongly. Male chimpanzees find middle-aged females almost as attractive as young ones as long as they are in season. This is obviously because the human habits of lifelong marriage and long, slow periods of child rearing are also unique. If a man is to devote his life to a wife, he must know that she has a potentially long reproductive life ahead of her. If he were to form occasional short-lived pair bonds throughout his life, it would not matter how young his mates were. We are, in other words, descended from men who chose young women as mates and so left more sons and daughters in the world than other men.[23]

THE LEGS THAT LAUNCHED A THOUSAND SHIPS

That many of the components of female beauty are clues to age, every woman and every cosmetic company well knows. But there is more to beauty than youth. The reasons that many youthful women are not beautiful are generally twofold: They are overweight or underweight, or their facial features do not fit our image of beauty. Beauty is a trinity of youth, figure, and face.

A pop song from the 1970s included the cruelly sexist line "nice legs, shame about the face." The importance of regular, symmetrical facial features is somewhat puzzling. Why should a man throw away a chance at mating with a young and fertile woman simply because she has a long nose or a double chin?

It is possible that facial features are a clue to genetic or nurtured quality, or to character and personality. Facial symmetry may well prove to be a clue to good genes or good health during development.[24] "The face is the most information-dense part of the body" is how Don Symons put it to me one day. And the less symmetrical a face, the less attractive. But asymmetry is not a common reason for ugliness; many people have perfectly symmetrical faces and yet are still ugly. The other noticeable feature of facial beauty is that the average face is more beautiful than any extreme. In 1883, Francis Galton discovered that merging the photographs of several women's faces produced a composite that is usually judged to be better looking than any of the individual faces that went into making it.[25] The experiment has been repeated recently with computer-merged photographs of female undergraduates: The more faces that go into the image, the more beautiful the woman appears.[26] Indeed, the faces of models are eminently forgettable. Despite seeing them on the covers of magazines every day, we learn to recognize few individuals. The faces of politicians, not known for their beauty, are much more memorable. Faces that are "full of character" are almost by definition nonaverage faces. The more average and unblemished the face, the more beautiful, but the less it tells you about its owner's character.

This attraction to the average—to a nose that is neither too long nor too short, to eyes that are not too close together nor too far apart, to a chin that is neither prominent nor receding, to lips that are full but not too full, to cheekbones that are prominent but not absurdly so, to a face that is the average, oval shape, neither too long nor too broad—crops up throughout literature as a theme of female beauty. It suggests to me that a Fisherian sexy-son—or rather, in this case, sexy-daughter—effect is at work. Given the importance of facial beauty, a man who chooses an ugly-faced mate will probably have daughters that marry late or marry second-rate husbands. Throughout human history men have fulfilled their ambitions through their daughters' looks. In societies with few other opportunities for social mobility, a great beauty could always marry above her station.[27] Of course women inherit their looks from their fathers as well as their mothers, so a woman should also prefer regular features in a man—and women mostly do.

All that the Fisher effect requires is for men to show a tendency to prefer the average face, and runaway selection will take over. Any man who deviates from the average preference has fewer or poorer grandchildren because his daughters are considered less beautiful than the average. It is a cruel, despotic fashion, one that enforces its pitiless logic at the expense of many a brilliant, kind, and accomplished woman who happens to be plain, and one that has ironically been made worse by the demographic transition to prescribed monogamy. In medieval Europe and in ancient Rome, powerful men took all the beauties into their harems, leaving a general shortage of women for the other men, so an ugly woman stood a better chance of eventually finding some man desperate enough to marry her. That may not sound very just, but justice is rarely the consequence of sexual selection.

PERSONALITIES

So much for what in women attracts men. What draws women to certain men? Male handsomeness is affected by the same trinity as female beauty—face, youth, and figure. But in study after study, women consistently agree that these factors matter less than personality and status. Men consistently place physical features above personality and status when considering women; women do not when considering men.[28]

The single exception is height. Tall men are universally considered more attractive by women than short men. In the world of dating agencies, the principle that a man must be taller than his date is so universal that it has been called "the cardinal principle of date selection." Out of 720 applications by couples for bank accounts, only one was from a couple in which the woman was taller than the man, and yet couples chosen at random from the population would show scores of such cases. People mate "assortatively" for height. Men seek shorter wives, and women seek taller husbands. This cannot be due only to the men. When shown drawings of men and women together and asked to write stories to go with them, even women who stated adamantly that the size of a man made no difference to them wrote stories about anxious or weak men more often when the man depicted was shorter than the woman. The laudatory metaphor "he's a big man" is found in many cultures. It has been calculated that every inch is worth $6,000 a year in salary in modern America.[29]

Bruce Ellis has summarized the evidence that personality is critical in men. In a monogamous society a woman often chooses a mate long before he has had a chance to become a "chief," and she must look for clues to his future potential rather than rely only on his past achievements. Poise, self assurance, optimism, efficiency, perseverance, courage, decisiveness, intelligence, ambition—these are the things that cause men to rise to the top of their professions. And not coincidentally, these are the things women find attractive. They are clues to future status. In one test of this truism, three scientists told their subjects stories about two different people of undefined gender taking part in a tennis match and doing equally well. One was portrayed as strong, competitive, dominant, and determined, the other as consistent, playing for fun rather than to win, easily intimidated by a stronger opponent, and uncompetitive. When asked to summarize the characters of these two people, women and men came up with similar descriptions. But whereas women said that the dominant one was more sexually attractive (if male), men did not find the dominant one more attractive (if female).[30]

Likewise, the same scientists videotaped an actor in two simulated interviews; in one he sat meekly in a chair near the door, with his head bowed, nodding at the interviewer, while in the other he was relaxed, leaning back and gesturing confidently. When shown the videos, women found the more dominant actor more desirable as a date and more sexually attractive, whereas men did not when the actor was female. Body language matters for male sex-iness.[31]

If women select mates on the basis of personality more than men do, this correlates with the fact noted in chapter 8, and well known to many couples, that women are better judges of char-acter. Good female judges of character left more descendants than bad. Good male judges did no better than bad male judges.

The importance of character may explain why Hollywood directors believe that the perfect box-office draw is a familiar, pop-ular male star and a little-known female beauty (and pay them accordingly). Male stars, such as Sean Connery and Mel Gibson, build their reputations gradually. Female stars, such as Julia Roberts and Sharon Stone, rocket to fame in a single movie. The recipe of the James Bond films was perfect: a new girl every time but the same old Bond. (Man, though less than some male mam-mals, exhibits the "Coolidge effect": a new female refreshes his libido. The effect is named after the famous story about President Calvin Coolidge and his wife being shown around a farm. Learning that a cockerel could have sex dozens of times a day, Mrs. Coolidge said: "Please tell that to the president." On being told, Mr. Coolidge asked, "Same hen every time?" "Oh, no, Mr. President. A different one each time." The president continued: "Tell that to Mrs. Coolidge.")[32]

The evidence that women do use direct clues of male status is overwhelming. American men who marry in a given year earn about one and a half times as much as men of the same age who do not. In a survey of two hundred tribal societies, two scientists con-firmed that the handsomeness of a man depends on his skills and prowess rather than on his appearance. Dominance in a man is uni-versally considered attractive by women. In Buss's study of thirty-seven societies, women put more value on men's financial prospects than vice versa. All in all, as Bruce Ellis put it in a recent review, "status and economic achievement are highly relevant barometers of male attractiveness, more so than physical attributes."[33]

What are the clues to status? Ellis suggests that clothes and ornaments provide one set of clues: an Armani suit, a Rolex watch, and a BMW are as blatantly revealing of rank as any admiral's sleeve stripes or Sioux chief's headdress. In a book that chronicled how fashion has always been, until recently, a matter of class emulation, Quentin Bell wrote: "The history of fashionable dress is tied to the competition between classes, in the first place the emulation of the aristocracy by the bourgeoisie and then the more extended competition which results from the ability of the proletariat to compete with the middle classes. . . . Implicit in the whole is a system of sartorial morality dependent upon pecuniary standards of value."[34]

Bobbi Low has surveyed hundreds of societies and come to the conclusion that male ornaments almost always relate to rank and status—maturity, seniority, physical prowess, ferocity, or ability to indulge in conspicuous consumption—whereas female ornaments tend to signal marital or pubertal status and sometimes husband's wealth. Certainly a Victorian duchess was emphasizing not her own wealth but her husband's in the class distinctions of her clothes. This applies as plainly in modern urban societies as it did in ancient tribal ones. Tom Wolfe was the first to comment on how the circular ornaments on the hoods of Mercedes-Benzes had become status symbols among Harlem drug dealers.

At this point some evolutionists seem dangerously close to arguing that women have evolved the ability to be impressed by BMWs. Yet BMWs have existed for only about one human generation. Either evolution is working absurdly fast, or there is something wrong. There are two ways to avoid this difficulty, one of which is popular at the University of Michigan, the other at Santa Barbara. The Michigan scientists say something like this: Women do not have an evolved ability to be impressed by BMWs, but they have an evolved ability to be flexible and to adapt to the social pressures of the society in which they grew up. The Santa Barbara scientists say: Behavior itself is rarely what has evolved; it is the underlying psychological attitude that evolves, and modern women possess a mental mechanism, evolved during the Pleistocene period, that enables them to read what correlates to status among men and find such clues desirable.

In a sense, both are saying the same thing. Women are impressed by signals of status, whatever those specific symbols are. Presumably at some point they learn the association between BMWs and wealth; it is not a difficult equation to solve.[35]

THE FASHION BUSINESS

We are back at a familiar paradox. Evolutionists and art historians agree that fashion is all about status. In their dress, women follow fashion more than men do. Yet women seek clues to status, which change with fashion, and men seek clues to fertility, which do not. Men should not care less what women wear as long as they are smooth-skinned, slim, young, healthy, and generally nubile. Women should care greatly about what men wear because it tells them a good deal about their background, their wealth, their social status, even their ambitions. So why do women follow clothes fashions more avidly than men?

I can think of several answers to this question. First, the theory is simply wrong, and men prefer status symbols, whereas women prefer bodies. Perhaps, but that flies in the face of an awful lot of robust evidence. Second, women's fashion is not about status after all. Third, modern Western societies have been in a two-century aberration from which they are just emerging. In Regency England, Louis XIV's France, medieval Christendom, ancient Greece, or among modern Yanomamö, men followed fashion as avidly as women. Men wore bright colors, flowing robes, jewels, rich materials, gorgeous uniforms, and gleaming, decorated armor. The damsels that knights rescued were no more fashionably attired than their paramours. Only in Victorian times did the deadly uniformity of the black frock coat and its dismal modern descendant, the gray suit, infect the male sex, and only in this century have women's hemlines gone up and down like yo-yos.

This suggests the fourth and most intriguing explanation, which is that women do care more about clothes and men do care less, but instead of influencing the other sex with their concerns, they influence their own. Each gender uses its own preferences to guide its own behavior. Experiments show that men think women care about physique much more than they actually do; women think men care about status cues much more than they actually do. So perhaps each sex simply acts out its instincts in the conviction that the other sex likes the same things as they do.

One experiment seems to support the idea that men and women mistake their own preferences for those of the opposite sex. April Fallon and Paul Rozin of the University of Pennsylvania showed four simple line drawings of male or female figures in swimsuits to nearly five hundred undergraduates. In each case the figures differed only in thinness. They asked the subjects to indicate their current figure, their ideal figure, the figure that they considered most attractive to the opposite sex, and the figure they thought most attractive in the opposite sex. Men's current, ideal, and attractive figures were almost identical; men are, on average, content with their figures. Women, as expected, were far heavier than what they thought most attractive to men, which was heavier still than their own ideal. But intriguingly, both sexes erred in their estimation of what the other sex most likes. Men think women like a heavier build than they do; women think men like women thinner than they do.[36]

However, such confusions cannot be the whole explanation of why women follow fashion because it does not work for other features of attraction. Women are far more concerned with their own youth than men despite the fact that they mostly do not themselves seek younger partners.

And yet the notion that fashion is about status revolts us in a democratic age. We pretend instead that fashion is actually about showing off a body to best advantage. New fashions are worn by gorgeous models, and perhaps women buy them because they subconsciously credit the beauty to the dress and not the model. Surveys reveal what everybody knows: Men are attracted by women in revealing, tight, or skimpy clothing; women are less attracted by such clothing on men. Most female fashions are more or less explicitly designed to enhance beauty; for example, a gigantic crinoline made a waist look small simply by contrast. A woman is careful to choose clothes that "suit" her particular figure or hair color. Moreover, since most men grow up seeing women dressed and spend their critical periods seeing clothed women, their ideals of beauty include images of clothed women as well as naked ones. Havelock Ellis recounted the story of a boy who, standing before a painting of the Judgment of Paris, was asked which goddess he thought was the most beautiful. He replied: "I can't tell because they haven't their clothes on."[37]

But the most characteristic feature of fashion, today at least, is its obsession with novelty. We have already seen how Bell thinks this comes about, as the trendsetters try to escape their vulgar imitators. Low thinks the key to women's fashion is novelty. "Any conspicuous display which signals the ability to read fashion trends" is a clue to a woman's status.[38] Being the first in fashion is certainly a status symbol among women. Without the ability to induce constant obsolescence, fashion designers would be a lot less rich than they are.

This brings us back to the shifting sands of cultural standards of beauty. Beauty cannot be commonplace in a monogamous species like man; it must stand out. Men are discriminating because they will get the chance to marry only one or perhaps two women, so they are always interested in the best they can get, never in the ordinary. In a crowd of women all wearing black, the single one in red would surely catch the eye of a man, whatever her figure or face was like.

The very word fashion used to mean something between conformity and custom, where now it means novelty and modernity. Remarking on painful corsets and the hypocrisy of low necklines in a puritanical society, Quentin Bell observed: "The case against the fashion is always a strong one; why is it then that it never results in an effective verdict? Why is it that both public opinion and formal regulations are invariably set at nought, while sartorial custom, which consists in laws that are imposed without formal sanctions, is obeyed with wonderful docility, and this despite the fact that its laws are unreasonable, arbitrary, and not infrequently cruel."[39]

I am left feeling that this puzzle is, in the present state of evolutionary and sociological thinking, insoluble. Fashion is change and obsolescence imposed on a pattern of tyrannical conformity. Fashion is about status, and yet the sex that is obsessed with fashion is trying to impress the sex that cares least about status.

THE FOLLY OF SEXUAL PERFECTIONISM

Whatever determines sexual attraction, the Red Queen is at work. If for most of human history beautiful women and dominant men had more children than their rivals—which they surely did because the dominant men chose beautiful wives, and together they lived off the toil of their rivals—then in each generation women became that little bit more beautiful and men that little bit more dominant. But their rivals did, too, being descendants of the same successful couples. So standards rose, too. A beautiful woman needed to shine still more brightly to stand out in the new firmament. And a dominant man needed to bully or scheme still more mercilessly to get his way. Our senses are easily dulled by the commonplace, however exceptional it may seem elsewhere or at other times. As Charles Darwin put it, "If all our women were to become as beautiful as the Venus de Medici, we should for a time be charmed; but we should soon wish for variety; and as soon as we had obtained variety, we should wish to see certain characters in our women a little exaggerated beyond the then existing common standard."[40] This, incidentally, is as concise a statement as could be made for why eugenics would never work. A page later Darwin describes the Jollof tribe of West Africa, famous for the beauty of its women; it deliberately sold its ugly women into slavery. Such Nazi eugenics would indeed gradually raise the level of beauty in the tribe, but the men's subjective standards of beauty would rise as fast. Since beauty is an entirely subjective concept, the Jollofs were doomed to perpetual disappointment.

The depressing part of Darwin's insight is that it shows how beauty cannot exist without ugliness. Sexual selection, Red Queen—style, is inevitably a cause of dissatisfaction, vain striving, and misery to individuals. All people are always looking for greater beauty or handsomeness than they find around them. This brings up yet another paradox. It is all very well to say that men want to marry beautiful women and women want to marry rich and powerful men, but most of us never get the chance. Modern society is monogamous, so most of the beautiful women are married to dominant men already. What happens to Mr. and Ms. Average? They do not remain celibate; they settle for something second best. In black grouse the females are perfectionist, the males indiscriminate. In a monogamous human society, neither sex can afford to be either perfectionist or indiscriminate. Mr. Average chooses a plain woman, and Ms. Average chooses a wimp. They temper their idealist preferences with realism. People end up married to their equals in attractiveness: The homecoming queen marries the football hero; the nerd marries the girl in glasses; the man with mediocre prospects marries the woman with mediocre looks. So pervasive is this habit that exceptions stand out a mile: "What on earth can she see in him?" we ask of a model's dull and unsuccessful husband, as if there must be some hidden clue to his worth that the rest of us have missed. "How did she manage to catch him?" we ask of a high-flying man married to an ugly woman.

The answer is that we each instinctively know our relative worth as surely as in Jane Austen's day people knew their place in the class system. Bruce Ellis showed how we manage this "assortative mating" pattern. He gave each of thirty students a numbered card to stick on their foreheads. Each could now see the others' numbers, but nobody knew his or her own. He told them to pair up with the highest number they could find. Immediately the person with 30 on her forehead was surrounded by a buzzing crowd, so she adjusted her expectations upward and refused to pair up with just anybody, settling eventually for somebody with a number in the high twenties. The person with number 1, meanwhile, after trying to persuade number 30 of his worth, then lowered his sights and went progressively down the scale, steadily discovering his low status, until he ended up taking the first person who would accept him, probably number 2.[41]

The game shows with uncomfortable realism how we measure our own relative desirability from others' reactions to us. Repeated rejection causes us to lower our sights; an unbroken string of successful seductions encourages us to aim a little higher. But it is worth it to get off the Red Queen's treadmill before you drop.

Chapter 10

THE INTELLECTUAL CHESS GAME

Were I (who to my cost already am

One of those strange prodigious Creatures Man.)

A Spirit free, to choose for my own share,

What Case of Flesh, and Blood, I pleased to weare,

I'd be a Dog, *a* Monkey, *or a* Bear.

Or anything but that vain Animal,

Who is so proud of being rational.

The senses are too gross, and he'll contrive

A Sixth, to contradict the other Five;

And before certain instinct, will preferr

Reason, which Fifty times for one does err.

—John Wilmot, Earl of Rochester

The time: 300,000 years ago. The place: the middle of the Pacific Ocean. The occasion: a conference of bottle-nosed dolphins to discuss the evolution of their own intelligence. The conference was being held over an area of about twelve square miles of ocean so that the participants could fish in between meetings; it was during the squid season. The sessions consisted of long soliloquies by invited speakers followed by a series of commentaries in Squeak, the language of Pacific bottle-noses. Squawk speakers from the Atlantic were able to hear memorized translations at night. The matter at issue was simple: Why did bottle-nosed dolphins have brains that were so much bigger than those of other animals? The bottle-nose brain was twice as large as that of many other dolphins. The first speaker argued that it was all a matter of language. Dolphins needed big brains to enable them to hold in their heads the concepts and the grammar with which to express themselves. The ensuing commentaries were merciless. The language theory solved nothing, said the commentators. Whales had complex language, and every dolphin knew how stupid whales were. Only the year before a group of bottle-noses had fooled an old humpback whale into attacking his best friend by sending out soliloquies about infidelity in humpback language. The second squeaker, a male, was more favorably received, for he argued that this was indeed the purpose of dolphin intelligence: to deceive. Are we not, he squeaked, the global masters of deception and manipulation? Do we not spend all our time scheming to outwit one another in the pursuit of female dolphins? Are we not the only species in which "triadic" interactions among alliances of individuals are known? The third speaker replied that this was all very laudable, but why us? Why bottle-nosed dolphins? Why not sharks or porpoises? There was a dolphin in the River Ganges whose brain weighed only five hundred grams. A bottle-nose brain weighed fifteen hundred grams. No, he replied, the answer lay plainly in the fact that of all the creatures on earth, bottle-nosed dolphins were the ones that had the most varied and flexible diet. They could eat squid or fish or . . . well, all sorts of different kinds of fish. That variety demanded flexibility, and flexibility demanded a big brain that could learn.

The final speaker of the day was scornful of all his predecessors. If social complexity was what required intelligence, why were none of the social animals on land intelligent? The speaker had heard stories of an ape species that was almost as big-brained as dolphins; indeed, for its body size it was even bigger. It lived in bands on the African savanna and used tools and hunted meat as well as gathered plants for food. It even had language of a sort, though with none of the richness of Squeak. It did not, he squeaked drolly, eat fish.[1]

THE APE THAT MADE IT

Around 18 million years ago there were tens of species of ape living in Africa and many others in Asia. Over the next 15 million years most of them became extinct. A Martian zoologist who arrived in Africa about 3 million years ago would probably have concluded that the apes were bound for the trash heap of history, an outdated model of animal made obsolescent by competition with the monkeys. Even if he noticed that there was one ape, a close relative of the chimpanzee, that walked on two feet, entirely upright, he would not have predicted much of a future for it.

For its size, midway between a chimpanzee and an orang-utan, the upright ape, known to science now as *Australopithecus afarensis* and to the world as "Lucy,"[2] had a "normal" brain size: about four hundred cubic centimeters—bigger than the modern chimp, smaller than the modern orangutan. Its posture was peculiarly humanlike, undoubtedly, but its head was not. Apart from its uncannily human legs and feet, we would not have had any trouble thinking of it as an ape. Yet over the next 3 million years the heads of its descendants exploded in size. Brain capacity doubled in the first 2 million years and almost doubled again in the next million, to reach the fourteen hundred cubic centimeters of modern people. The heads of chimps, gorillas, and orangutans stayed roughly the same. So did the other descendant of Lucy's species, the so-called robust australopithecines, or nutcracker people, who became specialist plant eaters.

What caused the sudden and spectacular expansion of that one ape's head, from which so much else flowed? Why did it happen to one ape and not another? What can account for the astonishing speed, and the accelerating speed of the change? These questions may seem to have nothing to do with the subject of this book, but the answer may lie with sex. If new theories are right, the evolution of man's big head was the result of a Red Queen sexual contest between individuals of the same gender.

On one level the evolution of big-headedness in man's ancestors is easily explained. Those that had big heads had more young than those that did not. The young, inheriting the big heads, therefore had bigger heads than their parents' generation. This process, moving in fits and starts, faster in some places than in others, eventually caused the trebling of the brain capacity of man. It could have happened no other way. But the intriguing thing is what made the big-brained people likely to have more children than the small-brained ones. After all, as a diverse array of observers from Charles Darwin to Lee Kwan Yew, the former prime minister of Singapore, have noted with regret, clever people are not noticeably more prolific breeders than stupid people.

A time-traveling Martian could go back and examine the three consecutive descendants of *Australopithecus*, *Homo habilis*, *Homo erectus*, and so-called archaic *Homo sapiens*. He would find a steady progression in brain size—that much we know from the fossils— and he would be able to tell us what the clever ones were using their bigger brains for. We can do something similar today simply by looking at what modern human beings use their brains for. The trouble is that every aspect of human intelligence you consider as uniquely human applies to the other apes as well. A vast chunk of our brains is used for visual perception; but it is hardly plausible that Lucy suddenly needed better visual perception than her distant cousins. Memory, hearing, smell, face recognition, self-awareness, manual dexterity—they all have more space in the human than in the chimp brain, yet it is hard to understand why any of them was more likely to cause Lucy to have more children than it was to cause a chimp to have more. We need some qualitative leap from ape to man, some difference of kind rather than degree that transformed the human mind in ways that for the first time made the biggest possible brain the best possible brain.

There was a time when it was easy to define what made humans different from other animals. Humans had learning; animals had instincts. Humans used tools; animals did not. Humans had language; animals did not. Humans had consciousness; animals did not. Humans had culture; animals did not. Humans had self-awareness; animals did not. Gradually these differences have been blurred or shown to be differences in degree rather than in kind. Snails learn. Finches use tools. Dolphins use language. Dogs are conscious. Orangutans recognize themselves in mirrors. Japanese macaques pass on cultural tricks. Elephants mourn their dead.

This is not to say that all animals are as good as humans at each of these tasks, but remember that humans were once no better than them and yet they came under sudden pressure to get better and better, while animals did not. A well-trained humanist is already scoffing at such sophistry. Only people can make tools as well as use them. Only people can use grammar as well as vocabulary. Only people can empathize as well as feel emotion. But this sounds uncannily like special pleading. I find the instinctive arrogance of the human sciences thoroughly unconvincing because so many of its bastions have already fallen to the champions of animals. Beaten back from position after position, the humanists simply pretend they never intended to hold them in the first place and redefine the retreat as tactical. Almost all discussions of consciousness assume a priori that it is a uniquely human feature when it is patently obvious to anybody who has ever kept a dog that the average dog can dream, feel sad or glad, and recognize individual people; to call it an unconscious automaton is perverse.

THE MYTH OF LEARNING

At this point the humanist usually retreats to his strongest bastion: learning. The human, he says, is uniquely flexible in his behavior, adapting to skyscrapers, deserts, coal mines, and tundra with equal ease. That is because he learns far more than animals and relies on instincts far less. Learning how the world is rather than simply arriving in it with a fully formed program for survival is a superior strategy, but it demands a bigger brain. Therefore, the bigger brain of the human reflects a shift away from instinct and toward learning.

Like just about everybody else who has ever thought about these things, I found such logic impeccable until I read a chapter in a book called *The Adapted Mind* by Leda Cosmides and John Tooby of the University of California at Santa Barbara.[3] They set out to challenge the conventional wisdom, which has dominated psychology and most other social sciences for many decades, that instinct and learning are opposite ends of a spectrum, that an animal that relies on instincts does not rely on learning and vice versa. This simply is not so. Learning implies plasticity, whereas instinct implies preparedness. So, for example, in learning the vocabulary of her native language, a child is almost infinitely plastic. She can learn that the word for a cow is *vache* or *cow* or any other word. And likewise in knowing that she must blink or duck when a ball approaches her face at speed, a child would not need to have plasticity at all. To have to learn such a reflex would be painful. So the blink reflex is prepared, and the vocabulary store in her brain is plastic.

But she did not learn that she needed a vocabulary store. She was born with it and with an acute curiosity to learn the names of things. More than that, when she learned the word *cup*, she knew without being told that it was a general name for any whole cup, not its contents or its handle and not the specific cup she saw first, but the whole class of objects called cups. Without these two innate instincts, the "whole object assumption" and the "taxonomic assumption," language would be a lot harder to learn. Children would often find themselves in the position of the apocryphal explorer who points at a never-before-seen animal and says to his local guide, "What's that?" The guide replies, "Kangaroo," which means in his language "I don't know."

In other words, it is hard to conceive how people can learn (be plastic) without sharing assumptions (being prepared). The old idea that plasticity and preparedness were opposites is plainly wrong. The psychologist William James argued a century ago that man had both more learning capacity *and more instincts*, rather than more learning and fewer instincts. He was ridiculed for this, but he was right.

Return to the example of language. The more scientists study language, the more they realize that hugely important aspects of it, such as grammar and the desire to speak in the first place, are not learned by imitation at all. Children simply develop language. Now this might seem crazy because a child reared in isolation would not, as James I of England hoped he might, simply grow up to speak Hebrew. How could he? Children must learn the vocabulary and the particular rules of inflection and syntax specific to their language. True, but almost all linguists now agree with Noam Chomsky that there is a "deep structure" that is universal to all languages and that is programmed into the brain rather than learned. Thus, the reason all grammars conform to a similar deep structure (for example, they use either word order or inflection to signify whether a noun is object or subject) is that all brains have the same "language organ."

Children plainly have a language organ in their brains ready and waiting to apply the rules. They infer the basic rules of grammar without instruction, a task that has been shown to be beyond the power of a computer unless the computer has been endowed with some prior knowledge.

From about the age of one and a half until soon after puberty children have a fascination with learning a language and are capable of learning several languages far more easily than adults can. They learn to talk irrespective of how much encouragement they are given. Children do not have to be taught grammar, at least not of living languages that they hear spoken; they divine it. They are constantly generalizing the rules they have learned in defiance of the examples they hear (such as "persons gived" rather than "people gave"). They are learning to talk in the same way that they are learning to see, by adding the plasticity of vocabulary to the preparedness of a brain that insists on applying rules. The brain has to be taught that large animals with udders are called cows. But to see a cow standing in a field, the visual part of the brain employs a series of sophisticated mathematical filters to the image that it receives from the eye—all unconscious, innate, and unteachable. In the same way, the language part of the brain knows without being taught that the word for a large animal with an udder is likely to behave grammatically like other nouns and not like verbs.[4]

The point is that nothing could be more "instinctive" than the predisposition to learn a language. It is virtually unteachable. It is hard-wired. It is not learned. It is—horrid thought—genetically determined. And yet nothing could be more plastic than the vocabulary and syntax to which that predisposition applies itself. The ability to learn a language, like almost all the other human brain functions, is an instinct for learning.

If I am right and people are just animals with more than usually trainable instincts, then it might seem that I am excusing instinctive behavior. When a man kills another man or tries to seduce a woman, he is just being true to his nature. What a bleak, amoral message. Surely there is a more natural basis for morality in the human psyche than that? The centuries-old debate between the followers of Rousseau and Hobbes—whether we are corrupted noble savages or civilized brutes—has missed the point. We are instinctive brutes, and some of our instincts are unsavory. Of course some instincts are very much more moral, and the vast human capacity for altruism and generosity—the glue that has always held society together—is just as natural as any selfishness.

Yet selfish instincts are there, too. Men are much more instinctively capable of murder and of sexual promiscuity than women, for example. But Hobbes's vindication means nothing because instincts combine with learning. None of our instincts is inevitable; none is insuperable. Morality is never based upon nature. It never assumes that people are angels or that the things it asks human beings to do come naturally. "Thou shalt not kill" is not a gentle reminder but a fierce injunction to men to overcome any instincts they may have or face punishment.

NURTURE IS NOT NECESSARILY THE OPPOSITE OF NATURE

The Jamesian notion that man has instincts to learn things at a stroke demolishes the whole dichotomy of learning versus instinct, nature versus nurture, genes versus environment, human nature versus human culture, innate versus acquired, and all the dualisms that have plagued the study of the mind ever since René Descartes. For if the brain consists of evolved mechanisms highly specific and intricately designed but flexible in content, then it is impossible to use the fact that a behavior is flexible as an indication that it is "cultural." The ability to use language is "genetic" in the sense that it is inherent in the genes' instructions for putting together a human body to include a detailed language-acquisition device. It is also "cultural" in the sense that the vocabulary and syntax of the language are arbitrary and learned. It is also developmental in the sense that the language acquisition device grows after birth and feeds off the examples it sees around it. Just because language is acquired after birth does not means that it is cultural. Teeth are also acquired after birth.

"There is no more a gene for aggression than there is for wisdom teeth," wrote Stephen Jay Gould, implying that behavior must be cultural and not "biological."[5] His facts are right, of course, but that is exactly why his implication is wrong. Wisdom teeth are not cultural artifacts; they are genetically determined even though they develop in late adolescence and even though there is not a single gene that says "grow wisdom teeth." By the term "a gene for aggression," Gould means that the difference between the aggressiveness of person A and person B would be due to a difference in gene X. But just as all sorts of environmental differences (such as nutrition and dentists) can cause A to have bigger wisdom teeth than B, so all sorts of genetic differences (affecting how the face grows, how the body absorbs calcium, how the sequence of teeth are ordered) can cause person A to have bigger wisdom teeth than person B. Exactly the same applies to aggression.

Somewhere in our education we unthinkingly absorb the idea that nature (genes) and nurture (environment) are opposites and that we must make a choice between them. If we choose environmentalism, then we are espousing a universal human nature that is as blank as a sheet of paper awaiting culture's pen, that humans are therefore perfectible and born equal. If we choose genes, then we espouse irreversible genetic differences between races and between individuals. We are fatalists and elitists. Who would not hope with all her heart that the geneticists were wrong?

Robin Fox, an anthropologist who has called this dilemma a quarrel between original sin and the perfectibility of man, portrayed the dogma of environmentalism thus:

> This Rousseauist tradition has a remarkably strong grip
> on the post-Renaissance occidental imagination. It is
> feared that without it we shall be prey to reactionary
> persuasion by assorted villains, from social Darwinists
> to eugenicists, fascists and new-right conservatives. To
> fend off this villainy, the argument goes, we must assert
> that man is either innately neutral (tabula rasa) or
> innately good and that bad circumstances are what make
> him behave wickedly.[6]

Although the notion of a tabula rasa goes back to John Locke, it was in this century that it reached the zenith of its intellectual hegemony. Reacting to the idiocies of social Darwinists and eugenicists, a series of thinkers first in sociology, then anthropology, and finally psychology shifted the burden of proof firmly away from nurture and onto nature. Until proved otherwise, man must be considered a creature of his culture, rather than culture a product of man's nature.

Emil Durkheim, the founder of sociology, set out in 1895 his assertion that social science must assume people are blank slates on which culture writes. Since then, if anything, this idea has hardened into three cast-iron assumptions: First, anything that varies between cultures must be culturally rather than biologically acquired; second, anything that develops rather than appears fully formed at birth must also be learned; third, anything genetically determined must be inflexible. No wonder social science is irredeemably wedded to the notion that nothing in human behavior is "innate," for things do vary greatly between cultures, do develop after birth, and are plainly flexible. Therefore, the mechanisms of the human mind cannot be innate. Everything must be cultural. The reason men find young women more sexually attractive than old women must be that their culture teaches them subtly to favor youth, not because their ancestors left more descendants if they had an innate preference for youth.[7]

Anthropology's turn was next. With the publication of Margaret Mead's *Coming of Age in Samoa* in 1928, the discipline was transformed. Mead asserted that sexual and cultural variety was effectively infinite and was therefore the product of nurture. She did little to prove nurture's predominance—indeed, what empirical evidence Mead did adduce was largely, it now seems, wishful thinking[8]—but she shifted the burden of proof. Mainstream anthropology remains to this day committed to the view that there is only a blank human nature.[9]

Psychology's conversion was more gradual. Freud believed in universal human mental attributes—such as the Oedipal complex. But his followers became obsessed with trying to explain everything according to individual early childhood influences, and Freudianism came to mean blaming one's early nurture for one's nature. Soon psychologists came to believe that even the mind of an adult was a general-purpose learning device. This approach reached its apogee in the behaviorism of B. F. Skinner. He argued that brains are simply devices for associating any cause with any effect.

By the 1950s, looking back at what Nazism had done in the name of nature, few biologists felt inclined to challenge what their human-science colleagues asserted. Yet uncomfortable facts were already appearing. Anthropologists had failed to find the diversity Mead had promised. Freudians had explained very little and altered even less by their appeals to early influences. Behaviorism could not account for the innate preferences of different species of animal to learn different things: Rats are better at running mazes than pigeons. Sociology's inability to explain or rectify the causes of delinquency was an embarrassment. In the 1970s a few brave "sociobiologists" began to ask why, if other animals had evolved natures, humans would be exempt. They were vilified by the social science establishment and told to go back to ant-watching. Yet the question they had asked has not gone away.[10]

The principal reason for the hostility to sociobiology was that it seemed to justify prejudice. Yet this was simply a confusion. Genetic theories of racism, or classism or any kind of *ism*, have nothing in common with the notion that there is a universal, instinctive human nature. Indeed, they are fundamentally opposed because one believes in universals and the other in racial or class particulars. Genetic differences have been assumed just because genes are involved. Why should that be the case? Is it not possible that the genes of two individuals are identical? The logos painted on the tails of two Boeing 747s depend on the airlines that own them, but the tails beneath are essentially the same: They were made in the same factory of the same metal. You do not assume because they are owned by different airlines that they were made in different factories. Why, then, must we assume because there are differences between the speech of the French and the English that they must have brains that are not influenced by genes at all? Their brains are the products of genes—not different genes, the same genes. There is a universal human language-acquisition device, just as there is a universal human kidney and a universal 747 tail structure.

Think, too, of the totalitarian implications of pure environmentalism. Stephen Jay Gould once caricatured the views of genetic determinists in this way: "If we are programmed to be what we are, then these traits are ineluctable. We may, at best, channel them, but we cannot change them."[11] He meant genetically programmed, but the same logic applies with even more force to environmental programming. Some years later Gould wrote: "Cultural determinism can be just as cruel in attributing severe congenital diseases—autism, for example—to psychobabble about too much parental love, or too little."[12]

If, indeed, we are the product of our nurture (and who can deny that many childhood influences are ineluctable—witness accent?), then we have been programmed by our various upbringings to be what we are and we cannot change it—rich man, poor man, beggar-man, thief. Environmental determinism of the sort most sociologists espouse is as cruel and horrific a creed as the biological determinism they attack. The truth is, fortunately, that we are an inextricable and flexible mixture of the two. To the extent that we are the product of the genes, they are all and always will be genes that develop and are calibrated by experience, as the eye learns to find edges or the mind learns its vocabulary. To the extent that we are products of the environment, it is an environment that our designed brains choose to learn from. We do not respond to the "royal jelly" that worker bees feed to certain grubs to turn them into queens. Nor does a bee learn that a mother's smile is a cause for happiness.

THE MENTAL PROGRAM

When, in the 1980s, artificial intelligence researchers joined the ranks of those searching for the mechanism of mind, they, too, began with behaviorist assumptions: that the human brain, like a computer, was an association device. They quickly discovered that a computer was only as good as its programs. You would not dream of trying to use a computer as a word processor unless you had a word-processing program. In the same way, to make a computer capable of object recognition or motion perception or medical diagnosis or chess, you had to program it with "knowledge." Even the "neural network" enthusiasts of the late 1980s quickly admitted that their claim to have found a general learning-by-association device was false: Neural networks depend crucially on being told what answer to reach or what pattern to find, or on being designed for a particular task, or on being given straightforward examples to learn from. The "connectionists," who placed such high hopes in neural networks, had stumbled straight into the traps that had caught the behaviorists a generation earlier. Untrained connectionist networks proved incapable even of learning the past tense in English.[13]

The alternative to connectionism, and to behaviorism before it, was the "cognitive" approach, which set out to discover the mind's internal mechanisms. This first flowered with Noam Chomsky's assertions in *Syntactic Structures*, a book published in 1957, that general-purpose association-learning devices simply could not solve the problem of inferring the rules of grammar from speech.[14] It needed a mechanism equipped with knowledge about what to look for. Linguists gradually came to accept Chomsky's argument. Those studying human vision, meanwhile, found it fruitful to pursue the "computational" approach advocated by David Marr, a young British scientist at MIT. Marr and Tomaso Poggio systematically laid bare the mathematical tricks that the brain was using to recognize solid objects in the image formed in the eye. For example, the retina of the eye is wired in such a way as to be especially sensitive to edges between contrasting dark and light parts of an image; optical illusions prove that people use such edges to delineate the boundaries of objects. These and other mechanisms in the brain are "innate" and highly specific to their task, but they are probably perfected by exposure to examples. No general-purpose induction here.[15]

Almost every scientist who studies language or perception now admits that the brain is equipped with mechanisms, which it did not "learn" from the culture but developed with exposure to the world; these mechanisms specialize in interpreting the signals that are perceived. Tooby and Cosmides argue that "higher" mental mechanisms are the same. There are specialized mechanisms in the mind that are "designed" by evolution to recognize faces, read emotions, be generous to one's children, fear snakes, be attracted to certain members of the opposite sex, infer mood, infer semantic meaning, acquire grammar, interpret social situations, perceive a suitable design of tool for a certain job, calculate social obligations, and so on. Each of these "modules" is equipped with some knowledge of the world necessary for doing such tasks, just as the human kidney is designed to filter the blood.

We have modules for learning to interpret facial expressions—parts of our brain learn that and nothing else. At ten weeks we assume that objects are solid, and therefore two objects cannot occupy the same space at the same time—an assumption that no amount of exposure to cartoon films will later undo. Babies express surprise when shown tricks that imply two objects can occupy the same place. At eighteen months babies assume there is no such thing as action at a distance—that object A cannot be moved by object B unless they touch. At the same age we show more interest in sorting tools according to their function than according to their color. And experiments show that, like cats, we assume any object capable of self-generated motion is an animal, which is something we only partially unlearn in our machine-infested world.[16]

That last is an example of how many of the instincts in our heads develop on the assumption that the world is that of the Pleistocene period, before cars. Infant New Yorkers find it far easier to acquire a fear of snakes than of cars, despite the far greater danger posed by the latter. Their brains are simply predisposed to fear snakes.

Fearing snakes and assuming that self-propelled motion is a sign of an animal are instincts that are probably as well developed in monkeys as in people. Nor is the unwillingness of adults to have sex with people with whom they have lived as children—the incest-avoidance instinct—peculiarly human. Lucy did not need a bigger brain for these things any more than a dog did.

The one thing Lucy did not need was to have to start from

scratch and learn the world afresh every generation. Culture could not teach her to detect edges in the visual field; it did not teach her the rules of grammar. It could have taught her to fear snakes, but why bother? Why not let her be born with a fear of snakes? It is not obvious to somebody with an evolutionary perspective quite why we must consider learning so valuable. If learning really did replace instincts rather than enhance and train them, then we would spend half our lives relearning things that monkeys know automatically, such as the fact that unfaithful mates can cuckold you. Why bother to learn them? Why not allow the Baldwin effect to turn them into instincts and spend slightly less time going through the laborious business of adolescence? If a bat had to learn to use its sonar navigation from its parents, rather than simply developing the ability as it grew, or a cuckoo had to learn the way to Africa in winter, rather than "knowing" before setting off, then there would be a lot more dead bats and lost cuckoos every generation. Nature chooses to equip bats with echo-location instincts and cuckoos with migration instincts because it is more efficient than making them learn. True, we learn a lot more than bats and cuckoos do. We learn mathematics and a vocabulary of tens of thousands of words and what people's characters are like. But this is because we have instincts to learn these things (with the possible exception of mathematics), not because we have fewer instincts than bats or cuckoos.

THE TOOLMAKER MYTH

Until the mid 1970s the question of why people needed big brains when other animals did not had only really been posed by the anthropologists and archaeologists who study the bones and tools of ancient human beings. Their answer, persuasively summarized by Kenneth Oakley in 1949 in a book called *Man the Toolmaker*, was that man was a tool user and toolmaker par excellence and that he developed a big brain for that purpose. Given the increasing sophistication of man's tools throughout his history, and the sudden leaps of technical skill that seemed to accompany each change in skull size—from *habilis* to *erectus*, from *erectus* to *sapiens*, from Neanderthal to modern—this made some sense. But there were two problems with it. First, during the 1960s the ability of animals, especially chimps, to make and use tools was discovered, which rather took the shine off *Homo habilis*'s somewhat basic tool kit. Second, there was a suspicious bias about the argument. Archaeologists study stone tools because that is what they find preserved. An archaeologist of a million years in the future would call ours the concrete age, with some justice, but he might never even know about books, newspapers, television broadcasts, the clothes industry, the oil business, even the car industry—all traces of which would have rusted away. He might assume that our civilization was characterized by hand-to-hand combat by naked people over concrete citadels. Perhaps, in like fashion, the Neolithic age was distinguished from the Paleolithic not by its tool kit but by the invention of language or marriage or nepotism or some such unfossilizable signature. Wood probably loomed larger than stone in people's lives, yet no wooden tools survive.[17]

Besides, the evidence from the tools, far from suggesting continuous human ingenuity, speaks of monumental and tedious conservatism. The first stone tools, the Oldowan technology of *Homo habilis*, which appeared about 2.5 million years ago in Ethiopia, were very simple indeed: roughly chipped rocks. They barely improved at all over the next million years, and far from experimenting, they became gradually more standardized. They were then replaced by the Acheulian technology of *Homo erectus*, which consisted of hand axes and teardrop-shaped stone devices. Again, nothing happened for a million years and more, until about 200,000 years ago when there was a sudden and dramatic expansion in the variety and virtuosity of tools at about the time that *Homo sapiens* appeared. From then on there was no looking back: Tools grew ever more varied and accomplished until the invention of metal. But it comes too late to explain big heads; heads had been swelling ever since 3 million years ago.[18]

Making the tools that *erectus* used is not especially hard.

Everybody could do it, presumably, which is why it was done all over Africa. There was no inventiveness or creativity going on. For a million years these people made the same dull hand axes, yet their brains were already grossly large by ape standards. Plainly, the instincts of manual dexterity, perception of shape, and reverse engineering from function to form were useful to these people, but it is highly implausible to account for the enlargement of the brain as driven entirely by an enlargement of these instincts.

The first rival to the toolmaking theory was "man the hunter." In the 1960s, starting with the work of Raymond Dart, there was much interest in the notion that man was the only ape to have taken up a meat diet and hunting as a way of life. Hunting, went the logic, required forethought, cunning, coordination, and the ability to learn skills such as where to find game and how to get close to it. All true, all utterly banal. Anybody who has ever seen a film of lions hunting zebra on the Serengeti will know how skillful lions are at each of the tasks mentioned above. They stalk, ambush, cooperate, and deceive their prey as carefully as any group of humans ever could. Lions do not need vast brains, so why should we? The fashion for man the hunter gave way to woman the gatherer, but similar arguments applied. It is simply unnecessary to be capable of philosophy and language to be able to dig tubers from the ground. Baboons do it just as well as women.[19]

Nonetheless, one of the most startling things to come out of the great studies of the !Kung San people of the Namib desert in the 1960s was the enormous accumulation of local lore that hunter-gatherer people possess—when and where to hunt for each kind of animal, how to read a spoor, where to find each kind of plant food, which kind of food is available after rain, which things are poisonous and which medicinal. Of the !Kung, Melvin Konner wrote, "Their knowledge of wild plants and animals is deep and thorough enough to astonish and inform professional botanists and zoologists."[20] Without this accumulated knowledge it would not have been possible for mankind to develop so rich and varied a diet, for the results of trial-and-error experiments would not have been cumulative but would have had to be relearned every generation. We would have been limited to fruit and antelope meat, not daring to try tubers, mushrooms, and the like. The astonishing symbiotic relationship between the African honey guide bird and people, in which the bird leads a man to a bees' nest and then eats what remains of the honey when he leaves, depends on the fact that people know because they have been told that honey guides lead them to honey. To accumulate and pass on this store of knowledge required a large memory and a large capacity for language. Hence the need for a large brain.

The argument is sound enough, but once more it applies with equal force to every omnivore on the African plains. Baboons must know where to forage at what time and whether to eat centipedes and snakes. Chimpanzees actually seek out a special plant whose leaves can cure them of worm infections, and they have cultural traditions about how to crack nuts. Any animal whose generations overlap and which lives in groups can accumulate a store of knowledge of natural history that is passed on merely by imitation. The explanation fails the test of applying only to humans.[21]

THE BABY APE

The humanist might be feeling a little frustrated by this line of argument. After all, we have big brains and we use them. The fact that lions and baboons have small ones and get by does not mean that we are not helped by our brains. We get by rather better than lions and baboons. We have built cities, and they have not. We invented agriculture, and they did not. We colonized ice-age Europe, and they did not. We can live in the desert and the rain forest; they are stuck on the savanna. Yet the argument still has considerable force because big brains do not come free. In human beings, 18 percent of the energy that we consume every day is spent in running the brain. That is a mighty costly ornament to stick on top of the body just in case it helps you invent agriculture, just as sex was a mighty costly habit to indulge in merely in case it led to innovation (chapter 2). The human brain is almost as costly an invention as sex, which implies that its advantage must be as immediate and as large as sex's was.

For this reason it is easy to reject the so-called neutral theory of the evolution of intelligence, which has been popularized in recent years mainly by Stephen Jay Gould.[22] The key to his argument is the concept of "neoteny"—the retention of juvenile features into adult life. It is a commonplace of human evolution that the transition from *Australopithecus* to *Homo* and from *Homo habilis* to *Homo erectus* and thence to *Homo sapiens* all involved prolonging and slowing the development of the body so that it still looked like a baby when it was already mature. The relatively large brain case and small jaw, the slender limbs, the hairless skin, the unrotated big toe, the thin bones, even the external female genitalia—we look like baby apes.[23]

The skull of a baby chimpanzee looks much more like the skull of an adult human being than either the skull of an adult chimpanzee or the skull of a baby human being. Turning an apeman into a man was a simple matter of changing the genes that affect the rate of development of adult characters, so that by the time we stop growing and start breeding, we still look rather like a baby. "Man is born and remains more immature and for a longer period than any other animal," wrote Ashley Montagu in 1961.[24]

The evidence for neoteny is extensive. Human teeth erupt through the jaw in a set order: the first molar at the age of six, compared with three for a chimp. This pattern is a good indication of all sorts of other things because the teeth must come at just the right moment relative to the growth of the jaw. Holly Smith, an anthropologist at the University of Michigan, found in twenty-one species of primate a close correlation between the age at which the first molar erupted and body weight, length of gestation, age at weaning, birth interval, sexual maturity, life span, and especially brain size. Because she knew the brain size of fossil hominids, she was able to predict that Lucy would have erupted her first molar at three and lived to forty, much like chimpanzees, whereas the average *Homo erectus* would have erupted his at nearly five and lived to fifty-two.[25]

Neoteny is not confined to man. It is also a characteristic of several kinds of domestic animals, especially dogs. Some dogs are sexually mature when they are still stuck in an early phase of wolf development: They have short snouts, floppy ears, and the sort of behavior that wolf pups show—retrieving for example. Other, such as sheepdogs, are stuck at a different phase: longer snouts, half-cocked ears, and chasing. Still others, such as German shepherds, have the full range of wolf hunting and attacking behaviors plus long snouts and cocked ears.[26]

But whereas dogs are truly neotenic, breeding at a young age and looking like wolf puppies, humans are peculiar. They look like infant apes, true, but they breed at an advanced age. The combination of a slow change in the shape of their head and a long period of youthfulness means that as adults they have astonishingly large brains for an ape. Indeed, the mechanism by which ape-men turned into men was clearly a genetic switch that simply slowed the developmental clock. Stephen Jay Gould argues that rather than seek an adaptive explanation of features like language, perhaps we should simply regard them as "accidental," though useful, by-products of neoteny's achievement of large brain size. If something as spectacular as language can be the product of simply a large brain plus culture, then there need be no specific explanation of why larger brains are required because their advantages are obvious.[27]

The argument is based on a false premise. As Chomsky and others have amply demonstrated, language is one of the most highly designed capabilities imaginable, and far from being a by-product of a big brain, it is a mechanism with a very specific pattern that develops in children without instruction. It also has obvious evolutionary advantages, as a moment's reflection will reveal. Without, for example, the trick of recursion (subordinate phrases) it becomes impossible to tell even the simplest story. In the words of Steve Pinker and Paul Bloom, "It makes a big difference whether a far-off region is reached by taking the trail that is in front of the large tree or the trail that the large tree is in front of. It makes a difference whether that region has animals that you can eat or animals that can eat you." Recursion could easily have helped a Pleistocene man survive or breed. Language, conclude Pinker and Bloom, "is a design imposed on neural circuitry as a response to evolutionary pressure."[28] It is not the whirring by-product of the mental machine.

The neoteny argument does have one advantage: It shows a possible reason why apes and baboons did not follow man down the path to ever bigger brains. It is possible that the neoteny mutation simply never arose in our primate cousins. Or, more intriguingly, as I shall explain later, the mutation may have arisen but never had a reason to spread.

GOSSIP'S GRIP

Those outside anthropology had never paid much obeisance to man the toolmaker or any other explanation for intelligence. For most people, the advantages of intelligence were obvious. It led to more learning and less instinct, which meant that behavior could be more flexible, which was rewarded by evolution. We have already seen how shot full of holes this argument is. Learning is a burden on the individual, in place of flexible instincts, and the two are not opposites in any case. Mankind is not the learning ape, he is the clever ape with more instincts and more open to experience. Not having seen this flaw in the logic, the disciplines that considered such matters, especially philosophy, always showed a strange lack of curiosity in the whole question of intelligence. Philosophers assume that intelligence and consciousness have obvious advantages and get on with the serious debate about what consciousness is. Before the 1970s there was very little evidence that any of them had even posed the obvious evolutionary question: Why is intelligence a good thing?

So the force with which the question was suddenly put in 1975 by two zoologists working independently had an enormous impact. Richard Alexander of the University of Michigan was one. In the tradition of the Red Queen, he expressed skepticism about whether what Charles Darwin had called "the hostile forces of nature" were a sufficiently challenging adversary for an intelligent mind. The point is that the challenges presented by stone tools or tubers are mostly predictable ones. Generation after generation of chipping a tool off a block of stone or knowing where to look for tubers calls for the same level of skill each time. With experience each gets easier. It is rather like learning to ride a bicycle; once you know how to do it, it comes naturally. Indeed, it becomes "unconscious," as if conscious effort were simply not needed every time. Likewise, *Homo erectus* did not need consciousness to know that you should stalk zebras upwind every time lest they scent you or that tubers grow beneath certain trees. It came as naturally to him as riding a bike does to us. Imagine playing chess against a computer that has only one opening gambit. It might be a good opening gambit, but once you know how to beat it, you can play the same response yourself, game after game. Of course, the whole point of chess is that your opponent can select one of many different ways to respond to each move you make.

It was logic like this that led Alexander to propose that the key feature of the human environment that rewarded intelligence was the presence of other human beings. Generation after generation, if your lineage is getting more intelligent, so is theirs. However fast you run, you stay in the same place relative to them. Humans became ecologically dominant by virtue of their technical skills, and that made humans the only enemy of humans (apart from parasites). "Only humans themselves could provide the necessary challenge to explain their own evolution," wrote Alexander.[29]

True enough, but Scottish midges and African elephants are "ecologically dominant" in the sense that they outnumber or outrank all potential enemies, yet neither has seen the need to develop the ability to understand the theory of relativity. In any case, where is the evidence that Lucy was ecologically dominant? By all accounts her species was an insignificant part of the fauna of the dry, wooded savanna where she lived.[30]

Independently, Nicholas Humphrey, a young Cambridge zoologist, came to a conclusion similar to Alexander's. Humphrey began an essay on the topic with the story of how Henry Ford once asked his representatives to find out which parts of the Model-T never went wrong. They came back with the answer that the kingpin had never gone wrong; so Ford ordered it made to an inferior specification to save money. "Nature," wrote Humphrey, "is surely at least as careful an economist as Henry Ford."[31]

Intelligence must therefore have a purpose; it cannot be an expensive luxury. Defining intelligence as the ability to "modify behavior on the basis of valid inference from evidence," Humphrey argued that the use of intelligence for practical invention was an easily demolished straw man. "Paradoxically, subsistence technology, rather than requiring intelligence, may actually become a substitute for it." The gorilla, Humphrey noted, is intelligent as animals go, yet it leads the most technically undemanding life imaginable. It eats the leaves that grow abundantly all around it. But the gorilla's life is dominated by social problems. The vast majority of its intellectual effort is expended on dominating, submitting to, reading the mood of, and affecting the lives of other gorillas.

Likewise, Robinson Crusoe's life on the desert island was technically fairly straightforward, says Humphrey. "It was the arrival of Man Friday on the scene that really made things difficult for Crusoe." Humphrey suggested that mankind uses his intellect mainly in social situations. "The game of social plot and counterplot cannot be played merely on the basis of accumulated knowledge, any more than a game of chess can." A person must calculate the consequences of his own behavior and calculate the likely behavior of others. For that he needs at least a glimpse of his own motives in order to guess the things that are going through others' minds in similar situations, and it was this need for self-knowledge that drove the increase in conscious awareness.[32]

As Horace Barlow of Cambridge University has pointed out, the things of which we are conscious are mostly the mental events that concern social actions: We remain unconscious of how we see, walk, hit a tennis ball, or write a word. Like a military hierarchy, consciousness operates on a "need to know" policy. "I can think of no exception to the rule that one is conscious of what it is possible to report to others and not conscious of what it is not possible to report."[33] John Crook, a psychologist with a special interest in Eastern philosophy, has made much the same point: "Attention therefore moves cognition into awareness, where it becomes subject to verbal formulation and reporting to others."[34]

What Humphrey and Alexander described was essentially a Red Queen chess game. The faster mankind ran—the more intelligent he became—the more he stayed in the same place because the people over whom he sought psychological dominion were his own relatives, the descendants of the more intelligent people from previous generations. As Pinker and Bloom put it, "Interacting with an organism of approximately equal mental abilities whose motives are at times outright [sic] malevolent makes formidable and ever-escalating demands on cognition."[35] If Tooby and Cosmides are right about mental modules, among the modules that were selected to increase in size by this intellectual chess tournament was the "theory of mind" module, the one that enables us to form an opinion about one another's thoughts, together with the means to express our own thoughts through the language modules.[36] There is plenty of good evidence for this idea when you look about you. Gossip is one of the most universal of human habits. No conversation between people who know each other well—fellow employees, fellow family members, old friends—ever lingers for long on any topic other than the behavior, ambitions, motives, frailties, and affairs of other absent—or present—members of the group. That is the reason the soap opera is the quintessentially effective way to entertain people.[37] Nor is this a Western habit. Konner wrote of his experience with !Kung San tribesmen:

> After two years with the San, I came to think of the Pleistocene epoch of human history (the 3 million years during which we evolved) as one interminable marathon encounter group. When we slept in a grass hut in one of their villages, there were many nights when its flimsy walls leaked charged exchanges from the circle around the fire, frank expressions of feeling and contention beginning when the dusk fires were lit and running on until dawn.[38]

Virtually all novels and plays are about the same subject, even when disguised as history or adventure. If you want to understand human motives, read Proust or Trollope or Tom Wolfe, not Freud or Piaget or Skinner. We are obsessed with one another's minds. "Our intuitive commonsense psychology far surpasses any scientific psychology in scope and accuracy," wrote Don Symons.[39] Horace Barlow points out that great literary minds are, almost by definition, great mind-reading minds. Shakespeare was a far better psychologist than Freud, and Jane Austen a far better sociologist than Durkheim. We are clever because we are—and to the extent that we are—natural psychologists.[40]

Indeed, novelists themselves saw this first. In George Eliot's *Felix Holt, the Radical*, she gives a concise summary of the Alexander-Humphrey theory:

> Fancy what a game of chess would be if all the chessmen had passions and intellects, more or less small and cunning; if you were not only uncertain about your adversary's men, but a little uncertain also about your own. . . . You would be especially likely to be beaten, if you depended arrogantly on your mathematical imagination, and regarded your passionate pieces with contempt. Yet this imaginary chess is easy compared with a game a man has to play against his fellowmen with other fellowmen for instruments.

The Alexander-Humphrey theory, which is widely known as the Machiavellian hypothesis,[41] sounds rather obvious, but it could never have been proposed in the 1960s before the "selfish" revolution in the study of behavior or by anybody steeped in the ways of social science, for it requires a cynical view of animal communication. Until the mid 1970s zoologists thought of communication in terms of information transfer: It was in the interests of both the communicator and the recipient that the message be clear, honest, and informative. But as Lord Macaulay put it,[42] "The object of oratory alone is not truth but persuasion." In 1978, Richard Dawkins and John Krebs pointed out that animals use communication principally to manipulate one another rather than to transfer information. A bird sings long and eloquently to persuade a female to mate with him or a rival to keep clear of his territory. If he were merely passing on information, he need not make the song so elaborate. Animal communication, said Dawkins and Krebs, is more like human advertising than like airline timetables. Even the most mutually beneficial communication, like that between a mother and a baby, is pure manipulation, as every mother who has been woken in the night by a desperate-sounding infant who merely wants company knows. Once scientists had begun thinking in this way, they looked at animal social life in an entirely new light.[43]

One of the most striking pieces of evidence for deception's role in communication comes from experiments that Leda Cosmides did when at Stanford University and that Gerd Gigerenzer and his colleagues did at Salzburg University. There is a simple logical puzzle called the Wason test, which people are bafflingly bad at. It consists of four cards placed on the table. Each card has a letter on one side and a number on the other. At present the cards read as follows: D, F, 3, 7. Your task is to turn over only those cards that you need to in order to prove the following rule to be true or false: *If a card has a D on one side, then it has a 3 on the other.*

When presented with this test, less than one-quarter of Stanford students got it right, an average performance. (The right answer, by the way, is D and 7.) But it has been known for years that people are much better at the Wason test if it is presented differently. For example, the problem can be set as follows: "You are a bouncer in a Boston bar, and you will lose your job unless you enforce the following law: If a person is drinking beer, then he must be over twenty years old." The cards now read: "drinking beer, drinking Coke, twenty-five years old, sixteen years old." Now three-quarters of the students get the right answer: Turn over the cards marked "drinking beer" and "sixteen years old." But the problem is logically identical to the first one. Perhaps the more familiar context of the Boston bar is what helps people do better, but other equally familiar examples elicit poor performance. The secret of why some Wason tests are easier than others has proved to be one of psychology's enduring enigmas.

Cosmides and Gigerenzer have solved the enigma. If the law to be enforced is not a social contract, the problem is difficult—however simple its logic; but if it is a social contract, like the beer-drinking example, then it is easy. In one of Gigerenzer's experiments, people were good at enforcing the rule "If you take a pension, then you must have worked here ten years" by wanting to know what was on the back of the cards "worked here eight years" and "got a pension"—so long as they were told they were the employer. But if told they were an employee and still set the same rule, they turned over the cards "worked here for twelve years" and "did not get a pension," as if looking for cheating employers—even though the logic clearly implies that cheating employers are not infringing the rule.

Through a long series of experiments Cosmides and Gigerenzer proved that people are simply not treating the puzzles as pieces of logic at all. They are treating them as social contracts and looking for cheats. The human mind may not be much suited to logic at all, they conclude, but is well suited to judging the fairness of social bargains and the sincerity of social offers. It is a mistrustful Machiavellian world.[44]

Richard Byrne and Andrew Whiten of the University of St. Andrews studied baboons in East Africa and witnessed an incident in which Paul, a young baboon, saw an adult female, Mel, find a large root. He looked around and then gave a sharp cry. The call summoned the baboon's mother, who "assumed" that Mel had just stolen the food from her young or threatened him in some way, and chased Mel away. Paul ate the root. This piece of social manipulation by the young baboon required some intelligence: a knowledge that its call would bring its mother, a guess at what the mother would "assume" had happened, and a prediction that it would lead to Paul's getting the food. It was also using intelligence to deceive. Byrne and Whiten went on to suggest that the habit of calculated deception is common in humans, occasional in chimpanzees, rare in baboons, and virtually unknown in other animals. Deceiving and detecting deception would then be the primary reason for intelligence. They suggest that the great apes acquired a unique ability to imagine alternative possible worlds as a means to deception.[45]

Robert Trivers has argued that to deceive others well, an

animal must deceive itself, and that self-deception's hallmark is a biased system of transfer from the conscious to the unconscious mind. Deception is therefore the reason for the invention of the *sub*conscious.[46]

Yet Byrne's and Whiten's account of the baboon incident goes right to the heart of what is wrong with the Machiavellian theory. It applies to every social species. For example, if you read any stories of life in a chimpanzee troop, the "plot" has a painful predictability about it to human ears. In Jane Goodall's account of the career of the successful male Goblin, we watch Goblin's precocious and confident rise in the hierarchy as he challenges and defeats first each of the females in the troop and then, one by one, the males: Humphrey, Jomeo, Sherry, Satan, and Evered:

> Only Figan [the alpha male] was exempt. Indeed, it was his relationship with Figan that enabled him to challenge these older and more experienced males: He almost never did so unless Figan was nearby.
>
> [To the human reader what comes next is startlingly obvious.]
>
> For some time we had been expecting Goblin to turn on Figan. Indeed, I am still puzzled as to why Figan, so socially adroit in all other ways, had not been able to predict the inevitable outcome of his sponsorship of Goblin.[47]

The plot has a few twists, but we are not surprised; Figan is soon toppled. Machiavelli at least warned his Prince to watch his back. Brutus and Cassius took great care to conceal their plot from Julius Caesar; they could never have pulled off the assassination if their open ambition had been so obvious. Not even the most power-blinded human dictator is taken by surprise as Figan was. Of course that only proves that people are cleverer than chimpanzees, which is no great surprise, but it starkly poses the question *why?* If Figan had had a bigger brain, he might have seen what was coming. So the evolutionary pressure that Nick Humphrey identified—to get better and better at solving social puzzles, reading minds, and predicting reactions—is all there in the chimps and baboons, too. As Geoffrey Miller, a psychologist at the University of Stanford, has put it, "All apes and monkeys show complex behavior replete with communication, manipulation, deception, and long-term relationships; selection for Machiavellian intelligence based on such social complexities should again predict much larger brains in other apes and monkeys than we observe."[48]

There have been several answers to this puzzle, none of which is entirely convincing. The first is Humphrey's own answer, which is that human society is more complex than ape society because it needs a "polytechnic school" in which young people can learn the practical skills of their species. This seems to me merely a retreat to the toolmaker theory. The second is the suggestion that alliance building among unrelated individuals is a key to success in human beings and that this complication vastly increases the rewards of intellect. To which comes the response: What about dolphins? There is growing evidence that dolphin society is based on shifting alliances of males and of females so that, for example, Richard Connor observed a pair of males that came across a small group of other males that had kidnapped a fertile female from her group. Instead of fighting them for the female, the pair went away and found some allies, came back, and with superior numbers stole the female from the first group.[49] Even in chimps the rise of a male to the alpha position and his tenure there is determined by his ability to command the loyalty of allies.[50] So the alliance theory once more seems too general to explain the sudden increase in human intelligence. Moreover, like most of these theories, it explains language, tactical thinking, social exchange, and the like, but it does not explain some of the things to which human beings devote much of their mental energy: music and humor, for example.

WITTINESS AND SEXINESS

At least the Machiavelli theory proposes an adversary for the human brain that is its equal, however clever it gets. Few of my readers will need reminding of the ruthlessness that human beings can show when in pursuit of self-interest. There is no such thing as being clever enough just as there is no such thing as being good enough at chess. Either you win or you do not. If winning pits you against a better opponent, as it does in the evolutionary tournament generation after generation, then the pressure to get better and better never lets up. The way the brains of human beings have gotten bigger at an accelerating pace implies that some such within-species arms race is at work.

So argues Geoffrey Miller. After laying bare the inadequacies of the conventional theories about intelligence, he takes a surprising turn.

> I suggest that the neocortex is not primarily or exclusively a device for toolmaking, bipedal walking, fire-using, warfare, hunting, gathering, or avoiding savanna predators. None of these postulated functions alone can explain its explosive development in our lineage and not in other closely related species. . . . The neocortex is largely a courtship device to attract and retain sexual mates: Its specific evolutionary function is to stimulate and entertain other people, and to assess the stimulation attempts of others.[51]

The only way, he suggests, that sufficient evolutionary pressure could suddenly and capriciously be sustained in one species to enlarge an organ far beyond its normal size is sexual selection. "Just as the peahen is satisfied with nothing less than a visually brilliant display of peacock plumage, I postulate that hominid males and females became satisfied with nothing less than psychologically brilliant, fascinating, articulate, entertaining companions." Miller's use of the peacock is deliberate. Wherever else in the animal kingdom we find greatly exaggerated and enlarged ornaments, we have been able to explain them by the runaway, sexy-son, Fisher effect of intense sexual selection (or the equally powerful Good-genes effect, as described in chapter 5). Sexual selection, as we have seen, is very different from natural selection in its effects, for it does not solve survival problems, it makes them worse. Female choice causes peacocks' tails to grow longer until they become a burden—then demands that they grow longer still. Miller used the wrong word: Peahens are never satisfied. And so, having found a force that produces exponential change in ornaments, it seems perverse not to consider it when trying to explain the exponential expansion of the brain.

Miller adduces some circumstantial evidence for his view. Surveys consistently place intelligence, sense of humor, creativity, and interesting personality above even such things as wealth and beauty in lists of desirable characteristics in both sexes.[52] Yet these characteristics fail entirely to predict youth, status, fertility, or parental ability, so evolutionists tend to ignore them—but there they are, right at the top of the list. Just as a peacock's tail is no guide to his ability as a father but despotic fashion punishes those who cease to respect it, so Miller suggests that men and women dare not step off the treadmill of selecting the wittiest, most creative and articulate person available with whom to mate. (Note that conventional "intelligence" as measured by examinations is not what he is talking about.)

Likewise, the manner in which sexual selection capriciously seizes upon preexisting perceptual biases fits with the fact that apes are by nature naturally "curious, playful, easily bored, and appreciative of simulation." Miller suggests that to keep a husband around long enough to help in raising children, women would have needed to be as varied and creative in their behavior as possible, which he calls the Scheherazade effect after the Arabian storyteller who entranced the Sultan with 1,001 tales so that he did not abandon her (and execute her) for another courtesan. The same would have applied to males who wanted to attract females, which Miller calls the Dionysus effect after the Greek god of dance, music, intoxication, and seduction. He might also have called it the Mick Jagger effect; he admitted to me one day that he could not understand what made strutting, middle-aged rock stars so attractive to women. In this respect Don Symons noted that tribal chiefs are both gifted orators and highly polygamous men.[53]

Miller notes that the bigger the brain became, the more necessary long-term pair bonds were. A human infant is born helpless and premature. If it were as advanced at birth as an ape, it would be twenty-one months in the womb.[54] But the human pelvis is simply incapable of bearing a child with a head that big, so it is born at nine months and treated like a helpless, external fetus for the next year, not even beginning to walk until it is at the age when it would expect to enter the world. This helplessness further enhances the pressure on women to keep men around to help feed them when encumbered with a child—the Scheherazade effect.

Miller finds that the most commonly voiced objection to the Scheherazade effect is that most people are not witty and creative but are dull and predictable. True enough, but compared to what? Our standards for what is considered entertaining have, if Miller is right, evolved as fast as our wit. "I think male readers may find it hard to imagine some four-foot-tall, half-hairy, flat-chested, hominid females being sexier than similar hominids," wrote Miller in a letter to me (referring to "Lucy"). "We're spoiled because sexual selection has already driven us so far that it's hard to appreciate how any point we've passed could have been considered an improvement. We are positively turned off by traits that half a million years ago would have been considered irresistibly sexy."[55]

Miller's theory draws attention to several facts that have remained unexplained in other theories, namely that dance, music, humor, and sexual foreplay are all features unique to human beings. Following the Tooby-Cosmides logic, we cannot argue that these are mere cultural habits foisted on us by "society." Plainly a desire to hear rhythmic tunes or to be made to laugh by wit develops innately. Following Miller we note that they are characterized by obsessions with novelty and virtuosity and much practiced by the young. From Beatlemania to Madonna (and back again to Orpheus), the sexual fascination of youth with musical creativity has been obvious. It is a human universal.

It is crucial for Miller's theory that human beings are especially selective about their mates. Indeed, among apes, people are unique in that both sexes are extremely choosy. A gorilla female is happy to be mated with whoever "owns" her harem. A gorilla male will mate with any estral female he can find. A chimp female is keen to mate with many different males in the troop. A chimp male will mate with any female in season. But women are highly selective about the men with whom they mate. So indeed are men. True, they are easily persuaded to go to bed with beautiful young women—but that is exactly the point. Most women are neither young nor beautiful, nor are they trying to seduce strange men. It is hard to overemphasize how unusual humans are in this respect. Males in some monogamous bird species such as pigeons and doves[56] do take care to select a female carefully, but in many other birds, the males are happy to have a fling with any passing female, as the evidence of sperm competition theory has demonstrated (chapter 7). Although he may prefer variety more than females do, man is a highly sexually selective male as males go.

Selectivity by one or the other sex is the prerequisite of sexual selection. And as I have argued in previous chapters, it is more than that. It is the almost invariant predictor of sexual selection. Fisher's runaway process for sexy sons and Zahavi-Hamilton's Good-genes effect simply cannot be avoided once one or the other sex is being selective. So we should actually expect some exaggeration of some feature or other in man as a simple consequence of sexual selection.[57]

Incidentally, Miller's argument draws attention to a little-appreciated aspect of sexual selection: It can affect both the selected sex and the selector. For example, among American blackbirds those species in which the female is large are also the species in which the male is much larger. The same is true of many mammals and birds. Among grouse, pheasants, seals, and deer, a greater ratio between male and female size occurs in the larger species. A recent analysis of this effect concludes that it is caused by sexual selection: The more polygamous the species, the more premium there is on large size in males; the more males are selected for large size, the more they inevitably leave large-size genes to their daughters as well as their sons. Genes can be "sex-linked" but usually only imperfectly or when there is a strong disadvantage to a daughter's inheriting the effect—as in the case of female birds and gaudy colors. Thus, sexual selection by males of females for large brains would result in larger brains for both sexes.[58]

OBSESSED WITH YOUTH

I believe that Miller's tale deserves a special twist from the neoteny theory (although he is not convinced). The neoteny theory is well established among anthropologists.. And the notion of human monogamous child rearing is well established among sociobiologists. Nobody has yet put the two together. If men began selecting mates that appeared youthful, then any gene that slowed the rate of development of adult characteristics in a woman would make her more attractive at a given age than a rival. Consequently, she would leave more descendants, who would inherit the same gene. Any neoteny gene would give the appearance of youthfulness. Neoteny, in other words, could be a consequence of sexual selection, and since neoteny is credited with increasing our intelligence (by enlarging the brain size at adulthood), it is to sexual selection that we should attribute our great intelligence.

The idea is hard to grasp at first, so a thought experiment may help. Imagine two primeval women: One develops at the normal rate, and the other has an extra neoteny gene so that she is hairless of body, large-brained, small-jawed, late maturing, and long-lived. At the age of twenty-five, both are widowed; each has had one child by her first husband. The men in the tribe have a preference for young women and twenty-five is not young, so neither stands much chance of getting a second husband. But there is one man who cannot find a wife. Given the alternatives, he chooses the younger-looking woman. She goes on to have three more children while her rival barely manages to rear the one she already had.

The details of the story do not matter. The point is that once males prefer youth, a gene for delaying the signs of aging would generally prosper at the expense of a normal gene, and a neoteny gene does exactly that. The gene would probably make the woman's sons appear neotenized as well as her daughters, for there is no reason that it should be specific to the female sex in its effects. The whole species would be driven into neoteny.

Christopher Badcock, a sociologist at the London School of Economics who unusually combines an interest in evolution and an interest in Freud, has proposed a similar idea. He suggested that neotenic (or, as he calls it, "paedomorphic") traits were favored by female choice rather than male choice. Younger males, he suggests, made more cooperative hunters, and therefore females who wanted meat picked younger-looking men. The principle is the same: Neotenic development is a consequence of a preference for it in one sex.[59]

This is not to deny that bigger brains themselves brought advantages in Machiavellian intelligence or language or seductiveness. Indeed, once these advantages became clear, men who were especially fussy about picking youthful-looking women would be most successful because they sometimes picked neotenic, big-brained women and therefore had more intelligent children. But it does suggest an escape from the question Why did it not happen to baboons?

However, Miller's sexual selection idea suffers from a near fatal flaw. Remember that it presupposes sexual choosiness by one or other sex. But what caused that choosiness? Presumably the cause was the fact that men took part in parental care, which gave women an incentive to confine probable paternity to one man and gave men an incentive to enter into a long-term relationship as long as he could be certain of paternity. Why then did men take part in parental care? Because by doing so they could increase the chances of rearing a child more than by trying to seek new partners. The reason for this was that children, unusual for ape infants, took a long time to mature, and men could help their wives during child rearing by hunting meat for them. Why did they take a long time to mature? Because they had big heads! The argument is circular.

That may not be fatal to it. Some of the best arguments, such as Fisher's theory of runaway sexual selection, are circular. The relationship between chickens and eggs is circular. Miller is actually rather proud of the theory's circularity because he believes we have learned from computer simulation that evolution is a process which pulls itself up by its bootstraps. There is no single cause and effect because effects can reinforce causes. If a bird finds itself to be good at cracking seeds, then it specializes in cracking seeds, which puts further pressure on its seed-cracking ability to evolve. Evolution is circular.

STALE MATE

It is a disquieting thought that our heads contain a neurological version of a peacock's tail—an ornament designed for sexual display whose virtuosity at everything from calculus to sculpture is perhaps just a side effect of the ability to charm. Disquieting and yet not altogether convincing. The sexual selection of the human mind is the most speculative and fragile of the many evolutionary theories discussed in this book, but it is also very much in the same vein as the others. I began this book by asking why all human beings were so similar and yet so different, suggesting that the answer lay in the unique alchemy of sex. An individual is unique because of the genetic variety that sexual reproduction generates in its perpetual chess tournament with disease. An individual is a member of a homogeneous species because of the incessant mixing of that variety in the pool of fellow human beings' genes. And I end with one of the strangest of the consequences of sex: that the choosiness of human beings in picking their mates has driven the human mind into a history of frenzied expansion for no reason except that wit, virtuosity, inventiveness, and individuality turn other people on. It is a somewhat less uplifting perspective on the purpose of humanity than the religious one, but it is also rather liberating. Be different.

THE
SELF-DOMESTICATED APE

Know then thyself, presume not God to scan,

The proper study of mankind is man.

Plac'd on this isthmus of a middle state,

A being darkly wise, and rudely great:

With too much knowledge for the sceptic side,

With too much weakness for the stoic's pride;

He hangs between; in doubt to act or rest,

In doubt to deem himself a god, or beast;

In doubt his mind or body to prefer;

Born but to die, and reas'ning but to err;

Alike in ignorance, his reason such,

Whether he thinks too little or too much.

—Alexander Pope, "An Essay on Man"

The study of human nature is at about the same stage as the study of the human genome, which is at about the same stage as the mapping of the world in the time of Herodotus. We know a few fragments in detail and some large parts in outline, but huge surprises still await us and errors abound. If we can free ourselves from the sterile dogmatic dispute about nature and nurture, we can gradually uncover the rest.

But just as Mercator could not get the relative sizes of Europe and Africa correct until he had the perspective that longitude and latitude provided, so the perspective of other animals is vital to the study of human nature. It is impossible to understand the social life of a phalarope or a sage grouse or an elephant seal or a chimpanzee in isolation. You can describe each in glorious detail, of course: They are respectively polyandrous, lekking, harem-defending, fission-fusion. But only with the perspective of evolution can you truly understand why. Only then can you see the part that different opportunities for parental investment, different habitats, different diets, and different historical baggage have played in determining their natures. It is the purest nonsense to abandon the perspective of comparisons with other animals just because of our hubristic belief that humans alone are learning creatures that reinvent themselves at whim. So I make no apologies for mixing animals and human beings together in this book.

Nor is the fact of civilization sufficient to rescue our parochial egotism. We are, it is true, as domesticated as any dog or cow, perhaps more so. We have bred out of ouselves all sorts of instincts that were probably features of our Pleistocene nature, in the same way that human beings have bred out of the cow many of the characteristics of the Pleistocene aurochs. But scratch a cow and you still find an aurochs underneath. A herd of dairy calves released into a forest would soon reinvent the polygamous herd in which males competed for status. Dogs left to their own devices still become territorial pack animals in which the senior animals monopolize breeding. Turned loose on an African savanna, a group of young Americans would not re-create the exact existence of their ancestors; indeed, they would probably starve, so dependent have we been for millennia on cultural traditions of where to find food and how to live. But nor would such people invent an entirely inhuman social arrangement. As every experiment in free-wheeling communities up to and including Rajneeshpuram in Oregon has proved, human communities always invent a hierarchy and always atomize into possessive sexual bonds.

Mankind is a self-domesticated animal; a mammal; an ape; a social ape; an ape in which the male takes the iniative in courtship and females usually leave the society of their birth; an ape in which men are predators, women herbivorous foragers; an ape in which males are relatively hierarchical, females relatively egalitarian; an ape in which males contribute unusually large amounts of investment in the upbringing of their offspring by provisioning their mates and their children with food, protection, and company; an ape in which monogamous pair bonds are the rule but many males have affairs and occasional males achieve polygamy; an ape in which females mated to low-ranking males often cuckold their husbands in order to gain access to the genes of higher-ranking males; an ape that has been subject to unusually intense mutual sexual selection so that many of the features of the female body (lips, breasts, waists) and the mind of both sexes (songs, competitive ambition, status seeking) are designed for use in competition for mates; an ape that has developed an extraordinary range of new instincts to learn by association, to communicate by speech, and to pass on traditions. But still an ape.

Half the ideas in this book are probably wrong. The history

of human science is not encouraging. Galton's eugenics, Freud's unconscious, Durkheim's sociology, Mead's culture-driven anthropology, Skinner's behaviorism, Piaget's early learning, and Wilson's sociobiology all appear in retrospect to be riddled with errors and false perspectives. No doubt the Red Queen's approach is just another chapter in this marred tale. No doubt its politicization and the vested interests ranged against it will do as much damage as was done to previous attempts to understand human nature. The Western cultural revolution that calls itself political correctness will no doubt stifle inquiries it does not like, such as those into the mental differences between men and women. I sometimes feel that we are fated never to understand ourselves because part of our nature is to turn every inquiry into an expression of our own nature: ambitious, illogical, manipulative, and religious. "Never literary attempt was more unfortunate than my Treatise of Human Nature. It fell dead-born from the Press," said David Hume.

But then I remember how much progress we have made since Hume and how much nearer to the goal of a complete understanding of human nature we are than ever before. We will never quite reach that goal, and it would perhaps be better if we never did. But as long as we can keep asking *why*, we have a noble purpose.

NOTES

Chapter 1. Human Nature

1. Dawkins 1991.
2. Weismann 1889.
3. Weismann 1889.
4. A few scientists argue that Chinese people are indeed descended from "Peking man," the local version of *Homo erectus*, but the evidence is now heavily against them.
5. Karl Marx, in *Criticism of the Gotha Program* (1875), was paraphrasing Michael Bakunin, who declared, when on trial after the failure of an anarchist rising at Lyons (1870): "From each according to his faculties, to each according to his needs."
6. Not all anthropologists would agree that all modern people are descendants of a race that was confined to Africa until 100,000 years ago, but most do.
7. Tooby and Cosmides 1990.
8. Mayr 1983; Dawkins 1986.
9. Hunter, Nur, and Werren 1993.
10. Dawkins 1991.
11. Dawkins 1986.
12. Tiger 1991.
13. See Edward Tenner's article "Revenge Theory" in *Harvard Magazine*, March–April 1991, for why this is so.
14. Wilson 1975.

Chapter 2. The Enigma

1. Bell 1982.
2. Weismann 1889.
3. Brooks 1988.

4. J. Maynard Smith, interview.

5. Levin 1988.

6. Weismann 1889.

7. Bell 1982.

8. Fisher 1930.

9. Muller 1932.

10. Crow and Kimura 1965.

11. Wynne Edwards 1962.

12. Darwin 1859.

13. Humphrey 1983.

14. Williams 1966.

15. Fisher 1930; Wright 1931; Haldane 1932.

16. Huxley 1942.

17. Hamilton 1964; Trivers 1971.

18. Ghiselin 1974, 1988.

19. Maynard Smith 1971.

20. Stebbins 1950; Maynard Smith 1978.

21. Jaenike 1978.

22. Gould and Lewontin 1979.

23. Williams 1975; Maynard Smith 1978.

24. Maynard Smith 1971.

25. Ghiselin 1988.

26. Bernstein, Hopf, and Michod 1988.

27. Bernstein 1983; Bernstein, Byerly, Hopf, and Michod 1985.

28. Maynard Smith 1988.

29. Tiersch, Beck, and Douglas 1991.

30. Bull and Charnov 1985; Bierzychudek 1987b; Kondrashov and Crow 1991; Perrod, Richerd, and Valero 1991.

31. Bernstein, Hopf, and Michod 1988.

32. Kondrashov 1988.

33. Flegg, Spencer, and Wood 1985.

34. Stearns 1978; Michod and Levin 1988.

35. Kirkpatrick and Jenkins 1989; Wiener, Feldman, and Otto 1992.

36. Muller 1964.

37. Bell 1988.

38. Muller's ratchet has recently been found at work in viruses; see Chao 1992; Chao, Tran, and Matthews 1992.

39. Crow 1988.

40. Kondrashov 1982.

41. M. Meselson, interview.

42. Kondrashov 1988.

43. Hamilton 1990.

44. C. Lively, interview.

Chapter 3. The Power of Parasites

1. Hurst, Hamilton, and Ladle 1992.

2. M. Meselson, interview.

3. Maynard Smith 1986.

4. Williams 1966, Williams 1975.

5. Maynard Smith 1971.

6. Williams and Mitton 1973.

7. Williams 1975.

8. Bell 1982.

9. Bell 1982.

10. Ghiselin 1974.

11. Darwin 1859.

12. Bell 1982.

13. Schmitt and Antonovics 1986; Ladle 1992.

14. Williams 1966.

15. Bierzychudek 1987.

16. Harvey 1978.

17. Burt and Bell 1987.

18. Eldredge and Gould 1972.

19. Williams 1975.

20. Carroll 1871.

21. Van Valen 1973; L. Van Valen, interview.

22. Zinsser 1934; McNeill 1976.

23. *Washington Post*, December 16, 1991.

24. Krause 1992.

25. Dawkins 1990.

26. Assuming thirty minutes per bacterial generation, there are 1,226,400 bacterial generations in a human lifetime of seventy years. In the 7 million years since we shared an ancestor with chimpanzees, there have been just over 200,000 "human" generations of thirty years each.

27. O'Connell 1989.

28. Dawkins and Krebs 1979.

29. Schall 1990; May and Anderson 1990.

30. Levy 1992.

31. Ray 1992.

32. Ray 1992; T. Ray, interview.

33. L. Hurst, interview.
34. Burt and Bell 1987.
35. Bell and Burt 1990.
36. Kelley 1985; Schmitt and Antonovics 1986; Bierzychudek 1987a.
37. Haldane 1949; Hamilton 1990.
38. Hamilton, Axelrod, and Tanese 1990; W. D. Hamilton, interview.
39. Haldane 1949; Clarke 1979.
40. Clay 1991.
41. Bremermann 1987.
42. Nowak 1992; Nowak and May 1992.
43. Hill, Allsopp, Kwiatkowski, Anstey, Twumasi, Rowe, Bennett, Brewster, McMichael, and Greenwood 1991.
44. Potts, Manning, and Wakeland 1991.
45. Haldane 1949.
46. Jayakar 1970; Hamilton 1990.
47. Jaenike 1978; Bell 1982; Bremermann 1980; Tooby 1982; Hamilton 1980.
48. Hamilton 1964; Hamilton 1967; Hamilton 1971.
49. Hamilton, Axelrod, and Tanese 1990.
50. Hamilton, Axelrod, and Tanese 1990.
51. W. D. Hamilton, interviews.
52. W. D. Hamilton, interview; A. Pomiankowski, interview.
53. Glesner and Tilman 1978; Bierzychudek 1987.
54. Daly and Wilson 1983.
55. Edmunds and Alstad 1978, 1981; Seger and Hamilton 1988.
56. Harvey 1978.
57. Gould 1978.
58. C. Lively, interview.
59. Lively 1987.
60. C. Lively, interview.
61. Lively, Craddock, and Vrijenhoek 1990.
62. Tooby 1982.
63. Bell 1987.
64. Hamilton 1990.
65. Hamilton 1990.
66. Bell and Maynard Smith 1987.
67. W. D. Hamilton, interview.
68. M. Meselson, interview.
69. R. Ladle, interview.
70. G. Bell, interview; A. Burt, interview; Felsentein 1988; W. Hamilton, interview; J. Maynard Smith, interview; G. Williams, interview.
71. Metzenberg 1990.

Chapter 4. Genetic Mutiny and Gender

1. Hardin 1968.
2. I make no apology for using the word *gender* when I mean sex (male or female); I know it is a word that originally referred only to grammatical categories, but meanings change and it is usefully unambiguous to have a word other than *sex* for males and females.
3. Cosmides and Tooby 1981.
4. Leigh 1990.
5. See Dawkins 1976, 1982, for the clearest exposition of this case.
6. Hickey 1982; Hickey and Rose 1988.
7. Doolittle and Sapienza 1980; Orgel and Crick 1980.
8. Dawkins 1986.
9. Nee and Maynard Smith 1990.
10. Mereschkovsky 1905; Margulis 1981; Margulis and Sagan 1986.
11. Beeman, Friesen, and Denell 1992.
12. Hewitt 1972; Hewitt 1976; Hewitt and East 1978; Shaw, Hewitt, and Anderson 1985; Bell and Burt 1990; Jones 1991.
13. D. Haig, interview.
14. Haig and Grafen 1991.
15. Charlesworth and Hartl 1978.
16. For a comprehensive review of meiotic drive see *American Naturalist*, vol. 137, pp. 281–456, "The Genetics and Evolutionary Biology of Meiotic Drive," a symposium organized by T. W. Lyttle, L. M. Sandler, T. Prout, and D. D. Perkins, 1991.
17. Haig and Grafen 1991.
18. D. Haig, interview; see also S. Spandrel (unpublished).
19. Hamilton 1967; Dawkins 1982; Bull 1983; Hurst 1992a; L. Hurst, interview.
20. Leigh 1977.
21. Cosmides and Tooby 1981.
22. Margulis 1981.
23. Cosmides and Tooby 1981; Hurst and Hamilton 1992.
24. Anderson 1992; Hurst 1991b; Hurst 1992b.
25. Werren, Skinner, and Huger 1986; Werren 1987; Hurst 1990; Hurst 1991c.
26. Mitchison 1990.
27. L. Hurst, interview; see also Parker, Baker, and Smith 1972 and Hoekstra 1987 for additional, but not rival, features of the evolution of anisogamy and two genders.
28. Frank 1989.
29. Gouyon and Couvet 1987; Frank 1989; Frank 1991; Hurst and Pomiankowski 1991.

30. Hurst 1991a.

31. Hurst and Hamilton 1992.

32. Hurst, Godfray, and Harvey 1990.

33. Hurst, Godfray, and Harvey 1990.

34. Olsen and Marsden 1954; Olsen 1956; Olsen and Buss 1967.

35. Lienhart and Vermelin 1946.

36. Hamilton 1967.

37. Cosmides and Tooby 1981.

38. Bull and Bulmer 1981; Frank 1990.

39. Bull and Bulmer 1981; J. J. Bull, interview.

40. Frank and Swingland 1988; Charnov 1982; Bull 1983; J. J. Bull, interview.

41. Warner, Robertson, and Leigh 1975.

42. Bull 1983; Bull 1987; Conover and Kynard 1981.

43. Dunn, Adams, and Smith 1990; Adams, Greenwood, and Naylor 1987.

44. Head, May, and Pendleton 1987.

45. J. J. Bull, interview.

46. Bull 1983; Werren 1991; Hunter, Nur, and Werren 1993.

47. Trivers and Willard 1973.

48. Trivers and Willard 1973.

49. The sex ratio of presidential children was first noticed by Laura Betzig and Samantha Weber of the University of Michigan.

50. Trivers and Willard 1973.

51. Austad and Sunquist 1986.

52. Clutton-Brock and Iason 1986; Clutton-Brock 1991; Huck, Labov, and Lisk 1986.

53. T. H. Clutton-Brock, interview.

54. Clutton-Brock, Albon, and Guinness 1984.

55. Symington 1987.

56. For baboons, see Altmann 1980; for macaques, see Silk 1983, Simpson and Simpson 1982, and Small and Hrdy 1986; for a general summary, see Van Schaik and Hrdy 1991; for howler monkeys, I rely on K. Glander, interview; for a skeptical view of this data, T. Hasegawa, correspondence.

57. Hrdy 1987.

58. Van Schaik and Hrdy 1991.

59. Goodall 1986.

60. Grant 1990; Betzig and Weber 1992.

61. Grant 1990; V. J. Grant, correspondence.

62. Bromwich 1989.

63. K. McWhirter: "The gender vendors." Independent newspaper, London 27 October 1991, pages 54–55.

64. B. Gledhill, interview.

65. For zebra finches, see Burley 1981; for red-cockaded woodpeckers, see Gowaty and Lennartz 1985; for bald eagles, see Bortolotti 1986; for other hawks, see Olsen and Cockburn 1991.

66. N. D. Kristof: "Asia, Vanishing Point for As Many As 100 Million Women." *International Herald Tribune*, 6 November 1991, page 1.

67. Rao 1986; Hrdy 1990.

68. M. Nordborg, interview.

69. Bromwich 1989.

70. James 1986; James 1989; W. H. James, interview.

71. Unterberger and Kirsch 1932.

72. Dawkins 1982.

73. A. C. Hurlbert, personal communication.

74. Fisher 1930; R. L. Trivers, interview.

75. Betzig 1992a.

76. Dickemann 1979; Boone 1988; Voland 1988; Judge and Hrdy 1988.

77. Hrdy 1987; Cronk 1989; Hrdy 1990.

78. Dickemann 1979.

79. Dickemann 1979; Kitcher 1985; Alexander 1988; Hrdy 1990.

80. S. B. Hrdy, interview.

81. Dickemann 1979.

Chapter 5. The Peacock's Tale

1. Troy and Elgar 1991.

2. Trivers 1972; see also Dawkins 1976.

3. Atmar 1991.

4. Darwin 1871.

5. Diamond 1991b.

6. Cronin 1992.

7. Marden 1992.

8. Baker 1985; Gotmark 1992.

9. Ridley, Rands, and Lelliott 1984.

10. Halliday 1983.

11. Hoglund and Robertson 1990.

12. Møller 1988.

13. Hoglund, Eriksson, and Lindell 1990.

14. Andersson 1982.

15. Cherry 1990.

16. Houde and Endler 1990.

17. Evans and Thomas 1992.

18. Fisher 1930.

19. Jones and Hunter 1993.

20. Ridley and Hill 1987.

21. Taylor and Williams 1982.

22. Boyce 1990.

23. Cronin 1992.

24. The best volumes on the two factions of sexual selection are Bradbury and Andersson 1987 and Cronin 1992.

25. O'Donald 1980; Lande 1981; Kirkpatrick 1982; see Arnold 1983.

26. Weatherhead and Robertson 1979.

27. Pomiankowski, Iwasa, and Nee 1991

28. Pomiankowski 1990.

29. Dugatkin 1992; Gibson and Hoglund 1992. Copying has also been proven in fallow deer: Balmford 1991.

30. Pomiankowski 1990; see also Trail 1990 for why capuchin birds and other monomorphic lekking species experience female-female competition.

31. Partridge 1980.

32. Balmford 1991.

33. Alatalo, Hoglund, and Lundberg 1991.

34. Hill 1990.

35. Diamond 1991a.

36. Zahavi 1975.

37. Dawkins 1976; Cronin 1992.

38. Andersson 1986; Pomiankowski 1987; Grafen 1990; Iwasa, Pomiankowski and Nee 1991.

39. Møller 1991.

40. Hamilton and Zuk 1982.

41. Ward 1988; Pruett-Jones, Pruett-Jones, and Jones 1990; Zuk 1991; Zuk 1992.

42. Low 1990.

43. Cronin 1992.

44. Møller 1990.

45. Hillgarth 1990; N. Hillgarth and M. Zuk, interview.

46. Kirkpatrick and Ryan 1991.

47. Boyce 1990; Spurrier, Boyce, and Manly 1991.

48. Thornhill and Sauer 1992.

49. Møller 1992.

50. Møller and Pomiankowski (in press); see also Balmford, Thomas, and Jones 1993; A. Pomiankowski, interview.

51. Maynard Smith 1991; see Cronin 1992 for a history of how people have repeatedly made the mistake of thinking choice must be conscious and active,

and that therefore it was unreasonable to expect female animals to choose their mates using "rational" criteria.

52. Zuk 1992.
53. Zuk, in press.
54. Zuk, Thornhill, Ligon, and Johnson 1990; Ligon, Thornhill, Zuk, and Johnson 1990.
55. Flinn 1992.
56. Daly and Wilson 1983.
57. Folstad and Karter 1992; Zuk 1992.
58. Zuk, in press.
59. Wederkind 1992.
60. Hamilton 1990b.
61. Kodric-Brown and Brown 1984.
62. Dawkins and Krebs 1978.
63. Dawkins and Guilford 1991.
64. Low, Alexander, and Noonan 1987.
65. T. Guilford, interview; B. Low, interview.
66. Ryan 1991; M. Ryan, interview.
67. Basolo 1990.
68. Green 1987.
69. Eberhard 1985.
70. Kramer 1990.
71. Enquist and Arak 1993.
72. Gilliard 1963.
73. Houde and Endler 1990; J. Endler, interview.
74. Kirkpatrick 1989.
75. Searcy 1992.
76. Burley 1981.
77. The hypnosis idea is my own: see Ridley 1981. But it receives some indirect support from later experiments on peacocks and other pheasants: See Rands, Ridley, and Lelliott 1984; Davison 1983; Ridley, Rands, and Lelliott 1984; Petrie, Halliday, and Sanders 1991.
78. Gould and Gould 1989.
79. Pomiankowski and Guilford 1990.
80. A. Pomiankowski, interview.

Chapter 6. Polygamy and the Nature of Men

1. Betzig 1986.
2. Brown 1991; Barkow, Cosmides, and Tooby 1992.

3. Crook and Crook 1988.

4. Betzig and Weber 1992.

5. Trivers 1972.

6. Bateman 1948.

7. Alexander 1974, 1979; Irons 1979.

8. Clutton-Brock and Vincent 1991; Gwynne 1991.

9. For a clear summary of the argument that paternal care leads to the female initiative in courtship, and the evidence for it, see my namesake's paper: Ridley (Mark) 1978.

10. Symons 1979; D. Symons, interview.

11. Symons 1979.

12. Symons 1979.

13. Tripp 1975; Symons 1979.

14. Maynard Smith and Price 1973.

15. Trivers 1971; Maynard Smith 1977; Emlen and Oring 1977.

16. Pleszczynska and Hansell 1980; Garson, Pleszczynska and Holm 1981. Incidentally, polygamy can mean having many mates of either sex; polygyny means specifically males having many female mates. Although polygyny is more precise, I have stuck with the more familiar words throughout this book: *polygamy* for males, *polyandry* for females.

17. L. Betzig, interview.

18. Borgehoff Mulder 1988, 1992; M. Borgehoff Mulder, interview.

19. "Polygamists emerge from secrecy seeking not just peace but respect" by Dirk Johnson. *New York Times*, 9 April 1991, page A22.

20. Green 1993.

21. Symons 1979 put it this way: "Heterosexual relations are structured to a substantial degree by the nature and interests of the human female."

22. Crook and Gartlan 1966; Jarman 1974; Clutton-Brock and Harvey 1977.

23. Avery and Ridley 1988; Vos 1979.

24. Smith 1984.

25. Foley and Lee 1989.

26. Foley 1987; Foley and Lee 1989; Leakey and Lewin 1992; Kingdon 1993.

27. Symons 1987; K. Hill, interview.

28. Alexander 1988; R. D. Alexander, interview.

29. Kaplan and Hill 1985b; Hewlett 1988.

30. Kaplan and Hill 1985a; Hill and Kaplan 1988; Hawkes 1992; Cosmides and Tooby 1992; K. Hawkes, interview.

31. Cashdan 1980; Cosmides and Tooby 1992.

32. N. Chagnon, interview; Cronk 1991.

33. Rosenberg and Birdzell 1986.

34. Goodall 1990.

35. Daly and Wilson 1983.
36. "Dolphin Courtship: Brutal, Cunning and Complex" by N. Angier, *New York Times*, 18 February 1992, p. C1.
37. Dickemann 1979.
38. Hartung 1982.
39. L. Betzig, interview.
40. Betzig 1986.
41. Betzig 1986.
42. Finley, quoted in Betzig 1992b; the Gibbon quote is from *The Decline and Fall of the Roman Empire*, volume 1, chapter 7.
43. Betzig 1992c.
44. Betzig 1992a.
45. Scruton 1986.
46. Brown and Hotra 1988.
47. D. E. Brown, interview.
48. Goodall 1986. However, old females are killed by the victors.
49. N. Chagnon, interview.
50. Chagnon 1968; Chagnon 1988.
51. I'm indebted to Archie Fraser for pointing out this parallel.
52. Chagnon 1968.
53. Smith 1984.
54. D. E. Brown, interview.

Chapter 7. Monogamy and the Nature of Women

1. Møller 1987; Birkhead and Møller 1992.
2. Murdock and White 1969; Fisher 1992 makes the interesting case that sexism, despotism, polygamy, and male "ownership" of wives were all invented along with the plow, which removed from women all their share in food winning; as women have come back into the work force in recent decades, so their say and status have improved.
3. Hrdy 1981; Hrdy 1986.
4. Bertram 1975; Hrdy 1979; Hausfater and Hrdy 1984. A remarkable experiment by Emlen, Demong, and Emlen 1989 greatly strengthened the contention that infanticide was an adaptive strategy. By removing territorial females, Emlen induced female jacanas—a role-reversed species—to kill the eggs of males with nests in their newly acquired territories.
5. Dunbar 1988.
6. Wrangham 1987; R. W. Wrangham, interview.

7. Goodall 1986, 1990; Hiraiwa-Hasegawa 1988; Yamamura, Hasegawa, and Ito 1990.

8. Daly and Wilson 1988.

9. Martin and May 1981.

10. Hasegawa and Hiraiwa-Hasegawa 1990; Diamond 1991b.

11. White 1992; Small 1992.

12. Short 1979.

13. Eberhard 1985; Hyde and Elgar 1992; Bellis, Baker, and Gage 1990; Baker and Bellis 1992.

14. Harcourt, Harvey, Larson, and Short 1981; Hyde and Elgar 1992.

15. Connor, Smolker, and Richards 1992.

16. Smith 1984. This explanation, that cool testicles are designed to increase the shelf life of stored sperm, fits the facts far better than the old notion that sperm must be manufactured in a cool organ or they will be deformed.

17. Harvey and May 1989.

18. Payne and Payne 1989.

19. Birkhead and Møller 1992.

20. Hamilton 1990b.

21. Westneat, Sherman, and Morton 1990; Birkhead and Møller 1992.

22. Potts, Manning, and Wakeland 1991.

23. Burley 1981.

24. Møller 1987.

25. Baker and Bellis 1989, 1992.

26. Birkhead and Møller 1992.

27. Hill and Kaplan 1988; K. Hill, interview.

28. K. Hill, interview.

29. Wilson and Daly 1992; R. W. Wrangham, interview.

30. Cherfas and Gribbin 1984; Flinn 1988.

31. Morris 1967.

32. Birkhead and Møller 1992.

33. Alexander and Noonan 1979.

34. The first authors to see it this way were Cherfas and Gribbin 1984.

35. Hrdy 1979; Symons 1979; Benshoof and Thornhill 1979; Diamond 1991b; Fisher 1992; Sillen-Tullberg and Møller 1993.

36. Korpimaki 1991.

37. Alatalo, Lundberg, and Stahlbrandt 1982. Recent research suggests that the wife, at least, knows what is happening. See Veiga 1992; Slagsvold, Amundsen, Dale, and Lampe 1992.

38. Veiga 1992.

39. Møller and Birkhead 1989.

40. Darwin 1803.

41. Wilson and Daly 1992.

42. Wilson and Daly 1992.

43. Thornhill and Thornhill 1983, 1989; Posner 1992.

44. Gaulin and Schlegel 1980; Wilson and Daly 1992; Regalski and Gaulin 1992.

45. A. Fraser, personal communication.

46. Malinowski 1927.

47. Wilson and Daly 1992,

48. French revolutionary law, quoted in translation by Wilson and Daly 1992.

49. Alexander 1974; Kurland 1979.

50. Betzig 1992a.

51. Voland 1988, 1992.

52. Boone 1988.

53. Darwin 1803.

54. Betzig 1992a.

55. Betzig 1992a.

56. Betzig 1992a.

57. Thornhill 1990.

58. Thornhill 1990.

59. Kitcher 1985; Vining 1986.

60. Perusse 1992.

61. W. Irons, interview; N. Polioudakis, interview.

Chapter 8. Sexing the Mind

1. Gaulin and Fitzgerald 1986; Jacobs, Gaulin, Sherry, and Hoffman 1990.

2. Konner 1982.

3. Darwin 1871.

4. Silverman and Eals 1992.

5. Maccoby and Jacklin 1974; Daly and Wilson 1983; Moir and Jessel 1991.

6. M. Bailey, interview.

7. Gaulin and Hoffman 1988.

8. Silverman and Eals 1992.

9. Wilson 1975; Kingdon 1993.

10. Daly and Wilson 1983.

11. Symons 1979.

12. Hudson and Jacot 1991.

13. Tannen 1990.

14. Gaulin and Hoffman 1988.

15. Maccoby and Jacklin 1974; Ehrhardt and Meyer-Bahlburg 1981; Rossi 1985; Moir and Jessel 1991.

16. Moir and Jessel 1991.
17. McGuinness 1979.
18. McGuinness 1979.
19. Imperato-McGinley, Peterson, Gautier, and Sturla 1979.
20. Daly and Wilson 1983; Moir and Jessel 1991.
21. Hoyenga and Hoyenga 1980.
22. Tannen 1990.
23. Tiger and Shepher 1977; Daly and Wilson 1983; Moir and Jessel 1991.
24. Fisher 1992.
25. Interviewed in the *Sunday Times* (London), 7 June 1992.
26. Dörner 1985, 1989; M. Bailey, interview; Le Vay 1992.
27. M. Bailey, interview; D. Hamer, interview.
28. Dickemann 1992.
29. Symons 1987.
30. Thornhill 1989a.
31. Buss 1989, 1992.
32. Ellis 1992.
33. Buss 1989, 1992.
34. Kenrick and Keefe 1989.
35. Ellis and Symons 1990.
36. Ellis and Symons 1990.
37. Symons 1987.
38. Mosher and Abramson 1977.
39. Ellis and Symons 1990.
40. Alatalo, Hoglund, and Lundberg 1991.
41. Fisher 1992.
42. Symons 1989.
43. Brown 1991.
44. Wilson 1978.
45. Tooby and Cosmides 1989.
46. Moir and Jessel 1991.

Chapter 9. *The Uses of Beauty*

1. M. Bailey, interview; D. Hamer, interview; F. Whitam, interview. Levay 1993.
2. Freud 1913.
3. Westermarck 1891.
4. Wolf 1966, 1970; Degler 1991.
5. Daly and Wilson 1983.

6. Shepher 1983.
7. Thornhill 1989b.
8. Thorpe 1954, 1961.
9. Marler and Tamura 1964.
10. Slater 1983.
11. Seid 1989.
12. *Washington Post*, 28 July 1992.
13. Frisch 1988; Anderson and Crawford 1992.
14. Smuts 1993.
15. Elder 1969; Buss 1992.
16. Ellis 1992.
17. Fisher 1930.
18. D. Singh, interview.
19. Low, Alexander, and Noonan 1987; Leakey and Lewin 1992; D. Singh, interview.
20. Ellis 1905.
21. The same idea—that fair hair is a sexually selected trait—has been put forward by Jonathan Kingdon recently; see Kingdon 1993.
22. Kingdon 1993.
23. This is a further reason that I am not convinced by Helen Fisher's (1992) theory that human pair bonds lasted about four years on average.
24. R. Thornhill, interview.
25. Galton 1883.
26. See "No Better Than Average" by M. Ridley, *Science* 257:328.
27. Dickemann 1979.
28. Buss 1992; Gould and Gould 1989.
29. Berscheid and Walster 1974; Gillis and Avis 1980; Ellis 1992; Shellberg 1992.
30. Sadalla, Kenrick, and Vershure 1987; Ellis 1992.
31. Daly and Wilson 1983.
32. Daly and Wilson 1983.
33. Ellis 1992. The other facts in this paragraph are from Trivers 1985; Ford and Beach 1951; Pratto, Sidanius, and Stallworth 1992; and Buss 1989.
34. Bell 1976.
35. Symons 1992; R. Alexander, interview.
36. Fallon and Rozin 1985.
37. Ellis 1905.
38. Low 1979.
39. Bell 1976.
40. Darwin 1871.
41. B. Ellis, interview.

Chapter 10. The Intellectual Chess Game

1. Connor, Smolker, and Richards (1992) argue that the social complexity of dolphin species roughly correlates with brain size. Bottle-nosed dolphins seem to be the most socially complex and the largest-brained species of all.
2. Johansen and Edey 1981.
3. Tooby and Cosmides 1992.
4. Bloom 1992; Pinker and Bloom 1992.
5. Gould 1981.
6. Fox 1991.
7. Durkheim 1895.
8. Brown 1991.
9. Mead 1928.
10. Wilson 1975.
11. Gould 1978.
12. Gould 1987.
13. Pinker and Bloom 1992.
14. Chomsky 1957.
15. Marr 1982; Hurlbert and Poggio 1988.
16. Tooby and Cosmides 1992.
17. Leakey and Lewin 1992.
18. Lewin 1984.
19. Dart 1954; Ardrey 1966.
20. Konner 1982.
21. R. Wrangham, interview.
22. Gould 1981.
23. Badcock 1991.
24. Montagu 1961.
25. Leakey and Lewin 1992.
26. Budiansky 1992.
27. S. J. Gould, reported in Pinker and Bloom 1992.
28. Pinker and Bloom 1992.
29. Alexander 1974, 1990.
30. Potts 1991.
31. Humphrey 1976.
32. Humphrey 1976, 1983.
33. Barlow, unpublished.
34. Crook 1991.
35. Pinker and Bloom 1992.
36. Tooby and Cosmides 1992.
37. Barlow 1990; Barkow 1992.

38. Konner 1982.
39. Symons 1987.
40. Barlow 1987.
41. Byrne and Whiten 1985, 1988, 1992.
42. Macaulay's works, vol. 11, "Essay on the Athenian Orators."
43. Dawkins and Krebs 1978.
44. Cosmides 1989; Cosmides and Tooby 1992; Gigerenzer and Hug (in press).
45. Byrne and Whiten 1985, 1988, 1992.
46. Trivers 1991.
47. Goodall 1986.
48. Miller 1992.
49. Connor, Smolker, and Richards 1992.
50. De Waal 1982.
51. Miller 1992.
52. Buss 1989.
53. Symons 1979; G. Miller, interview.
54. Leakey and Lewin 1992.
55. G. Miller, correspondence.
56. Erickson and Zenone 1976.
57. Miller 1992; see also Miller and Todd 1990.
58. Webster 1992.
59. Badcock 1991.

BIBLIOGRAPHY

Adams, J., Greenwood, P., and Naylor, P. 1987. Evolutionary aspects of environmental sex determination. *International Journal of Invertebrate Reproductive Development* 11:123–136.

Alatalo, R. V., Hoglund, J., and Lundberg, A. 1991. Lekking in the black grouse—a test of male viability. *Nature* 352:155–156.

Alatalo, R. V., Lundberg, A., and Stahlbrandt, K. 1982. Why do pied flycatcher females mate with already mated males? *Animal Behaviour* 30:585–593.

Alexander, R. D. 1974. The evolution of social behavior. *Annual Review of Ecology and Systematics* 5:325–383.

Alexander, R. D. 1979. *Darwinism and Human Affairs*. University of Washington Press, Seattle.

Alexander, R. D. 1988. Evolutionary approaches to human behavior: what does the future hold? In *Human Reproductive Behavior* (ed. L. Betzig, M. Borgehoff Mulder, and P. Turke). Pp. 317–341. Cambridge University Press, Cambridge.

Alexander, R. D. 1990. *How Did Humans Evolve? Reflections on the Uniquely Unique Species*. Museum of Zoology, University of Michigan, Special Publication No. 1.

Alexander, R. D., and Noonan, K. M. 1979. Concealment of ovulation, parental care, and human social evolution. In *Evolutionary Biology and Human Social Behavior* (ed. N. Chagnon and W. Irons). Pp. 436–453. Duxbury, North Scituate, Massachusetts.

Altmann, J. 1980. *Baboon Mothers and Infants*. Harvard University Press, Cambridge, Massachusetts.

Anderson, A. 1992. The evolution of sexes. *Science* 257:324–326.

Anderson, J. L., and Crawford, C. B. 1992. Modeling costs and benefits of adolescent weight control as a mechanism for reproductive suppression. *Human Nature* 3:299–334.

Andersson, M. 1982. Female choice selects for extreme tail length in a widow bird. *Nature* 299:818–820.

Andersson, M. 1986. Evolution of condition-dependent sex ornaments and mating preferences: sexual selection based on viability differences. *Evolution* 40:804–816.

Ardrey, R. 1966. *The Territorial Imperative.* Atheneum, New York.

Arnold, S. J. 1983. Sexual selection: the interface of theory and empiricism. In *Mate Choice* (ed. P. Bateson). Pp. 67–107. Cambridge University Press, Cambridge.

Atmar, W. 1991. On the role of males. *Animal Behaviour* 41:195–205.

Austad, S., and Sunquist, M. E. 1986. Sex-ratio manipulation in the common opossum. *Nature* 324:58–60.

Avery, M. I., and Ridley, M. W. 1988. Gamebird mating systems. In *The Ecology and Management of Gamebirds* (ed. P. J. Hudson and M. R. W. Rands). Blackwell, Oxford.

Badcock, C. 1991. *Evolution and Individual Behavior: An Introduction to Human Sociobiology.* Blackwell, Oxford.

Baker, R. R. 1985. Bird coloration: in defence of unprofitable prey. *Animal Behaviour* 33:1387–1388.

Baker, R. R., and Bellis, M. A. 1989. Number of sperm in human ejaculates varies in accordance with sperm competition. *Animal Behaviour* 37:867–869.

Baker, R. R., and Bellis, M. A. 1992. Human sperm competition: infidelity, the female orgasm and "kamikaze" sperm. Paper delivered to the fourth annual meeting of the Human Behavior and Evolution Society, Albuquerque, New Mexico, 22–26 July 1992.

Balmford, A. 1991. Mate choice on leks. *Trends in Ecology and Evolution* 6:87–92.

Balmford, A., Thomas, A. L. R., and Jones, I. L. 1993. Aerodynamics and the evolution of long tails in birds. *Nature* 361:628–631.

Barkow, J. H. 1992. Beneath new culture is old psychology: gossip and social stratification. In *The Adapted Mind* (ed. J. H. Barkow, L. Cosmides, and J. Tooby). Pp. 627–637. Oxford University Press, New York.

Barkow, J. H., Cosmides, L., and Tooby, J. (eds.). 1992. *The Adapted Mind.* Oxford University Press, New York.

Barlow, H. 1987. The biological role of consciousness. In *Mindwaves* (ed. C. Blakemore and S. Greenfield). Pp. 361–374. Blackwell, Oxford.

Barlow, H. 1990. The mechanical mind. *Annual Review of Neuroscience* 13:15–24.

Barlow, H. (unpublished). The inevitability of consciousness. Chapter draft.

Basolo, A. L. 1990. Female preference predates the evolution of the sword in swordtail fish. *Science* 250:808–810.

Bateman, A. J. 1948. Intrasexual selection in *Drosophila. Heredity* 2:349–368.

Beeman, R. W., Friesen, K. S., and Denell, R. E. 1992. Maternal-effect selfish genes in flour beetles. *Science* 256:89–92.

Bell, G. 1982. *The Masterpiece of Nature.* Croom-Helm, London.

Bell, G. 1987. Two theories of sex and variation. In *The Evolution of Sex and Its Consequences* (ed. S. C. Stearns). Pp. 117–133. Birkhauser, Basel.

Bell, G. 1988. *Sex and Death in Protozoa: The History of an Obsession.* Cambridge University Press, Cambridge.

Bell, G., and Burt, A. 1990. B-chromosomes: germ-line parasites which induce changes in host recombination. *Parasitology* 100:S19–S26.

Bell, G., and Maynard Smith, J. 1987. Short-term selection for recombination among mutually antagonistic species. *Nature* 328:66–68.

Bell, Q. 1976. *On Human Finery.* (Second edition.) Hogarth Press, London.

Bellis, M. A., Baker, R. R., and Gage, M. J. G. 1990. Variation in rat ejaculates consistent with the kamikaze-sperm hypothesis. *Journal of Mammalogy* 71:479–480.

Benshoof, L., and Thornhill, R. 1979. The evolution of monogamy and concealed ovulation in humans. *Journal of Social and Biological Structures* 2:95–106.

Bernstein, H. 1983. Recombinational repair may be an important function of sexual reproduction. *Bioscience* 33:326–331.

Bernstein, H., Byerly, H. C., Hopf, F. A., and Michod, R. E. 1985. Genetic damage, mutation and the evolution of sex. *Science* 229:1277–1281.

Bernstein, H., Hopf, F. A., and Michod, R. E. 1988. Is meiotic recombination an adaptation for repairing DNA, producing genetic variation, or both? In *The Evolution of Sex* (ed. R. E. Michod and B. R. Levin). Pp. 139–160. Sinauer Associates, Sunderland, Massachusetts.

Berscheid, E., and Walster, E. 1974. Physical Attractiveness. In *Advances in Experimental Social Psychology* (ed. L. Berkowitz), Vol. 7. Academic Press, New York.

Bertram, B. C. R. 1975. Social factors influencing reproduction in wild lions. *Journal of Zoology* 177:463–482.

Betzig, L. L. 1986. *Despotism and Differential Reproduction: A Darwinian View of History.* Aldine, Hawthorne, New York.

Betzig, L. L. 1992a. Medieval monogamy. In *Darwinian Approaches to the Past* (ed. S. Mithen and H. Maschner). Plenum, New York.

Betzig, L. L. 1992b. Roman polygyny. *Ethology and Sociobiology* 13:309–349.

Betzig, L. L. 1992c. Roman monogamy. *Ethology and Sociobiology* 13:351–383.

Betzig, L. L., and Weber, S. 1992. Polygyny in American politics. *Politics and Life Sciences* 12, no. 1.

Bierzychudek, P. 1987a. Resolving the paradox of sexual reproduction: a review of experimental tests. In *The Evolution of Sex and Its Consequences* (ed. S. C. Stearns). Pp. 163–174. Birkhauser, Basel.

Bierzychudek, P. 1987b. Patterns in plant parthenogenesis. In *The Evolution of Sex and Its Consequences* (ed. S. C. Stearns). Pp. 197–217. Birkhauser, Basel.

Birkhead, T. R., and Møller, A. P. 1992. *Sperm Competition in Birds.* Academic Press, London.

Bloom, P. 1992. Language as a biological adaptation. Paper delivered to the fourth annual meeting of the Human Behavior and Evolution Society, Albuquerque, New Mexico, 22–26 July 1992.

Boone, J. 1988. Parental investment, social subordination and population processes among the 15th and 16th century Portuguese nobility. In *Human Reproductive Behavior* (ed. L. Betzig, M. Borgehoff Mulder, and P. Turke). Pp. 201–219. Cambridge University Press, Cambridge.

Borgehoff Mulder, M. 1988. Is the polygyny threshold model relevant to humans? Kipsigis evidence. In *Mating Patterns* (ed. C. G. N. Mascie-Taylor and A. J. Boyce). Pp. 209–230. Cambridge University Press, Cambridge.

Borgehoff Mulder, M. 1992. Women's strategies in polygynous marriage. *Human Nature* 3:45–70.

Bortolotti, G. R. 1986. Influence of sibling competition on nestling sex ratios of sexually dimorphic birds. *American Naturalist* 127:495–507.

Boyce, M. S. The Red Queen visits sage grouse leks. *American Zoologist* 30:263–270.

Bradbury, J. W., and Andersson, M. B. (eds.) 1987. *Sexual Selection: Testing the Alternatives.* Dahlem Workshop report. *Life Sciences* 39. John Wiley, Chichester.

Bremermann, H. J. 1980. Sex and polymorphism as strategies in host-pathogen interactions. *Journal of Theoretical Biology* 87:671–702.

Bremermann, H. J. 1987. The adaptive significance of sexuality. In *The Evolution of Sex and Its Consequences* (ed. S. C. Stearns). Pp. 135–161. Birkhauser, Basel.

Bromwich, P. 1989. The sex ratio and ways of manipulating it. *Progress in Obstetrics and Gynaecology* 7:217–231.

Brooks, L. 1988. The evolution of recombination rates. In *The Evolution of Sex* (ed. R. E. Michod and B. R. Levin). Pp. 87–105. Sinauer, Massachusetts.

Brown, D. E. 1991. *Human Universals.* McGraw-Hill, New York.

Brown, D. E., and Hotra, D. 1988. Are presciptively monogamous societies effectively monogamous? In *Human Reproductive Behavior* (ed. L. Betzig, M. Borgehoff Mulder, and P. Turke). Pp. 153–160. Cambridge University Press, Cambridge.

Budiansky, S. 1992. *The Covenant of the Wild: Why Animals Chose Domestication.* William Morrow, New York.

Bull, J. J. 1983. *Evolution of Sex-Determining Mechanisms.* Benjamin/Cummings, Menlo Park, California.

Bull, J. J. 1987. Sex-determining mechanisms: an evolutionary perspective. In *The Evolution of Sex and Its Consequences.* (ed. S. C. Stearns). Pp. 93–115. Birkhauser, Basel.

Bull, J. J., and Bulmer, M. G. 1981. The evolution of XY females in mammals. *Heredity* 47:347–365.

Bull, J. J., and Charnov, E. L. 1985. On irreversible evolution. *Evolution* 39:1149–1155.

Burley, N. 1981. Sex ratio manipulation and selection for attractiveness. *Science* 211:721–722.

Burt, A., and Bell, G. 1987. Mammalian chiasma frequencies as a test of two theories of recombination. *Nature* 326:803–805.

Buss, D. 1989. Sex differences in human mate preferences: evolutionary hypotheses tested in 37 cultures. *Behavioral and Brain Sciences* 12:1–49.

Buss, D. 1992. Mate preference mechanisms: consequences for partner choice and intrasexual competition. In *The Adapted Mind* (ed. J. H. Barkow, L. Cosmides, and J. Tooby). Pp. 249–266. Oxford University Press, New York.

Byrne, R. W., and Whiten, A. 1985. Tactical deception of familiar individuals in baboons. *Animal Behaviour* 33:669–673.

Byrne, R. W., and Whiten, A. (eds.). 1988. *Machiavellian Intelligence: Social Expertise and the Evolution of Intellect in Monkeys, Apes and Humans.* Clarendon Press, Oxford.

Byrne, R. W., and Whiten, A. 1992. Cognitive evolution in primates: evidence from tactical deception. *Man* 27:609–627.

Carroll, L. 1871. *Through the Looking-Glass and What Alice Found There.* Macmillan, London.

Cashdan, E. 1980. Egalitarianism among hunters and gatherers. *American Anthropologist* 82:116–120.

Chagnon, N. A. 1968. *Yanomamö: The Fierce People.* Holt, Rinehart and Winston, New York.

Chagnon, N. A. 1988. Life histories, blood revenge and warfare in a tribal population. *Science* 239:935–992.

Chagnon, N. A., and Irons, W. (eds.). 1979. *Evolutionary Biology and Human Social Behavior: An Anthropological Perspective.* Duxbury, North Scituate, Massachusetts.

Chao, L. 1992. Evolution of sex in RNA viruses. *Trends in Ecology and Evolution* 7:147–151.

Chao, L., Tran, T., and Matthews, C. 1992. Muller's ratchet and the advantage of sex in the virus phi-6. *Evolution* 46:289–299.

Charlesworth, B., and Hartl, D. L. 1978. Population dynamics of the segregation distorter polymorphism of *Drosophila melanogaster. Genetics* 89:171–192.

Charnov, E. L. 1982. *The Theory of Sex Allocation.* Princeton University Press, Princeton.

Cherfas, J., and Gribbin, J. 1984. *The Redundant Male.* Pantheon, New York.

Cherry, M. I. 1990. Tail length and female choice. *Trends in Ecology and Evolution* 5:349–350.

Chomsky, N. 1957. *Syntactic Structures.* Mouton, The Hague.

Clarke, B. C. 1979. The evolution of genetic diversity. *Proceedings of the Royal Society of London* B 205:453–474.

Clay, K. 1991. Parasitic castration of plants by fungi. *Trends in Ecology and Evolution* 6:162–166.

Clutton-Brock, T. H. 1991. *The Evolution of Parental Care.* Princeton University Press, Princeton.

Clutton-Brock, T. H., and Albon, S. D., and Guinness, F. E. 1984. Maternal dominance, breeding success and birth sex ratios in red deer. *Nature* 308:358–360.

Clutton-Brock, T. H., and Harvey, P. H. 1977. Primate ecology and social organization. *Journal of Zoology* 183:1–39.

Clutton-Brock, T. H., and Iason, G. R. 1986. Sex ratio variation in mammals. *Quarterly Review of Biology* 61:339–374.

Clutton-Brock, T. H., and Vincent, A. C. J. 1991. Sexual selection and the potential reproductive rates of males and females. *Nature* 351:58–60.

Connor, R. C., Smolker, R. A., and Richards, A. F. 1992. Two levels of alliance formation among male bottlenose dolphins (*Tursiops* sp.). *Proceedings of the National Academy of Sciences, USA* 89:987–990.

Conover, D. O., and Kynard, B. E. 1981. Environmental sex determination: interaction of temperature and genotype in a fish. *Science* 213:577–579.

Cosmides, L. M. 1989. The logic of social exchange: has natural selection shaped how humans reason? Studies with the Wason selection task. *Cognition* 31:187–276.

Cosmides, L. M., and Tooby, J. 1981. Cytoplasmic inheritance and intragenomic conflict. *Journal of Theoretical Biology* 89:83–129.

Cosmides, L. M., and Tooby, J. 1992. Cognitive adaptations for social exchange. In *The Adapted Mind* (ed. J. H. Barkow, L. Cosmides, and J. Tooby). Pp. 163–228. Oxford University Press, New York.

Cronin, H. 1992. *The Ant and the Peacock*. Cambridge University Press, Cambridge.

Cronk, L. 1989. Low socioeconomic status and female-biased parental investment. The Mukogodo example. *American Anthropologist* 9:414–429.

Cronk, L. 1991. Wealth, status and reproductive success among the Mukogodo of Kenya. *American Anthropologist* 93:345–360.

Crook, J. H. 1991. Consciousness and the ecology of meaning: new findings and old philosophies. In *Man and Beast Revisited* (ed. M. H. Robinson and L. Tiger). Pp. 203–223. Smithsonian, Washington, DC.

Crook, J. H., and Crook, S. J. 1988. Tibetan polyandry: problems of adaptation and fitness. In *Human Reproductive Behavior* (ed. L. Betzig, M. Borgehoff Mulder, and P. Turke). Pp. 97–114. Cambridge University Press, Cambridge.

Crook, J. H., and Gartlan, J. S. 1966. Evolution of primate societies. *Nature* 210:1200–1203.

Crow, J. F. 1988. The importance of recombination. In *The Evolution of Sex* (ed. R. E. Michod and B. R. Levin). Pp. 56–73. Sinauer, Massachusetts.

Crow, J. F., and Kimura, M. 1965. Evolution in sexual and asexual populations. *American Naturalist* 99:439–450.

Daly, M., and Wilson, M. 1983. *Sex, Evolution and Behavior*. (Second edition.) Wadsworth, Belmont, California.

Daly, M., and Wilson, M. 1988. *Homicide*. Aldine, Hawthorne, New York.

Dart, R. 1954. The predatory transition from ape to man. *International Anthropological and Linguistic Review* 1:201–213.

Darwin, E. 1803. *The Temple of Nature; or, The Origin of Society.* J. Johnson, London.

Darwin, C. 1859. *On the Origin of Species by Means of Natural Selection, or the Preservation of Favoured Races in the Struggle for Life.* John Murray, London.

Darwin, C. 1871. *The Descent of Man and Selection in Relation to Sex.* John Murray, London.

Davison, G. W. H. 1983. The eyes have it: ocelli in a rain forest pheasant. *Animal Behaviour* 31:1037–1042.

Dawkins, M., and Guilford, T. 1991. The corruption of honest signalling. *Animal Behaviour* 41:865–873.

Dawkins, R. 1976. *The Selfish Gene.* Oxford University Press, Oxford.

Dawkins, R. 1982. *The Extended Phenotype.* Oxford University Press, Oxford.

Dawkins, R. 1986. *The Blind Watchmaker.* Longmans, London.

Dawkins, R. 1990. Parasites, desiderata lists and the paradox of the organism. *Parasitology* 100:S63–S73.

Dawkins, R. 1991. Darwin triumphant: Darwinism as a universal truth. In *Man and Beast Revisited* (ed. M. H. Robinson and L. Tiger). Pp. 23–39. Smithsonian, Washington, DC.

Dawkins, R., and Krebs, J. R. 1978. Animal signals: information or manipulation. In *Behavioural Ecology* (ed. J. R. Krebs and N. B. Davies). Pp. 282–309. Blackwell, Oxford.

Dawkins, R., and Krebs, J. R. 1979. Arms races between and within species. *Proceedings of the Royal Society of London* B 205:489–511.

Degler, C. N. 1991. *In Search of Human Nature.* Oxford University Press, Oxford.

de Waal, F. 1982. *Chimpanzee Politics.* Jonathan Cape, London.

Diamond, J. M. 1991a. Borrowed sexual ornaments. *Nature* 349:105.

Diamond, J. M. 1991b. *The Rise and Fall of the Third Chimpanzee.* Radius, London.

Dickemann, M. 1979. Female infanticide and reproductive strategies of stratified human societies. In *Evolutionary Biology and Human Social Behavior* (ed. N. Chagnon and W. Irons). Pp. 321–367. Duxbury, North Scituate, Massachusetts.

Dickemann, M. 1992. Phylogenetic fallacies and sexual oppression. *Human Nature* 3:71–87.

Doolittle, W. F., and Sapienza, C. 1980. Selfish genes, the phenotype paradigm and genome evolution. *Nature* 284:601–603.

Dorner, G. 1985. Sex-specific gonadotrophin secretion, sexual orientation and gender role behaviour. *Endokrinologie* 86:1–6.

Dorner, G. 1989. Hormone-dependent brain development and neuroendocrine prophylaxis. *Experimental and Clinical Endocrinology* 94:4–22.

Dugatkin, L. 1992. Sexual selection and imitation: females copy the mate choice of others. *American Naturalist* 139:1384–1389.

Dunbar, R. I. M. 1988. *Primate Social Systems.* Croom Helm, London.

Dunn, A. M., Adams, J., and Smith, J. E. 1990. Intersexes in a shrimp: a possible disadvantage of environmental sex determination. *Evolution* 44:1875–1878.

Durkheim, E. 1895/1962. *The Rules of the Sociological Method.* Free Press, Glencoe, Illinois.

Eberhard, W. G. 1985. *Sexual Selection and Animal Genitalia.* Harvard University Press, Cambridge, Massachusetts.

Edmunds, G. F., and Alstad, D. N. 1978. Coevolution in insect herbivores and conifers. *Science* 199:941–945.

Edmunds, G. F., and Alstad, D. N. 1981. Responses of black pine leaf scales to host plant variability. In *Insect Life-History Patterns: Habitat and Geographic Variation* (ed. R. F. Denno and H. Dingle). Springer Verlag, New York.

Ehrhardt, A. A., and Meyer-Bahlburg, H. F. L. 1981. Effects of parental sex hormones on gender-related behavior. *Science* 211:1312–1314.

Elder, G. H. 1969. Appearance and education in marriage mobility. *American Sociological Review* 34:519–533.

Eldredge, N., and Gould, S. J. 1972. Punctuated equilibria: an alternative to phyletic gradualism. In *Models in Paleobiology* (ed. T. J. M. Schopf). Pp. 82–115. Freeman Cooper, San Francisco.

Ellis, B. J. 1992. The evolution of sexual attraction: evaluative mechanisms in women. In *The Adapted Mind* (ed. J. H. Barkow, L. Cosmides, and J. Tooby). Pp. 267–288. Oxford University Press, New York.

Ellis, B. J., and Symons, D. 1990. Sex differences in sexual fantasy: an evolutionary psychological approach. *Journal of Sex Research* 27:527–555.

Ellis, H. 1905. *Studies in the Psychology of Sex.* F. A. Davis, New York.

Emlen, S. T., Demong, N. J., and Emlen, D. J. 1989. Experimental induction of infanticide in female wattled jacanas. *The Auk* 106:1–7.

Emlen, S. T., and Oring, L. W. 1977. Ecology, sexual selection and the evolution of mating systems. *Science* 197:215–223.

Enquist, M., and Arak, A. 1993. Selection of exaggerated male traits by female aesthetic senses. *Nature* 361:446–448.

Erickson, C. J., and Zenone, P. G. 1976. Courtship differences in male ring doves: avoidance of cuckoldry? *Science* 192:1353–1354.

Evans, M. R., and Thomas, A. L. R. 1992. Aerodynamic and mechanical effects of elongated tails in the scarlet-tufted malachite sunbird: measuring the cost of a handicap. *Animal Behaviour* 43:337–347.

Fallon, A. E., and Rozin, P. 1985. Sex differences in perception of desirable body shape. *Journal of Abnormal Psychology* 94:102–105.

Felsenstein, J. 1988. Sex and the evolution of recombination. In *The Evolution of Sex* (ed. R. E. Michod and B. R. Levin). Pp. 74–86. Sinauer, Massachusetts.

Fisher, H. E. 1992. *Anatomy of Love: The Natural History of Monogamy, Adultery and Divorce.* Norton, New York.

Fisher, R. A. 1930. *The Genetical Theory of Natural Selection.* Clarendon Press, Oxford.

Flegg, P. B., Spencer, D. M., and Wood, D. A. 1985. *The Biology and Technology of the Cultivated Mushroom.* John Wiley, Chichester.

Flinn, M. V. 1988. Mate guarding in a Caribbean village. *Ethology and Sociobiology* 9:1–28.

Flinn, M. V. 1992. Evolution and function of the human stress response. Paper delivered to the fourth annual meeting of the Human Behavior and Evolution Society, Albuquerque, New Mexico, 22–26 July 1992.

Foley, R. A. 1987. *Another Unique Species.* Longmans, London.

Foley, R. A., and Lee, P. C. 1989. Finite social space, evolutionary pathways and reconstructing hominid behaviour. *Science* 243:901–905.

Folstad, I., and Karter A. J. 1992. Parasites, bright males and the immunocompetence handicap. *American Naturalist* 139:603–622.

Ford, C. S, and Beach, F. A. 1951. *Patterns of Sexual Behavior.* Harper and Row, New York.

Fox, R. 1991. Aggression then and now. In *Man and Beast Revisited* (ed. M. H. Robinson and L. Tiger). Pp. 81–93. Smithsonian, Washington, D.C.

Frank, S. A. 1989. The evolutionary dynamics of cytoplasmic male sterility. *American Naturalist* 133:345–376.

Frank, S. A. 1990. Sex allocation theory for birds and mammals. *Annual Review of Ecology and Systematics* 21:13–55.

Frank, S. A. 1991. Divergence of meiotic drive suppression systems as an explanation for sex-biased hybrid sterility and inviability. *Evolution* 45:262–267.

Frank, S. A., and Swingland, I. R. 1988. Sex ratio under conditional sex expression. *Journal of Theoretical Biology* 135:415–418.

Freud, S. 1913. *Totem and Taboo.* Vintage Books, New York.

Frisch, R. E. 1988. Fatness and fertlity. *Scientific American* 258:70–77.

Galton, F. 1883. *Inquiries into the Human Faculty and Its Development.* Macmillan, London.

Garson, P. J., Pleszczynska, W. K., and Holm, C. H. 1981. The "polygyny threshold" model: a reassessment. *Canadian Journal of Zoology* 59:902–910.

Gaulin, S. J. C., and Fitzgerald, R. W. 1986. Sex differences in spatial ability: an evolutionary hypothesis and test. *American Naturalist* 127:74–88.

Gaulin, S. J. C., and Hoffman, G. E. 1988. Evolution and development of sex differences in spatial ability. In *Human Reproductive Behavior* (ed. L. Betzig, M. Borgehoff Mulder, and P. Turke). Pp. 129–152. Cambridge University Press, Cambridge.

Gaulin, S. J. C., and Schlegel, A. 1980. Paternal confidence and paternal invest-ment: a cross-cultural test of a sociobiological hypothesis. *Ethology and Sociobiology* 1:301–309.

Ghiselin, M. T. 1974. *The Economy of Nature and the Evolution of Sex.* University of California Press, Berkeley.

Ghiselin, M. T. 1988. The evolution of sex: a history of competing points of view. In *The Evolution of Sex* (ed. R. E. Michod and B. R. Levin). Pp. 7–23. Sinauer, Massachusetts.

Gibson, R. M., and Hoglund, J. 1992. Copying and sexual selection. *Trends in Ecology and Evolution* 7:229–231.

Gigerenzer, G., and Hug, K. (in press). *Reasoning About Social Contracts: Cheating and Perspective Change.* Institut für Psychologie, Universitat Salzburg, Austria.

Gilliard, E. T. 1963. The evolution of bower birds. *Scientific American* 209:38–46.

Gillis, J. S., and Avis, W. E. 1980. The male-taller norm in mate selection. *Personality and Social Psychology Bulletin* 6:396–401.

Glesner, R. R., and Tilman, D. 1978. Sexuality and the components of environ-mental uncertainty: clues from geographical parthenogenesis in terrestrial ani-mals. *American Naturalist* 112:659–673.

Goodall, J. 1986. *The Chimpanzees of Gombe.* Belknap, Cambridge, Massachusetts.

Goodall, J. 1990. *Through a Window.* Weidenfeld and Nicolson, London.

Gotmark, F. 1992. Anti-predator effect of conspicuous plumage in a male bird. *Animal Behaviour* 44:51–55.

Gould, J. L., and Gould, C. G. 1989. *Sexual Selection.* Scientific American, New York.

Gould, S. J. 1978. *Ever Since Darwin: Reflections in Natural History.* Andre Deutsch, London.

Gould, S. J. 1981. *The Mismeasure of Man.* Norton, New York.

Gould, S. J. 1987. *An Urchin in the Storm: Essays About Books and Ideas.* Norton, New York.

Gould, S. J., and Lewontin, R. C. 1979. The spandrels of San Marco and the Pan-glossian paradigm: a critique of the adaptationist program. *Proceedings of the Royal Society of London* B 205:581–598.

Gouyon, P-H., and Couvet, D. 1987. A conflict between two sexes, females and hermaphrodites. In *The Evolution of Sex and Its Consequences* (ed. S.C. Stearns). Pp. 243–261. Birkhauser, Basel.

Gowaty, P., and Lennartz, M. R. 1985. Sex ratios of nestling and fledgling red-cockaded woodpeckers (*Picoides borealis*). *American Naturalist* 126:347–353.

Grafen, A. 1990. Biological signals as handicaps. *Journal of Theoretical Biology* 144:517–546.

Grant, V. J. 1990. Maternal personality and sex of infant. *British Journal of Medical Psychology* 63:261–266.

Green, M. 1987. Scent marking in the Himalayan musk deer *(Moschus chrysogaster)*. *Journal of Zoology* 1987:721–737.

Green, R. 1993. *Sexual Science and the Law*. Harvard University Press, Cambridge, Massachusetts.

Gwynne, D. T. 1991. Sexual competition among females: what causes courtship-role reversal? *Trends in Ecology and Evolution* 6:118–121.

Haig, D., and Grafen, A. 1991. Genetic scrambling as a defence against meiotic drive. *Journal of Theoretical Biology* 153:531–558.

Haldane, J. B. S. 1932. *The Causes of Evolution*. Longmans, London.

Haldane, J. B. S. 1949. Disease and evolution. In *Symposium sui fattori ecologi e genetici della specializione negli animali, Supplemento a la Ricerca Scientifica Anno 19th*. Pp. 68–75.

Halliday, T. R. 1983. The study of mate choice. In *Mate Choice* (ed. P. Bateson). Cambridge University Press, Cambridge.

Hamilton, W. D. 1964. The genetical evolution of social behaviour. *Journal of Theoretical Biology* 7:1–52.

Hamilton, W. D. 1967. Extraordinary sex ratios. *Science* 156:477–488.

Hamilton, W. D. 1971. Geometry for the selfish herd. *Journal of Theoretical Biology* 31:295–311.

Hamilton, W. D. 1980. Sex versus non-sex versus parasite. *Oikos* 35:282–290.

Hamilton, W. D. 1990a. Memes of Haldane and Jayakar in a theory of sex. *Journal of Genetics* 69:17–32.

Hamilton, W. D. 1990b. Mate choice near and far. *American Zoologist* 30:341–351.

Hamilton, W. D., Axelrod, R., and Tanese, R. 1990. Sexual reproduction as an adaptation to resist parasites (a review). *Proceedings of the National Academy of Sciences of the USA* 87:3566–3573.

Hamilton, W. D., and Zuk, M. 1982. Heritable true fitness and bright birds: a role for parasites? *Science* 218:384–387.

Harcourt, A. H., Harvey, P. H., Larson, S. G., and Short, R. V. 1981. Testis weight, body weight and breeding system in primates. *Nature* 293:55–57.

Hardin, G. 1968. The tragedy of the commons. *Science* 162:1243–1248.

Hartung, J. 1982. Polygyny and the inheritance of wealth. *Current Anthropology* 23:1–12.

Harvey, H. T. 1978. *The Sequoias of Yosemite National Park*. Yosemite Natural History Association, Yosemite, California.

Harvey, P. H., and May, R. M. 1989. Out for the sperm count. *Nature* 337:508–509.

Hasegawa, T., and Hiraiwa-Hasegawa, M. 1990. Sperm competition and mating behavior. In *The Chimpanzees of the Mahale Mountains: Sexual Life-History Strategies* (ed. T. Nishida). Pp. 115–132. University of Tokyo Press, Tokyo.

Hausfater, G., and Hrdy, S. B. 1984. *Infanticide: Comparative and Evolutionary Perspectives*. Aldine, Hawthorne, New York.

Hawkes, K. 1992. Why hunter-gatherers work. Paper delivered to the fourth annual meeting of the Human Behavior and Evolution Society, Albuquerque, New Mexico, 22–26 July 1992.

Head, G., May, R. M., and Pendleton, L. 1987. Environmental determination of sex in reptiles. *Nature* 329:198–199.

Hewitt, G. M. 1972. The structure and role of B-chromosomes in the mottled grasshopper. *Chromosomes Today* 3:208–222.

Hewitt, G. M. 1976. Meiotic drive for B-chromosomes in the primary oocytes of *Myrmeleotettix maculatus* (Orthoptera: Acrididae). *Chromosoma* 56:381–391.

Hewitt, G. M., and East, T. M. 1978. Effects of B-chromosomes on development in grasshopper embryos. *Heredity* 41:347–356.

Hewlett, B. S. 1988. Sexual selection and paternal investment among Aka pygmies. In *Human Reproductive Behavior* (ed. L. Betzig, M. Borgehoff Mulder, and P. Turke). Pp. 263–275. Cambridge University Press, Cambridge.

Hickey, D. A. 1982. Selfish DNA: a sexually transmitted nuclear parasite. *Genetics* 101:519–531.

Hickey, D. A., and Rose, M. R. 1988. The role of gene transfer in the evolution of eukaryotic sex. In *The Evolution of Sex* (ed. R. E. Michod and B. R. Levin). Pp. 161–175. Sinauer, Massachusetts.

Hill, A., Allsopp, C. E. M., Kwiatkowski, D., Anstey, N. M., Twumasi, P. T., Rowe, P. A., Bennett, S., Brewster, D., McMichael, A. J., and Greenwood, B. M. 1991. Common West African HLA antigens are associated with protection from severe malaria. *Nature* 352:595–600.

Hill, G. E. 1990. Plumage coloration is a sexually selected indicator of male quality. *Nature* 350:337–339.

Hill, K., and Kaplan, H. 1988. Tradeoffs in male and female reproductive strategies among the Ache. In *Human Reproductive Behavior* (ed. L. Betzig, M. Borgehoff Mulder, and P. Turke). Pp. 277–305. Cambridge University Press, Cambridge.

Hillgarth, N. 1990. Parasites and female choice in the ring-necked pheasant. *American Zoologist* 30:227–233.

Hiraiwa-Hasegawa, M. 1988. Adaptive significance of infanticide in primates. *Trends in Ecology and Evolution* 3:102–105.

Hoekstra, R. F. 1987. The evolution of sexes. In *The Evolution of Sex and Its Consequences* (ed. S. C. Stearns). Pp. 59–92. Birkhauser, Basel.

Hoglund, J., Eriksson, M., and Lindell, L. E. 1990. Females of the lek-breeding great snipe, *Gallinago media*, prefer males with white tails. *Animal Behaviour* 40:23–32.

Hoglund, J., and Robertson, J. G. M. 1990. Female preferences, male decision rules and the evolution of leks in the great snipe, *Gallinago media*. *Animal Behaviour* 40:15–22.

Houde, A. E., and Endler, J. A. 1990. Correlated evolution of female mating preferences and male color patterns in the guppy *Poecilia reticulata*. *Science* 248:1405–1408.

Hoyenga, K. B., and Hoyenga, K. 1980. *Sex Differences*. Little Brown, Boston.

Hrdy, S. B. 1979. Infanticide among animals: a review, classification and examination of the implications for the reproductive strategies of females. *Ethology and Sociobiology* 1:13–40.

Hrdy, S. B. 1981. *The Woman That Never Evolved*. Harvard University Press, Cambridge, Massachusetts.

Hrdy, S. B. 1986. Empathy, polyandry, and the myth of the coy female. In *Feminist Approaches to Science* (ed. R. Bleier). Pergamon, New York.

Hrdy, S. B. 1987. Sex-biased parental investment among primates and other mammals: a critical re-evaluation of the Trivers-Willard hypothesis. In *Child Abuse and Neglect: Bio-social Dimensions* (ed. R. Gelles and J. Lancaster). Pp. 97–147. Aldine, Hawthorne, New York.

Hrdy, S. B. 1990. Sex bias in nature and in history: a late 1980s re-examination of the "biological origins" argument. *Yearbook of Physical Anthropology* 33:25–37.

Huck, U. W., Labov, J. D., and Lisk, R. D. 1986. Food-restricting young hamsters (*Mesocricetus auratus*) affects sex ratio and growth of subsequent offspring. *Biology of Reproduction* 35:592–598.

Hudson, L., and Jacot, B. 1991. *The Way Men Think*. Yale University Press, New Haven.

Humphrey, N. K. 1976. The social function of intellect. In *Growing Points in Ethology* (ed. P. P. G. Bateson and R. A. Hinde). Pp. 303–318. Cambridge University Press, Cambridge.

Humphrey, N. K. 1983. *Consciousness Regained: Chapters in the Development of Mind*. Oxford University Press, Oxford.

Hunter, M. S., Nur, U., and Werren, J. H. 1993. Origin of males by genome loss in an autoparasitoid wasp. *Heredity* 70:162–171.

Hurlbert, A. C., and Poggio, T. 1988. Making machines (and artificial intelligence) see. *Daedalus* 117:213–239.

Hurst, L. D. 1990. Parasite diversity and the evolution of diploidy, multicellularity and anisogamy. *Journal of Theoretical Biology* 144:429–443.

Hurst, L. D. 1991a. The evolution of cytoplasmic incompatibitility or when spite can be successful. *Journal of Theoretical Biology* 148:269–277.

Hurst, L. D. 1991b. Sex, slime and selfish genes. *Nature* 354:23–24.

Hurst, L. D. 1991c. The incidences and evolution of cytoplasmic male killers. *Proceedings of the Royal Society* B 244:91–99.

Hurst, L. D. 1992a. Is Stellate a relict meiotic driver? *Genetics* 130:229–230.

Hurst, L. D. 1992b. Intragenomic conflict as an evolutionary force. *Proceedings of the Royal Society* B 248:135–148.

Hurst, L. D., Godfray, H. C. J., and Harvey, P. H. 1990. Antibiotics cure asexuality. *Nature* 346:510–511.

Hurst, L. D., and Hamilton, W. D. 1992. Cytoplasmic fusion and the nature of sexes. *Proceedings of the Royal Society* B 247:189–207.

Hurst, L. D., and Pomiankowski, A. 1991. Causes of sex ratio bias may account for unisexual sterility in hybrids: a new explanation of Haldane's rule and related phenomena. *Genetics* 128:841–858.

Hurst, L. D., Hamilton, W. D., and Ladle, R. J. 1992. Covert sex. *Trends in Ecology and Evolution* 7:144–145.

Huxley, J. 1942. *Evolution: The Modern Synthesis.* George Allen and Unwin, London.

Hyde, L. M., and Elgar, M. A. 1992. Why do hopping mice have such tiny testes? *Trends in Ecology and Evolution* 7:359–360.

Imperato-McGinley, J., Peterson, R. E., Gautier, T., and Sturla, E. 1979. Androgens and the evolution of male gender identity among male pseudohermaphrodites with 5-alpha-reductase deficiency. *New England Journal of Medicine* 300:1233–1237.

Irons, W. Natural selection, adaptation and human social behavior. In *Evolutionary Biology and Human Social Behavior* (ed. N. Chagnon and W. Irons). Pp. 4–39. Duxbury, North Scituate, Massachusetts.

Iwasa, Y., Pomiankowski, A., and Nee, S. 1991. The evolution of costly mate preferences. II: The handicap principle. *Evolution* 45:1431–1442.

Jacobs, L. F., Gaulin, S. J. C., Sherry, D., and Hoffman, G. E. 1990. Evolution of spatial cognition: sex-specific patterns of spatial behavior predict hippocampal size. *Proceedings of the National Academy of Sciences USA* 87:6349–6352.

Jaenike, J. 1978. A hypothesis to account for the maintenance of sex within populations. *Evolutionary Theory* 3:191–194.

James, W. H. 1986. Hormonal control of the sex ratio. *Journal of Theoretical Biology* 118:427–441.

James, W. H. 1989. Parental hormone levels and mammalian sex ratios at birth. *Journal of Theoretical Biology* 139:59–67.

Jarman, P. J. 1974. The social organization of antelope in relation to their ecology. *Behaviour* 48:215–267.

Jayakar, S. 1970. A mathematical model for interaction of gene frequencies in a parasite and its host. *Theoretical Population Biology* 1:140–164.

Johansen, D. C., and Edey, M. 1981. *Lucy: The Beginnings of Mankind.* Simon and Schuster, New York.

Jones, I. L., and Hunter, F. M. 1993. Mutual sexual selection in a monogamous seabird. *Nature* 362:238–239.

Jones, R. N. 1991. B-chromosome drive. *American Naturalist* 137:430–442.

Judge, D. S., and Hrdy, S. B. 1988. Bias and equality in American legacies. Paper presented at eighty-seventh annual meeting of the American Anthropological Association. Phoenix, Arizona. November 1988.

Kaplan, H., and Hill, K. 1985a. Hunting ability and reproductive success among male Ache foragers. *Current Anthropology* 26:131–133.

Kaplan, H., and Hill, K. 1985b. Food sharing among Aché foragers: test of explanatory hypotheses. *Current Anthropology* 26:223–245.

Kelley, S. E. 1985. The mechanism of sib competition for the maintenance of sex in *Anthoxanthum odoratum*. Ph.D. thesis, Duke University, Durham, North Carolina.

Kenrick, D. T., and Keefe, R. C. 1989. Time to integrate sociobiology and social psychology. *Behavioral and Brain Sciences* 12:24–26.

Kingdon, J. 1993. *Self-Made Man and His Undoing*. Simon & Schuster, New York.

King-Hele, D. 1977. *Doctor of Revolution: The Life and Genius of Erasmus Darwin*. Faber and Faber, London.

Kirkpatrick, M. 1982. Sexual selection and the evolution of female choice. *Evolution* 36:1–12.

Kirkpatrick, M. 1989. Is bigger always better? *Nature* 337:116–117.

Kirkpatrick, M., and Jenkins, C. 1989. Genetic segregation and the maintenance of sexual reproduction. *Nature* 339:300–301.

Kirkpatrick, M., and Ryan, M. J. 1991. The evolution of mating preferences and the paradox of the lek. *Nature* 350:33–38.

Kitcher, P. 1985. *Vaulting Ambition: Sociobiology and the Quest for Human Nature*. MIT Press, Cambridge, Massachusetts.

Kodric-Brown, A., and Brown, J. H. 1984. Truth in advertising: the kind of traits favored by sexual selection. *American Naturalist* 124:309–323.

Kondrashov, A. S. 1982. Selection against harmful mutations in large sexual and asexual populations. *Genetic Research* 40:325–332.

Kondrashov, A. S. 1988. Deleterious mutations and the evolution of sexual reproduction. *Nature* 336:435–440.

Kondrashov, A. S. 1991. Haploidy or diploidy: which is better? *Nature* 351:314–315.

Konner, M. 1982. *The Tangled Wing: Biological Constraints on the Human Spirit*. Holt, Rinehart and Winston, New York.

Korpimaki, E. 1991. Poor reproductive success of polygynously mated female Tengmalm's owls: are better options available? *Animal Behaviour* 41:37–47.

Kramer, B. 1990. Sexual signals in electric fishes. *Trends in Ecology and Evolution* 5:247–249.

Krause, R. M. 1992. The origin of plagues: old and new. *Science* 257:1073–1078.

Kurland, J. A. 1979. Matrilines: the primate sisterhood and the human avunculate. In *Evolutionary Biology and Human Social Behavior* (ed. N. Chagnon and W. Irons). Pp. 145–180. Duxbury, North Scituate, Massachusetts.

Ladle, R. J. 1992. Parasites and sex: catching the Red Queen. *Trends in Ecology and Evolution* 7:405–408.

Lande, R. 1981. Models of speciation by sexual selection on polygenic traits. *Proceedings of the National Academy of Sciences* 78:3721–3725.

Leakey, R., and Lewin, R. 1992. *Origins Reconsidered: In Search of What Makes Us Human.* Little Brown, London.

Leigh, E. G. 1977. How does selection reconcile individual advantage with the good of the group? *Proceedings of the National Academy of Sciences* 74:4542–4546.

Leigh, E. G. 1990. Fisher, Wright, Haldane and the resurgence of Darwinism. Introduction to the Princeton Science Library edition of *The Causes of Evolution* by J. B. S. Haldane.

Le Vay, S. 1992. *Born That Way? The Biological Basis of Homosexuality.* Channel Four, London.

Le Vay, S. 1993. *The Sexual Brain.* MIT Press, Cambridge, Massachusetts.

Levin, B. R. 1988. The evolution of sex in bacteria. In *The Evolution of Sex* (ed. R. E. Michod and B. R. Levin). Pp. 194–211. Sinauer, Massachusetts.

Levy, S. 1992. *Artificial Life: The Quest for a New Creation.* Jonathan Cape, London.

Lewin, R. 1984. *Human Evolution: An Illustrated Introduction.* Blackwell Scientific Publications, Oxford.

Lienhart, R., and Vermelin, H. 1946. Observation d'une famille humaine a descendance exclusivement féminine. Essai d'interpretation de ce phénomène. *Comptes rendus de science de la société de biologie de Nancy et de ses filiales de Paris.* 140:537–540.

Ligon, J. D., Thornhill, R., Zuk, M., and Johnson, K. 1990. Male-male competition: ornamentation and the role of testosterone in sexual selection in red junglefowl. *Animal Behaviour* 40:367–373.

Lively, C. J., Craddock, C., and Vrijenhoek, R. C. 1990. Red Queen hypothesis supported by parasitism in sexual and clonal fish. *Nature* 344:864–866.

Lively, C. M. 1987. Evidence from a New Zealand snail for the maintenance of sex by parasitism. *Nature* 328:519–521.

Low, B. S. 1979. Sexual selection and human ornamentation. In *Evolutionary Biology and Human Social Behavior* (ed. N. Chagnon and W. Irons). Pp. 462–487. Duxbury, North Scituate, Massachusetts.

Low, B. S. 1990. Marriage systems and pathogen stress in human societies. *American Zoologist* 30:325–340.

Low, B. S., Alexander, R. D., and Noonan, K. M. 1987. Human hips, breasts and buttocks: is fat deceptive? *Ethology and Sociobiology* 8:249–257.

Maccoby, E. E., and Jacklin, C. N. 1974. *The Psychology of Sex Differences.* Stanford University Press, Palo Alto, California.

Malinowski, B. 1927. *Sex and Repression in Savage Society.* World Press, Cleveland.

Marden, J. H. 1992. Newton's second law of butterflies. *Natural History* 1/92:54–61.

Margulis, L. 1981. *Symbiosis in Cell Evolution.* W. H. Freeman, San Francisco.

Margulis, L., and Sagan, D. 1986. *Origins of Sex: Three Billion Years of Genetic Recombination.* Yale University Press, New Haven, Connecticut.

Marler, P. R., and Tamura, M. 1964. Culturally transmitted patterns of vocal behavior in sparrows. *Science* 146:1483–1486.

Marr, D. 1982. *Vision*. Freeman Cooper, San Francisco.

Martin, R. D., and May, R. M. 1981. Outward signs of breeding. *Nature* 293:7–9.

May, R. M., and Anderson R. M. 1990. Parasite-host co-evolution. *Parasitology* 100:S89–S101.

Maynard Smith, J. 1971. What use is sex? *Journal of Theoretical Biology* 30:319–335.

Maynard Smith, J. 1977. Parental investment—a prospective analysis. *Animal Behaviour* 25:1–9.

Maynard Smith, J. 1978. *The Evolution of Sex*. Cambridge University Press, Cambridge.

Maynard Smith, J. 1986. Contemplating life without sex. *Nature* 324:300–301.

Maynard Smith, J. 1988. The evolution of recombination. In *The Evolution of Sex* (ed. R. E. Michod and B. R. Levin). Pp. 106–125. Sinauer, Massachusetts.

Maynard Smith, J. 1991. Theories of sexual selection. *Trends in Ecology and Evolution* 6:146–151.

Maynard Smith, J., and Price, G. R. 1973. The logic of animal conflict. *Nature* 246:15–18.

Mayr, E. 1983. How to carry out the adaptationist program. *American Naturalist* 121:324–334.

McGuiness, D. 1979. How schools discriminate against boys. *Human Nature*, February 1979: 82–88.

McNeill, W. H. 1976. *Plagues and Peoples*. Anchor Press/Doubleday, New York.

Mead, M. 1928. *Coming of Age in Samoa*. William Morrow, New York.

Mereschkovsky, C. 1905. Le plante considérée comme une complex symbiotique. *Bulletin Société Science Naturelle, Ouest* 6:17–98.

Metzenberg, R. L. 1990. The role of similarity and difference in fungal mating. *Genetics* 125:457–462.

Michod, R. E., and Levin, B. R. (eds.). 1988. *The Evolution of Sex*. Sinauer, Sunderland, Massachusetts.

Miller, G. F. 1992. Sexual selection for protean expressiveness: a new model of hominid encephalization. Paper delivered to the fourth annual meeting of the Human Behavior and Evolution Society, Albuquerque, New Mexico, 22–26 July 1992.

Miller, G. F., and Todd, P. M. 1990. Exploring adaptive agency. I: Theory and methods for simulating the evolution of learning. In *Proceedings of the 1990 Connectionist Models Summer School* (ed. Touretzky, D. S., Elman, J. L., Sejnowski, T. J., and Hinton, G. E.). Pp. 65–80. Morgan Kauffmann, San Mateo, California.

Mitchison, N. A. 1990. The evolution of acquired immunity to parasites. *Parasitology* 100:S27–S34.

Moir, A., and Jessel, D. 1991. *Brain Sex: The Real Difference Between Men and Women.* Lyle Stuart, New York.

Møller, A. P. 1987. Intruders and defenders on avian breeding territories: the effect of sperm competition. *Oikos* 48:47–54.

Møller, A. P. 1988. Female choice selects for male sexual tail ornaments in the monogamous swallow. *Nature* 332:640–642.

Møller, A. P. 1990. Effects of a haematophagous mite on secondary sexual tail ornaments in the barn swallow *(Hirundo rustica)*: a test of the Hamilton and Zuk hypothesis. *Evolution* 44:771–784.

Møller, A. P. 1991. Sexual selection in the monogamous barn swallow *(Hirundo rustica)*. I. Determinants of tail ornament size. *Evolution* 45:1823–1836.

Møller, A. P. 1992. Female preference for symmetrical male sexual ornaments. *Nature* 357:238–240.

Møller, A. P., and Birkhead, T. R. 1989. Copulation behaviour in mammals: evidence that sperm competition is widespread. *Biological Journal of the Linnean Society* 38:119–131.

Møller, A. P., and Pomiankowski, A. (in press). Fluctuating asymmetry and sexual selection. *Genetica.*

Montagu, A. 1961. Neonatal and infant immaturity in man. *Journal of the American Medical Association* 178:56–57.

Morris, D. 1967. *The Naked Ape.* Dell, New York.

Mosher, D. L., and Abramson, P. R. 1977. Subjective sexual arousal to films of masturbation. *Journal of Consulting and Clinical Psychology* 45:796–807.

Muller, H. J. 1932. Some genetic aspects of sex. *American Naturalist* 66:118–138.

Muller, H. J. 1964. The relation of recombination to mutational advance. *Mutation Research* 1:2–9.

Murdock, G. P., and White, D. R. 1969. Standard cross-cultural sample. *Ethnology* 8:329–369.

Nee, S., and Maynard Smith, J. 1990. The evolutionary biology of molecular parasites. *Parasitology* 100:S5–S18.

Nowak, M. A. 1992. Variability of HIV infections. *Journal of Theoretical Biology* 155:1–20.

Nowak, M. A., and May, R. M. 1992. Coexistence and competition in HIV infections. *Journal of Theoretical Biology* 159:329–342.

O'Connell, R. L. *Of Arms and Men: A History of War, Weapons and Aggression.* Oxford University Press, Oxford.

O'Donald, P. 1980. *Genetic Models of Sexual Selection.* Cambridge University Press, Cambridge.

Olsen, M. W. 1956. Fowl pox vaccine associated with parthenogenesis in chicken and turkey eggs. *Science* 124:1078–1079.

Olsen, M. W., and Buss, E. G. 1967. Role of genetic factors and fowl pox virus in parthenogenesis in turkey eggs. *Genetics* 56:727–732.

Olsen, P. D., and Cockburn, A. 1991. Female-biased sex allocation in peregrine falcons and other raptors. *Behavioral Ecology and Sociobiology* 28:417–423.

Olsen, M. W., and Marsden, S. J. 1954. Natural parthenogenesis of turkey eggs. *Science* 120:545–546.

Orgel, L. E., and Crick, F. H. C. 1980. Selfish DNA: the ultimate parasite. *Nature* 284:604–607.

Parker, G. A., Baker R. R., and Smith, V. G. F. 1972. The origin and evolution of gamete dimorphism and the male-female phenomenon. *Journal of Theoretical Biology* 36:529–533.

Partridge, L. 1980. Mate choice increases a component of offspring fitness in fruit flies. *Nature* 283:290–291.

Payne, R. B., and Payne, L. L. 1989. Heritability estimates and behavior observations: extra-pair mating in indigo buntings. *Animal Behaviour* 38:457–467.

Perrot, V., Richerd, S., and Valero, M. 1991. Transition from haploidy to diploidy. *Nature* 351:315–317.

Perusse, D. (1992). Cultural and reproductive success in industrial societies: testing the relationship at the proximate and ultimate levels. *Behavioral and Brain Sciences*.

Petrie, M., Halliday, T., and Sanders, C. 1991. Peahens prefer peacocks with elaborate trains. *Animal Behaviour* 41:323–331.

Pinker, S., and Bloom, P. 1992. Natural language and natural selection. In *The Adapted Mind* (ed. J. H. Barkow, L. Cosmides, and J. Tooby). Pp. 405–447. Oxford University Press, New York.

Pleszczynska, W., and Hansell, R. I. C. 1980. Polygyny and decision theory: testing of a model in lark buntings (*Calamospiza melanocorys*). *American Naturalist* 116:821–830.

Pomiankowski, A. 1987. The costs of choice in sexual selection. *Journal of Theoretical Biology* 128:195–218.

Pomiankowski, A. 1990. How to find the top male. *Nature* 347:616–617.

Pomiankowski, A., and Guilford, T. 1990. Mating calls. *Nature* 344:495–496.

Pomiankowski, A., Iwasa, Y., and Nee, S. 1991. The evolution of costly mate preferences. I: Fisher and biased mutation. *Evolution* 45:1422–1430.

Posner, R. A. 1992. *Sex and Reason.* Harvard University Press, Cambridge, Massachusetts.

Potts, R. 1991. Untying the knot: evolution of early human behavior. In *Man and Beast Revisited* (ed. M. H. Robinson and L. Tiger). Pp. 41–59. Smithsonian, Washington, D.C.

Potts, W. K., Manning, C. J., and Wakeland E. K. 1991. Mating patterns in semi-natural populations of mice influenced by MHC genotype. *Nature* 352:619–621.

Pratto, F., Sidanius, J., and Stallworth, L. M. 1992. Sexual selection and the sexual and ethnic basis of social hierarchy. In *Social Stratification and Socioeconomic Inequality: A Comparative Analysis* (ed. J. Ellis). Praeger, New York.

Pruett-Jones, S. G., Pruett-Jones, M. A., and Jones, H. I. 1990. Parasites and sexual selection in birds of paradise. *American Zoologist* 30:287–298.

Rands, M. R. W., Ridley, M. W., and Lelliott, A. D. 1984. The social organization of feral peafowl. *Animal Behaviour* 32:830–835.

Rao, R. 1986. Move to stop sex-test abortion. *Nature* 324:202.

Ray, T. 1992. Evolution and optimisation of digital organisms. Unpublished manuscript, University of Delaware.

Regalski, J. M., and Gaulin, S. J. C. 1992. Whom are Mexican babies said to resemble? Monitoring and fostering paternal confidence in the Yucatán. Paper delivered to the fourth annual meeting of the Human Behavior and Evolution Society, Albuquerque, New Mexico, 22–26 July 1992.

Ridley, M. 1978. Paternal Care. *Animal Behaviour* 26:904–932.

Ridley, M. W. 1981. How did the peacock get his tail? *New Scientist* 91:398–401.

Ridley, M. W., and Hill, D. A. 1987. Social organization in the pheasant *(Phasianus colchicus):* harem formation, mate selection and the role of mate guarding. *Journal of Zoology* 211:619–630.

Ridley, M. W., Rands, M. R. W., and Lelliott, A. D. 1984. The courtship display of feral peafowl. *Journal of the World Pheasant Association* 9:20–40.

Rosenberg, N., and Birdzell, L. E. 1986. *How the West Grew Rich: The Economic Transformation of the Industrial World.* Basic Books, New York.

Rossi, A. S. (ed.). 1985. *Gender and the Life Course.* Aldine, Hawthorne, New York.

Ryan, M. J. 1991. Sexual selection and communication in frogs. *Trends in Ecology and Evolution* 6:351–355.

Sadalla, E. K., Kenrick, D. T., and Vershure, B. 1987. Dominance and heterosexual attraction. *Journal of Personality and Social Psychology* 52:730–738.

Schall, J. J. 1990. Virulence of lizard malaria: the evolutionary ecology of an ancient parasite-host association. *Parasitology* 100:S35–S52.

Schmitt, J., and Antonovics, J. 1986. Experimental studies of the evolutionary significance of sexual reproduction. IV. Effect of neighbor relatedness and aphid infestation on seedling performance. *Evolution* 40:830–836.

Scruton, R. 1986. *Sexual Desire: a Philosophical Investigation.* Weidenfeld and Nicolson, London.

Searcy, W. A. 1992. Song repertoire and mate choice in birds. *American Zoologist* 32:71–80.

Seger, J., and Hamilton, W. D. 1988. Parasites and sex. In *The Evolution of Sex* (ed. R. E. Michod and B. R. Levin). Pp. 139–160. Sinauer, Sunderland, Massachusetts.

Seid, R. P. 1989. *Never Too Thin: Why Women Are at War with Their Bodies.* Columbia University Press, New York.

Shaw, M. W., Hewitt, G. M., and Anderson, D. A. 1985. Polymorphism in the rates of meiotic drive acting on the B-chromosome of *Myrmeleotettix maculatus. Heredity* 55:61–68.

Shellberg, T. 1992. Tall bishops and genuflection genes. Paper delivered to the fourth annual meeting of the Human Behavior and Evolution Society, Albuquerque, New Mexico, 22–26 July 1992.

Shepher, J. 1983. *Incest: A Biosocial View.* Academic Press, Orlando, Florida.

Short, R. V. 1979. Sexual selection and its component parts, somatic and genital selection, as illustrated by man and the great apes. *Advances in the Study of Behaviour* 9:131–158.

Silk, J. B. 1983. Local resource competition and facultative adjustment of sex ratios in relation to competitive abilities. *American Naturalist* 121:56–66.

Sillen-Tullberg, B., and Møller, A. P. 1993. The relationship between concealed ovulation and mating systems in anthropoid primates: a phylogenetic analysis. *American Naturalist* 141:1–25.

Silverman, I., and Eals, M. 1992. Sex differences in spatial abilities: evolutionary theory and data. In *The Adapted Mind* (ed. J. H. Barkow, L. Cosmides, and J. Tooby). Pp. 533–549. Oxford University Press, New York.

Simpson, M. J. A., and Simpson, A. E. 1982. Birth sex ratios and social rank in rhesus monkey mothers. *Nature* 300:440–441.

Slagsvold, T., Amundsen, T., Dale, S., and Lampe, H. 1992. Female-female aggression explains polyterritoriality in male pied flycatchers. *Animal Behaviour* 43:397–407.

Slater, P. J. B. 1983. The Buzby phenomenon: thrushes and telephones. *Animal Behaviour* 31:308–309.

Small, M. F. 1992. What's love got to do with it? *Discover* 13:46–51.

Small, M. F., and Hrdy, S. B. 1986. Secondary sex ratios by maternal rank, parity and age in captive rhesus macaques (*Macaca mulatta*). *American Journal of Primatology* 11:359–365.

Smith, R. L. 1984. Human sperm competition. In *Sperm Competition and the Evolution of Animal Mating Systems* (ed. R. L. Smith). Pp. 601–659. Academic Press, Orlando, Florida.

Smuts, R. W. 1993 (in press). Fat, sex, class, adaptive flexibility and cultural change. *Ethology and Sociobiology.*

Spandrel, S. (unpublished). How the genome learnt Mendelian genetics, or you scratch my back, I'll stab yours.

Spurrier, M. F., Boyce. M. S., and Manly, B. F. J. 1991. Effects of parasites on mate choice by captive sage grouse. In *Ecology, Behavior and Evolution of Bird-Parasite Interactions* (ed. J. E. Loye and M. Zuk). Pp. 389–398. Oxford University Press, Oxford.

Stearns, S. C. (ed.). 1987. *The Evolution of Sex and Its Consequences.* Birkhauser, Basel.

Stebbins, G. L. 1950. *Variation and Evolution in Plants.* Columbia University Press, New York.

Symington, M. M. 1987. Sex ratio and maternal rank in wild spider monkeys: when daughters disperse. *Behavioral Ecology and Sociobiology* 20:421–425.

Symons. D. 1979. *The Evolution of Human Sexuality.* Oxford University Press, Oxford.

Symons, D. 1987. An evolutionary approach: can Darwin's view of life shed light on human sexuality? In *Theories of Human Sexuality* (ed. J. H. Geer and W. O'Donohue). Pp. 91–125. Plenum Press, New York.

Symons, D. 1989. The psychology of human mate preferences. *Behavioral and Brain Sciences* 12:34–35.

Symons, D. 1992. On the use and misuse of Darwinism in the study of human behavior. In *The Adapted Mind* (ed. J. H. Barkow, L. Cosmides, and J. Tooby). Pp. 137–159. Oxford University Press, New York.

Tannen, D. 1990. *You Just Don't Understand: Women and Men in Conversation.* William Morrow, New York.

Taylor, P. D., and Williams G. C. 1982. The lek paradox is not resolved. *Theoretical Population Biology* 22:392–409.

Thornhill, N. W. 1989a. Characteristics of female desirability: facultative standards of beauty. *Behavioral and Brain Sciences* 12:35–36.

Thornhill, N. W. 1989b. The evolutionary significance of incest rules. *Ethology and Sociobiology* 11:113–129.

Thornhill, N. W. 1990. The comparative method of evolutionary biology in the study of the societies of history. *International Journal of Contemporary Sociology* 27:7–27.

Thornhill, R., and Thornhill, N. W. 1983. Human rape: an evolutionary analysis. *Ethology and Sociobiology* 4:137–183.

Thornhill, R., and Thornhill, N. W. 1989. The evolution of psychological pain. In *Sociobiology and Social Sciences* (ed. R. J. Bell and N. J. Bell). Pp. 73–103. Texas Tech University Press, Lubbock.

Thornhill, R., and Sauer, P. 1992. Genetic sire effects on the fighting ability of sons and daughters and mating susccess of sons in a scorpion fly. *Animal Behaviour* 43:255–264.

Thorpe, W. H. 1954. The process of song-learning in the chaffinch as studied by means of the sound spectrograph. *Nature* 173:465–469.

Thorpe, W. H. 1961. *Bird Song: The Biology of Vocal Communication in Birds.* Cambridge University Press, Cambridge.

Tiersch, E. R., Beck, M. L., and Douglas, M. 1991. ZZW autotriploidy in a blue and yellow macaw. *Genetica* 84:209–212.

Tiger, L. 1991. Human nature and the psycho-industrial complex. In *Man and Beast*

Revisited (ed. M. H. Robinson and L. Tiger). Pp. 23–40. Smithsonian, Washington, DC.

Tiger, L., and Shepher, J. 1977. *Women in the Kibbutz.* Penguin, London.

Tooby, J. 1982. Pathogens, polymorphism and the evolution of sex. *Journal of Theoretical Biology* 97:557–576.

Tooby, J., and Cosmides, L. M. 1989. The innate versus the manifest: how universal does a universal have to be? *Behavioral and Brain Sciences* 12:36–37.

Tooby, J., and Cosmides, L. M. 1990. On the universality of human nature and the uniqueness of the individual: the role of genetics and adaptation. *Journal of Personality* 58:17–67.

Tooby, J., and Cosmides, L. M. 1992. The psychological foundations of culture. In *The Adapted Mind* (ed. J. H. Barkow, L. Cosmides, and J. Tooby). Pp. 19–136. Oxford University Press, New York.

Traill, P.W. 1990. Why should lek breeders be monomorphic? *Evolution* 44:1837–1852.

Tripp. C. A. 1975. *The Homosexual Matrix.* Signet, New York.

Trivers, R. L. 1971. The evolution of reciprocal altruism. *Quarterly Review of Biology* 46:35–57.

Trivers, R. L. 1972. Parental investment and sexual selection. In *Sexual Selection and the Descent of Man* (ed. B. Campbell). Pp. 136–179. Atherton, Chicago.

Trivers, R. L. 1985. *Social Evolution.* Benjamin/Cummings, Menlo Park, California.

Trivers, R. L. 1991. Deceit and self-deception: the relationship between communication and consciousness. In *Man and Beast Revisited* (ed. M. H. Robinson and L. Tiger). Pp. 175–191. Smithsonian, Washington, D.C.

Trivers, R. L., and Willard, D. 1973. Natural selection of parental ability to vary the sex ratio of offspring. *Science* 179:90–91.

Troy, S., and Elgar, M. A. 1991. Brush turkey incubation mounds: mate attraction in a promiscuous mating system. *Trends in Ecology and Evolution* 6:202–203.

Unterberger, F., and Kirsch, W. 1932. Bericht über versuche zür Beeinflussung des Geschlechtsverhältnisses bei Kaninchen nach Unterberger. *Monatsschrift für Geburtshilfe und Gynäkologie* 91:17–27.

van Schaik, C. P., and Hrdy, S. B. 1991. Intensity of local resource competition shapes the relationship between maternal rank and sex ratios at birth in cercopithecine primates. *American Naturalist* 138:1555–1562.

Van Valen, L. 1973. A new evolutionary law. *Evolutionary Theory* 1:1–30.

Veiga, J. 1992. Why are house sparrows predominantly monogamous? A test of hypotheses. *Animal Behaviour* 43:361–370.

Vining, D. R. 1986. Social versus reproductive success: the central theoretical problem of human sociobiology. *Behavioral and Brain Sciences* 9:167–187.

Voland, E. 1988. Differential infant and child mortality in evolutionary perspective: data from late 17th to 19th century Ostfriesland (Germany). In *Human*

Reproductive Behavior (ed. L. Betzig, M. Borgehoff Mulder, and P. Turke). Pp. 253–261. Cambridge University Press, Cambridge.

Voland, E. 1992. Historical demography and human behavioral ecology. Paper delivered to the fourth annual meeting of the Human Behavior and Evolution Society, Albuquerque, New Mexico, 22–26 July 1992.

Vos, G. J. de 1979. Adaptedness of arena behaviour in Black Grouse (*Tetrao tetrix*) and other grouse species (Tetraoninae). *Behaviour* 68:277–314.

Wallace, A. R. 1889. *Darwinism*. Macmillan, London.

Ward, P. I. 1988. Sexual dichromatism and parasitism in British and Irish freshwater fish. *Animal Behaviour* 36:1210–1215.

Warner, R. R., Robertson, D. R., and Leigh, E. G. 1975. Sex change and sexual selection. *Science* 190:633–638.

Weatherhead, P. L., and Robertson, R. J. 1979. Offspring quality and the polygyny threshold: "the sexy son hypothesis." *American Naturalist* 113:201–208.

Webster, M. S. 1992. Sexual dimorphism, mating system and body size in New World blackbirds (Icterinae). *Evolution* 46:1621–1641.

Wederkind, C. 1992. Detailed information about parasites revealed by sexual ornamentation. *Proceedings of the Royal Society* B 247:169–174.

Weismann, A. 1889. *Essays upon Heredity and Kindred Biological Problems*. Translated by E. B. Poulton, S. Schonland, and A. E. Shipley. Clarendon Press, Oxford.

Werren, J. H. 1987. The coevolution of autosomal and cytoplasmic sex ratio factors. *Journal of Theoretical Biology* 124:317–334.

Werren, J. H. 1991. The paternal-sex-ratio chromosome of *Nasonia*. *American Naturalist* 137:392–402.

Werren, J. H., Skinner, S. W., and Huger, A. M. 1986. Male-killing bacteria in a parasitic wasp. *Science* 231:990–992.

Westermarck, E. A. 1891. *The History of Human Marriage*. Macmillan, New York.

Westneat, D. F., Sherman, P. W., and Morton, M. L. 1990. The ecology of extra-pair copulations in birds. *Current Ornithology* 7:331–369.

White, F. 1992. Eros of the apes. *BBC Wildlife Magazine*, August 1992:39–47.

Wiener, P., Feldman, M. W., and Otto, S. P. 1992. On genetic segregation and the evolution of sex. *Evolution* 46:775–782.

Williams, G. C. 1966. *Adaptation and Natural Selection: A Critique of Some Current Evolutionary Thought*. Princeton University Press, Princeton.

Williams, G. C. 1975. Sex and evolution. In *Monographs in Population Biology*. Princeton University Press, Princeton.

Williams, G. C., and Mitton, J. B. 1973. Why reproduce sexually? *Journal of Theoretical Biology* 39:545–554.

Wilson, E. O. 1975. *Sociobiology: The New Synthesis*. Harvard University Press, Cambridge, Massachusetts.

Wilson, E. O. 1978. *On Human Nature*. Harvard University Press, Cambridge, Massachusetts.

Wilson, M., and Daly, M. 1992. The man who mistook his wife for a chattel. In *The Adapted Mind* (ed. J. H. Barkow, L. Cosmides, and J. Tooby). Pp. 289–322. Oxford University Press, New York.

Wolf, A. P. 1966. Childhood association, sexual attraction and the incest taboo: a Chinese case. *American Anthropologist* 68:883–898.

Wolf, A. P. 1970. Childhood association and sexual attraction: a further test of the Westermarck hypothesis. *American Anthropologist* 70:864–874.

Wrangham, R. W. 1987. The significance of African apes for recontructing human social evolution. In *The Evolution of Human Behavior: Primate Models* (ed. W. G. Kinzey). Pp. 51–71. SUNY Press, New York.

Wright, S. 1931. Evolution in Mendelian populations. *Genetics* 16:97–159.

Wynne-Edwards, V. C. 1962. *Animal Dispersion in Relation to Social Behaviour*. Oliver and Boyd, London.

Yamamura, N., Hasegawa, T., and Ito, Y. 1990. Why mothers do not resist infanticide: a cost-benefit genetic model. *Evolution* 44:1346–1357.

Zahavi, A. 1975. Mate selection—a selection for a handicap. *Journal of Theoretical Biology* 53:205–214.

Zinsser, H. 1934. *Rats, Lice and History*. Macmillan, London.

Zuk, M. 1991. Parasites and bright birds: new data and a new prediction. In *Ecology, Behavior and Evolution of Bird-Parasite Interactions* (ed. J. E. Loye and M. Zuk). Pp. 317–327. Oxford University Press, Oxford.

Zuk, M. 1992. The role of parasites in sexual selection: current evidence and future directions. *Advances in the Study of Behavior* 21:39–68.

Zuk, M. (in press). Immunology and the evolution of behavior. In *Behavioral Mechanisms in Evolutionary Biology* (ed. L. Real). University of Chicago Press, Chicago.

Zuk, M., Thornhill, R., Ligon, J. D., and Johnson, K. 1990. Parasites and mate choice in red junglefowl. *American Zoologist* 30:235–244.

INDEX

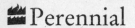 Perennial

ecco

Books by Matt Ridley:

NATURE VIA NURTURE: *Genes, Experience, and What Makes Us Human*
ISBN 0-06-000678-1 (New in hardcover from HarperCollins*Publishers*)
ISBN 0-06-054446-5 (audio cassette) • ISBN 0-06-054447-3 (audio CD)

Now that we finally understand genes, Matt Ridley urges us to put to rest the emotional 100-year-old war between nature and nurture.

"Bracingly intelligent, lucid, balanced—witty, too. . . . A scrupulous and charming look at our modern understanding of genes and experience."
—Dr. Oliver Sacks

GENOME: *The Autobiography of a Species in 23 Chapters*
ISBN 0-06-093290-2 (paperback)
New York Times Bestseller and Editors' Choice *New York Times Book Review*

By picking one newly discovered gene from each of the 23 human chromosomes and telling its story, Matt Ridley recounts the history of our species and its ancestors from the dawn of life to the brink of future medicine.

"A fascinating tour of the human genome. . . . If you want to catch a glimpse of the biotech century that is now dawning, *Genome* is an excellent place to start."
—*Wall Street Journal*

THE RED QUEEN: *Sex and the Evolution of Human Nature*
ISBN 0-06-055657-9 (paperback)

Matt Ridley answers dozens of the riddles of human nature and culture, and compels us to rethink everything from sexism to the endurance of romantic love.

"A terrific book, witty and lucid, and brimming with provocative conjectures."
—*Wall Street Journal*

THE BEST AMERICAN SCIENCE WRITING 2002
Edited by Matt Ridley
ISBN 0-06-621162-X (hardcover) • ISBN 0-06-093650-9 (paperback)

The third in an annual series dedicated to collecting the best science writing from the most prominent thinkers on the most current topics.

"The entire spectrum of science is covered with literary acumen here." —*Booklist*

Want to receive notice of author events and new books by Matt Ridley?
Sign up for Matt Ridley's AuthorTracker at www.AuthorTracker.com

---

Available wherever books are sold, or call 1-800-331-3761 to order.